D0765481

Buried Glory

Buried Glory

PORTRAITS OF SOVIET SCIENTISTS

Istvan Hargittai

OXFORD
UNIVERSITY PRESS

Oxford University Press is a department of the University of Oxford.
It furthers the University's objective of excellence in research, scholarship,
and education by publishing worldwide.

Oxford New York
Auckland Cape Town Dar es Salaam Hong Kong Karachi
Kuala Lumpur Madrid Melbourne Mexico City Nairobi
New Delhi Shanghai Taipei Toronto

With offices in
Argentina Austria Brazil Chile Czech Republic France Greece
Guatemala Hungary Italy Japan Poland Portugal Singapore
South Korea Switzerland Thailand Turkey Ukraine Vietnam

Published in the United States of America by
Oxford University Press
198 Madison Avenue, New York, NY 10016

Library of Congress Cataloging-in-Publication Data
Hargittai, Istvan, author.
Buried glory : portraits of Soviet scientists / Istvan Hargittai.
pages; cm
Includes bibliographical references and index.
ISBN 978-0-19-998559-3 (alkaline paper) 1. Scientists—Soviet Union—Biography.
2. Science—Soviet Union. I. Title.
Q141.H263 2013
509.2'247—dc23
2013005419
978-0-19-998559-3

9-8-7-6-5-4-3-2-1
Printed in the United States of America
on acid-free paper

I dedicate this book to the memory of my father,

Jenő Wilhelm (1901–1942). He was a Budapest lawyer, the first member of his family who acquired higher education. He coauthored a book about unfair competition. In 1942, anti-Jewish legislation in Hungary sent him—under most humiliating conditions in a so-called labor-service unit—to the Eastern Front. He was ordered to sweep mine fields with his bare hands, and was blown apart. His remains rest in a mass grave in Western Russia.

ALSO BY THE AUTHOR

Drive and Curiosity: What Fuels the Passion for Science (Prometheus, 2011)

Judging Edward Teller: A Closer Look at One of the Most Influential Scientists of the Twentieth Century (Prometheus, 2010)

With Magdolna Hargittai, *Symmetry through the Eyes of a Chemist*, Third Edition (Springer, 2009; 2010)

With Magdolna Hargittai, *Visual Symmetry* (World Scientific, 2009)

The DNA Doctor: Candid Conversations with James D. Watson (World Scientific, 2007)

The Martians of Science: Five Physicists Who Changed the Twentieth Century (Oxford University Press, 2006; 2008)

Our Lives: Encounters of a Scientist (Akadémiai Kiadó, 2004)

The Road to Stockholm: Nobel Prizes, Science, and Scientists (Oxford University Press, 2002; 2003)

With Magdolna Hargittai and Balazs Hargittai, *Candid Science I–VI: Conversations with Famous Scientists* (Imperial College Press, 2000–2006)

With Magdolna Hargittai, *In Our Own Image: Personal Symmetry in Discovery* (Plenum/Kluwer, 2000; Springer, 2012)

With Magdolna Hargittai, *Symmetry: A Unifying Concept* (Shelter Publications, 1994)

With R. J. Gillespie, *The VSEPR Model of Molecular Geometry* (Allyn & Bacon, 1991; Dover Publications, 2012)

CONTENTS

PREFACE

During the Cold War in the period 1945–1991, the balance of power between the two superpowers—the United States and the Soviet Union, including their allies—maintained a tenuous peace. Each of the two superpowers possessed enough destructive weaponry to annihilate the other many times over. In many other aspects the two superpowers were vastly different. The United States was more technologically advanced, whereas the Soviet Union had a troubled economy and a backward infrastructure. The sophisticated weaponry of the Soviet Union was to a large extent due to the outstanding achievements of its scientists and the communist regime's ability to concentrate its limited resources on selected tasks. The Soviet Union could not have become a superpower without a strong scientific background that relied on some traditions in science dating back to czarist Russia, but whose foundations were created from the beginning of Soviet power and developed in the 1920s and 1930s.

To be a scientist was one of the most privileged professions in the Soviet Union. It was a magnet for talent in view of the very restricted possibilities where gifted young people could aspire for a career. In contrast, the current Russia pays diminishing attention to science. One of the vice presidents of the Russian Academy of Sciences recently noted, "The demise of the Soviet Union hurt Russian science very much."[1]

Although communist ideology advocated a classless society, there was strong stratification in Soviet society. This is conspicuously demonstrated by the differentiation of its burial places. They had a hierarchy, with the Lenin Mausoleum on the Red Square in Moscow at its top—between 1953 and 1961 it was Lenin's *and* Stalin's mausoleum. Behind the mausoleum are buried the next echelon of Soviet leaders, each represented by a bust. Stalin is one of them; his bust was erected as late as 1970, indicating how hesitant post-Stalin Soviet leaderships were in condemning one of the bloodiest dictators in world history. When I last saw these graves (in June 2011), of the twelve, only Stalin's was covered with fresh flowers. Many of the most distinguished Soviet (and some international) politicians, military leaders, and communist revolutionaries are interred in the Kremlin wall, among them Sergei Korolev, the chief Soviet rocket constructor, and Igor Kurchatov, the nuclear czar.

The Novodevichy Cemetery in Moscow is Russia's most distinguished cemetery.[2] It is located close to the Moscow River and the center of the city.

In addition to political and military leaders, buried here are many of the topmost representatives of Soviet (as well as pre-Soviet and post-Soviet Russian) intelligentsia: writers, artists, scientists. A walk in the cemetery reveals such a plethora of great scientists of the Soviet decades that in itself is a manifestation of the importance that science played in the Soviet regime.

When I was a master's degree student in Moscow in the first half of the 1960s, I heard a lot about the Novodevichy Cemetery, but could not visit it. For years, it was closed "temporarily" for reconstruction. When I visited Moscow in the early 1980s, it was possible to visit the Novodevichy, but at the entrance the police took away all cameras, which they returned when the visitor left the cemetery. I was very much taken by the modern tombstone recently erected over the former Soviet leader Nikita Khrushchev's grave—the work of sculptor Ernst Neizvestny.[3] To me, its alternating black and white stones were an expression of antisymmetry. It fit Khrushchev's image eminently: he was one of Stalin's close associates and party to his crimes, yet he unmasked his master after his death. I lamented the ban on taking photographs in the Novodevichy, and soon I received color slides of the tombstone from unknown donors. I included the image in a forthcoming publication on symmetry.[4]

Nowadays, it is possible to visit Novodevichy freely, and on the occasion of my rare visits to Moscow, I never miss the opportunity to return there. Quite a few scientists whom I knew personally are buried there, some in their own right, others because they inherited the right from their families. Walking along the alleys of this beautiful memorial place, the thought is always with me that during the Soviet era there was an extraordinary accumulation of talent in science. They were the very men (there were hardly any women among them) who made it possible for the Soviet Union to become a superpower.

The expression "Buried Glory" in the title indicates that the scientist heroes in this book and their achievements belong to the past. Most have been unknown in the West, and while their activities have become a little better known lately, their personalities have remained mostly in obscurity. It is an interesting question why the Soviet scientists—with the notable exception of Andrei Sakharov—are so little known? This is in spite of their decisive contributions to the Soviet might. Western historiography tends to ignore them, even those that helped create the Soviet hydrogen bombs. The Soviet space program fared a little better due to the spectacular and easy-to-perceive successes, such as the first Sputnik in 1957 and the first manned flight by Yurii Gagarin in 1961.

Only one of the heroes of the twelve chapters in this book is still alive. Nine of them are buried in the Novodevichy Cemetery. Seven of them received the Nobel Prize, and the rest were also at that level of scientific achievement.

The coverage in this book is far from comprehensive, and it is not a history of Soviet science either. The primary goal of the book is to bring the personalities and the lives of a select set of Soviet scientists into human proximity. I hope that my selection of the sample of scientists introduced in this book will be adequate to create a good impression of what it meant to be creating in science under Soviet conditions and will also lead to an appreciation of the achievements of scientists who lived and labored in Soviet times.

Buried Glory

Introduction

Science in the Soviet Union always occupied a privileged position, but at times it was a very risky position, too. In the 1920s, up to the early 1930s, there was a tremendous boost in science in the new Soviet Union, and gifted young people joined the profession in large numbers. Considerable segments of the population that under the czarist regime could not have dreamed of higher education were now encouraged to pursue it, and the most gifted members were lured into scientific research. The Soviet government realized the importance of interactions with the rest of the world and arranged study trips for scientists to Western Europe. International visitors were welcomed at Soviet meetings and laboratories. The goal was to re-establish old contacts and establish new ones in order to lift the level of science in the Soviet Union to an international standard.

This openness changed from the early 1930s as a result of several factors coming together. Among them was the demand of the Soviet state to involve scientists increasingly in solving problems related to the economy and especially for the military. It began innocuously enough; thus, for example, there was a debate at the Ukrainian Institute of Physical Technology about whether or not the profile of the Institute should be changed to address practical problems. Whereas the initial discussions appeared to be open and democratic, they soon led to a sharp division between the representatives of the two views. There were subsequent repercussions targeting those who opposed the changes. Applications of science, especially for the military, necessitated making much of it classified. This was compounded by Stalin's growing paranoia about everything foreign, and he introduced an isolationist policy that characterized the Soviet Union during its later existence. After Stalin had eliminated those whom he considered rivals for power, he consolidated his monopoly and let his security organs loose to perform mass murders and deportations. Among the victims of this terror that included a considerable portion of the leadership of the Soviet Communist Party and the Red Army, there were a number of talented scientists: physicists, biologists, and others. Another period of terror flared up after World War II, in the period 1948–1953, during the last years of Stalin's life.

We note two distinguishing features between the two periods of terror. One was that in the second half of the 1930s, there was no branch of science where scientists were immune to persecution. On the other hand, in the post–World War II period, at the last minute, physics was exempted. This was due to the recognition of the importance of the nuclear weapons program. The question was whether ideological purity was more important than the atomic and hydrogen bombs. Physics could not have provided the latter without the application of the theory of relativity and quantum mechanics that were both considered the products of bourgeois ideologies.

The other distinguishing feature was that the prewar terror was not colored by anti-Semitism; whereas during the last years of Stalin's life, the Terror coincided with increasing anti-Semitic tendencies. The attacks on Jews culminated in the arrest and indictment of a group of distinguished medical doctors and professors, mostly Jewish, who were falsely accused of planning the assassination of Stalin and other Soviet leaders. The punishment and deportation of the entire Jewish population of the European areas of the Soviet Union had already been carefully choreographed under Stalin. Upon his death, these activities were abandoned, but covert anti-Semitism remained a characteristic feature of Soviet internal policy throughout the rest of the existence of the Soviet Union.

In the late 1930s, Stalin's henchman, Nikolai Ezhov, indiscriminately murdered and exiled people. His successor, Lavrentii Beria, was no less brutal and ruthless, but he was more "rational." Upon replacing Ezhov in 1939, he instituted an amnesty in which hundreds of thousands of prisoners were freed, constituting about one-third of those incarcerated. Beria wanted to make the prison camps economically viable. From the mid-1940s, when he was made responsible for the nuclear weapons program, he gave precedence to performance over political and ideological considerations.

In both the American and the Soviet nuclear programs, at least in the top echelon, there were many Jewish scientists. For the American program, this could be explained by the influx of refugee physicists from Germany, Hungary, Italy, and elsewhere prior to World War II. By the time they had become eligible for defense work, the atomic bomb project was available for them. This project was initiated later than other defense-related projects that also involved scientists, such as the radar project. The refugee physicists were also the first to recognize the danger that Germany might acquire an atomic bomb; hence they had pushed for an American counterprogram before others recognized its necessity. By then, most of the leading nonrefugee American scientists had found assignments in other programs. On the Soviet side, the collapse of the czarist regime also meant the elimination of the severe anti-Jewish restrictions in higher education; young, ambitious Jews flocked to centers of learning and research, and the emerging field of nuclear science attracted many of their best minds.

Against this background of the privileged situation of physics, the disadvantaged circumstances of other sciences were conspicuous during the last years of Stalin's reign. Among them, biology suffered the severest damage at the hands of the unscientific czar of biology and agricultural sciences, Trofim Lysenko. Stalin personally edited Lysenko's presentation to the 1948 meeting of the agricultural academy in which Lysenko launched an all-out attack on modern biology. This was part of the dedicated effort by Stalin and his regime to prevent Western influence from penetrating Soviet society. The areas of computerization, automation, and other technology-driven disciplines, often referred to under the umbrella term "cybernetics" (today, the term "informatics" is more applicable), were also singled out for attack. Cybernetics was considered to symbolize everything evil that an imperialistic-bourgeois society could offer. The devastating conditions of biology and cybernetics in the Soviet Union impacted not only the well-being of Soviet society but Soviet military capabilities as well, especially in the long run. This would be conspicuous when it became clear that the Soviet Union would not be able to match the American Strategic Defense Initiative (SDI), regardless of whether or not the program was feasible and whether or not much less expensive and sophisticated countermeasures could have neutralized it.

As mentioned above, the development of bombs saved physics from the ideological attacks. Ironically, practical considerations contributed—alas, with the opposite outcome—to Lysenko's support by Stalin and later by Khrushchev. He promised to lift the productivity of Soviet agriculture if Soviet agriculture would follow his teachings. The results were disastrous, both for biology in the Soviet Union and for Soviet agriculture. It was the physicists who would come to the rescue of biology, hesitantly at first but gradually with enhanced determination. Igor Tamm, Andrei Sakharov, Petr Kapitza, Igor Kurchatov, and others appeared increasingly determined to fight for modern biology. Sakharov's determination stemmed from his study of the possible biological consequences of nuclear testing. For Tamm, it was his general concern for science that mattered. On December 10, 1958, the day of the Nobel Prize award ceremony in Stockholm, Tamm, in his two-minute banquet speech, managed to highlight the importance of modern biology (see chapter 1).

Tamm had been Sakharov's mentor, and had strongly influenced the formation of his pupil's philosophy and outlook on the world. Both men condemned the way the Soviet leadership evoked anti-Semitism when the leadership found it convenient and relied on such occasions on the worst instincts of parts of the population. In his *Memoirs*, Sakharov quoted Tamm saying that there is "one foolproof way of telling if someone belongs to the Russian intelligentsia. A true Russian *intelligent* is never an anti-Semite. If he is infected with that virus, then he's something else, something terrible and dangerous."[1] The anti-Semitic character of the Soviet system remained in effect throughout the existence of the Soviet Union. There were limitations on the number of Jewish students that could be accepted at certain institutions of higher education, and some institutions did not accept any Jews at all.[2] It thus happened that Jewish professors taught a student body that Jewish students could not have joined. It also happened that when scientists were recruited for important, urgent projects, it was unofficially declared that there would be no limitation on the hiring of Jewish employees—this was another proof that under "normal" circumstances, there were severe limitations on Jews. It was also an open secret that a considerable portion (estimated at 25% to 30%) of the members of the Soviet Academy of Sciences would never vote to accept a Jewish nominee. It was a testimony to the professional strength of Jewish academicians that they were nonetheless elected.

The membership of the Soviet Academy of Sciences consisted of various strata. Some members were truly great scientists. Others were obvious representatives of the Communist Party. Yet others were elected as a result of complicated compromises between various groups of the members. Still, to a certain degree, the Academy of Sciences represented a hope for democratic action amid the hopelessness of a totalitarian regime. In rare, isolated cases, the majority of the membership defeated party nominees, and in a few cases it elected new members in spite of contrary party directives. By and large, however, the Communist Party kept the Academy under tight control, along with everything else in Soviet society, to the end of the existence of the Soviet Union. Most of the "heroes" of this book became members of the Science Academy, and they did so based on their scientific achievements. Their elections in some cases were facilitated and accelerated by their stellar participation in the development of Soviet nuclear power.

All scientists selected for inclusion in this book were top scientists, but there were many more that could have been selected just as well; beyond their excellence, my choice was subjective. There were a few whom I would have liked to add but decided not to lest the project grow too large. I hesitated over whether or not to add Igor Kurchatov, and here I would like to share my dilemma with my readers. He was an outstanding physicist, and it was to a great extent due to his activities that the Soviet Union caught up quite fast with the United States in nuclear matters. He was forty years old when he was made head of the Soviet nuclear program. I decided against his inclusion after I had read the information that was readily available about him, and found it an impossible task to realistically evaluate his personality and performance in the framework of the present project.

Kurchatov was the primary recipient of the intelligence pouring into Moscow about the American and other nuclear bomb developments. Some other leaders of the Soviet atomic bomb project were also receiving intelligence data, but nobody had the direct access to these classified reports that Kurchatov did. It is understandable that the Soviets guarded most rigorously the sources of such information and even the fact that they were the results of espionage activity. This is why Kurchatov could not reveal the sources of the information that he was obliged to share with his colleagues in order to utilize the information. His colleagues were amazed by the trustworthiness of this information, the solutions that proved almost invariably correct, and must have supposed that the data came either from Kurchatov's own research or from secret Soviet laboratories. If he was a conscientious scientist—and indications are that he was; to this day he is much revered among the Russian scientific community—his role of presenting scientific information without its sources must have tormented him. Some day—and it may never come—it might be possible to delineate his personal achievements from what he learned from espionage. Sorting out this problem was beyond the scope of the present project.

Yulii Khariton, to whom a chapter in this book is devoted, was also on the receiving end of intelligence, but in his case it was easy to delineate his scientific achievements before he joined the weapons program from his later activities as scientific chief of the secret Soviet nuclear laboratory, Arzamas-16.

My choices of the scientists for this book were directed by my interest and background in the physical sciences, although biology would have been another fascinating area of inquiry about great personalities. There, my first choices would have been the martyr Nikolai Vavilov and the scientist who figured prominently among those who brought modern biology back to the Soviet Union, Vladimir Engelhardt. As it is, they both appear only briefly in the book.

In this book, a dozen or so Soviet scientists are presented through impressionistic portraits. I happened to know personally nearly half of the scientists introduced here, and I have met family members of some whom I did not know. Some of these family members represented more than one of the heroes in this book. Most of the great scientists belonged to a privileged circle of Soviet society, and it was not unusual that their children married the children of other members of this circle.

The scientists presented are grouped in three sections; nuclear physics, low-temperature physics, and chemistry. Even in the chemistry group, most of the scientists were physicists or received training in physics. Twentieth-century physics and chemistry, especially in the first half of the century, overlapped a great deal. The labels physical chemistry and chemical physics demonstrate this blending of previously more-distinct disciplines. The principal discoveries of Nikolai Semenov, Yulii Khariton, Boris Belousov and Anatol Zhabotinsky, and Aleksandr Nesmeyanov were chemical in character even if only Belousov and Nesmeyanov could be labeled as bona fide chemists. The classifications "nuclear physics" and "low-temperature physics" here may also seem arbitrary since there was remarkable overlap in the activities of the physicists appearing in these pages. They did not recognize any of the artificial boundaries that the usual classification systems our educational and research establishments force onto science for practical considerations. The overlap among the activities of these scientists and their related backgrounds made some overlap among the chapters in this book inevitable. I tried to keep it within limits, but found that some overlap was useful to facilitate smooth reading. In discussing, for example, Lev Landau's incarceration and the actions taken by Petr Kapitza to secure his liberation, I look at the story from Landau's and Kapitza's perspectives in their respective chapters.

I believe that the portraits in the following pages provide a realistic and instructive picture of some of the best minds in twentieth-century science. In their sum, their stories bring us also closer to understanding not only these exceptional contributors to twentieth-century science, but also to understanding what to a great extent continues to be an enigma, the Soviet Union.

To facilitate geographical orientation, three maps are presented here.

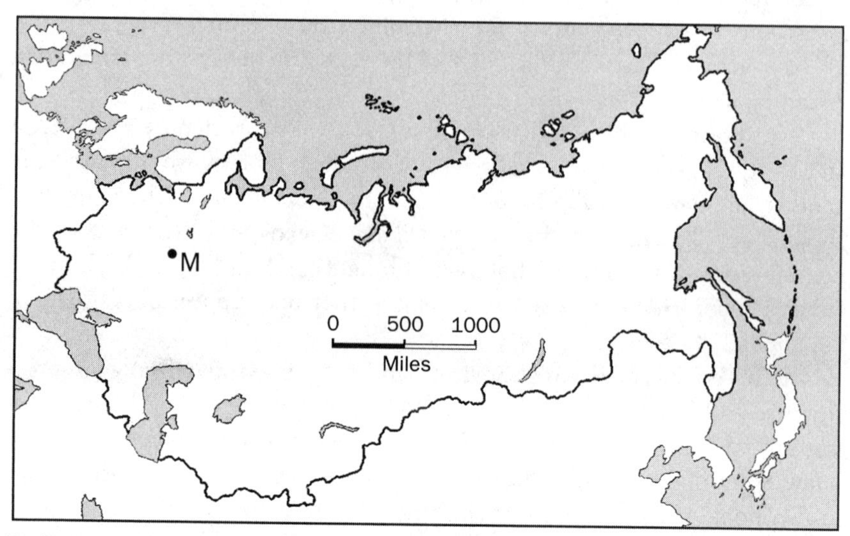

The Soviet Union, 1945–1991.

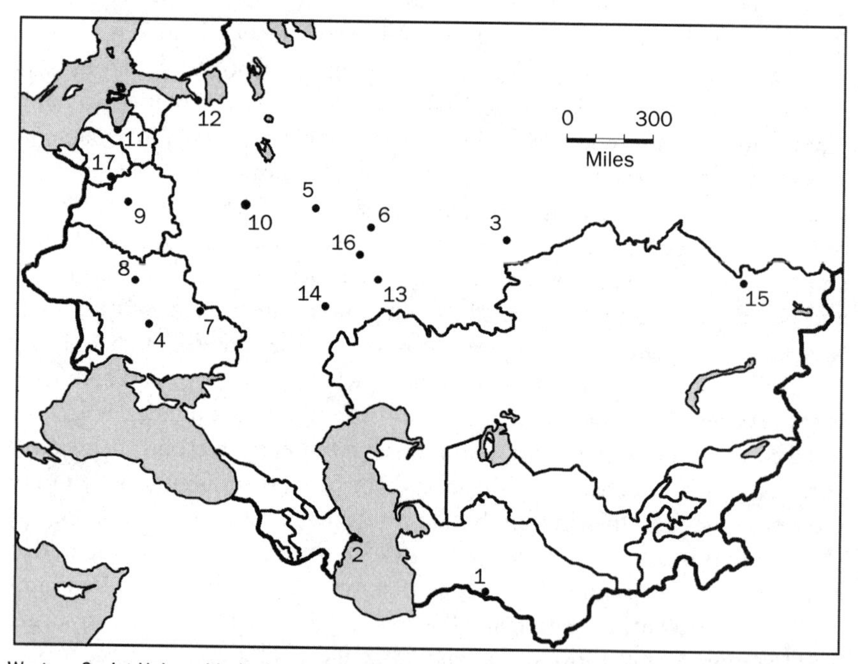

Western Soviet Union with the borders of the member republics and the locations figuring in the text marked (the names of the respective republics are in parentheses, except for Russia), 1–Ashkhabad (Turkmenistan), 2–Baku (Azerbaijan), 3–Chelyabinsk, 4–Elizavetgrad (today, Kirovograd, Ukraine), 5–Nizhnii Novgorod, in Soviet times, Gorky, 6–Kazan, 7–Kharkov (Ukraine), 8–Kiev (Ukraine), 9–Minsk (Belarus), 10–Moscow, 11–Riga (Latvia), 12–St. Petersburg, 13–Samara, 14–Saratov, 15–Semipalatinsk (Kazakhstan), 16–Ulyanovsk, 17–Vilnius (Lithuania).

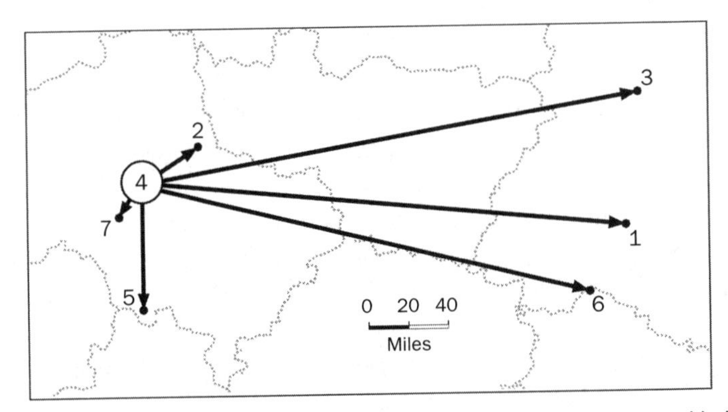

Moscow, Gorky—Nizhnii Novgorod and a few additional locations figuring in the text with distances (in miles) from Moscow as the crow flies, 1—Arzamas (244 miles), 2—Chernogolovka (35), 3—Nizhnii Novgorod (250), 4—Moscow, 5—Pushchino (65), 6—Sarov (231), 7—Troitsk (22)

Note on the scientific degrees in the Soviet/Russian system. University graduates who become teachers, engineers, and researchers receive a Diploma that can be considered equivalent to a master's degree, or somewhere between a master's degree and a bachelor's degree. Diplomas issued by Moscow State University and other leading institutions, for example, can certainly be considered equivalent to a master's. The Candidate of Science (C.Sc.) degree can be considered equivalent to a PhD degree. Just as the levels of American PhDs greatly vary depending on the granting institution, this is so in the Soviet and Russian system as well, although the ultimate granting agency is the State Commission of Accreditation, whose existence is supposed to guarantee a nationwide uniform level. The Doctor of Science (D.Sc.) degree follows the C.Sc. degree. The work for the C.Sc. degree is usually conducted under the supervision of a mentor; the work for the D.Sc. degree is usually independent research assisted by graduate students working for their lower degree. There is no equivalent to the D.Sc. degree in the US system, whereas in Russia this degree is usually a prerequisite for a professorial appointment. In this respect it is similar to the Habilitation in the German system. Membership in the Soviet, now Russian, Academy of Sciences can be considered to be part of the system of scientific degrees. It is a two-tier system. The first is the "corresponding membership" and the higher one is the "full membership." The titles are Corresponding Member of the Soviet (Russian) Academy of Sciences and Academician, respectively. The full membership does not necessarily follow the corresponding membership, and there are no rules stipulating how long one may be a corresponding member. Becoming an academician was and continues to be in the pinnacle of science—this was not a Soviet category, but a continuation of practice used in czarist Russia. In Soviet society, there was no aristocracy, and the communist higher echelon could be admired or despised. The scientific elite was revered.

***

Note on the transliteration of Russian names. The general rules of transliteration have been followed, but precedence was given to spellings that were already well known in the English-language literature. If the name had appeared in different versions, the most frequently used version was used. Variations may be, for example, Alexander, Aleksander, and Aleksandr or Alexey, Alexei, Aleksey, and Aleksei. Surnames have usually fewer variations in the literature, for example, Kitaigorodskii and Kitaigorodsky. I tried to be consistent throughout the text; exceptions include when original literature is being quoted, and when in the reference the name is given as in the original even if the spelling is different from the one used in the main text. Similar considerations applied to geographical names, leading to some inconsistency but retaining self-consistency. Thus, Gorky is used throughout rather than Gorkii, and Nizhnii Novgorod rather than Nizhny Novgorod.

PART ONE

Nuclear Physicists

Igor Tamm's grave in the Novodevichy Cemetery.

Source: Photograph by and courtesy of the author.

Igor Tamm's portrait.

Source: Courtesy of Valentina Berezovskaya, Moscow.

1

Igor Tamm

EXEMPLARY CONSISTENCY

Igor E. Tamm (1895–1971) came from a part-German, part-Cossack family. He grew up in turbulent times in Russia and his parents sent him to Scotland to study and to be away from politics. When Tamm returned he became a theoretical physicist. He was disillusioned with politics though he remained a Marxist all of his life.

He achieved great success in theoretical physics, including his Nobel-Prize-winning theory of the Vavilov-Cherenkov radiation. During the war he was engaged in defense-related research, but his greatest contribution to Soviet might came after the war when he was a leading member of the secret project on nuclear weapons.

He placed his principles above his career and benefits. It was a miracle that he survived unscathed both the trappings of a corrupt society and Stalin's terror. His life served as an example for his peers.

Igor Evgenevich Tamm was born July 8, 1895, in Vladivostok.* When he was three years old, his family moved to the town of their roots, Elizavetgrad in the Ukraine, in the center of the triangle of Kiev, Odessa, and Dnepropetrovsk. Today, the town is called Kirovograd. At the time of his birth, Tamm's family was in the Far East, because his engineer father worked on building the Trans-Siberian railway. Upon the family's return to Elizavetgrad, he was put in charge of the waterworks, electric power station, and the trolleys of the city. Elizavetgrad nurtured talent in both science and the arts. The world-renowned American chemist Moses Gomberg came from this city, and the Hungarian composer and performer Franz Liszt gave his last concert in Russia there.

Tamm was born into the Russian Empire of the ruthless Czar Nikolai II. He heard a lot about politics; he was ten years old at the time of the

* He was born on June 25 according to the Julian calendar. In 1918, the Soviets adopted the Gregorian calendar; here, I give the dates according to the Gregorian calendar, except in a few special cases where I find it helpful to give both.

Russian-Japanese war. During his high school years he became a dedicated socialist. His parents were worried, and upon his graduation from high school they sent him away for one year to Edinburgh, Scotland. In 1914, just as World War I was breaking out, he returned and entered Moscow State University to major in physics and mathematics. When the war broke out, there was universal patriotic euphoria, but Tamm opposed the war, and joined those who felt similarly. They considered revolution to be the only way for Russia to progress. Still, Tamm wanted to help his slightly older colleagues who had been conscripted and were on the front. In 1915, he joined the Russian Red Cross organization and volunteered to serve as a medic in the Russian Army, a "med brother" analogous to "med sisters."[1] After one year he returned to Moscow to continue his studies. It was also in 1915 that he joined the Social Democratic Party. In 1917, he married Natalia Shuiskaya.

In 1917, two revolutions occurred in Russia. The first was the bourgeois revolution in February. It prompted Tamm to become an orator and a prolific author of newspaper articles and of leaflets. He consistently argued against the war. He ran for representative of Elizavetgrad for the First All-Russian Congress of Workers and Soldiers, which convened in Petrograd, June 16–July 7 (June 3–24), 1917. At this time he was so dedicated to politics that during the election campaign he wrote: "I would give half of my life for this," that is, to be a representative at this congress.[2] Vladimir Lenin's communist revolution happened on November 7 (October 25); subsequently known as the Great October Socialist Revolution, it was celebrated every November 7 in the Soviet Union and throughout the communist world.

The socialists, called also social democrats, comprised many fractions and included Lenin and his followers. Lenin and his comrades won the majority in a crucial vote, and from then on, they called themselves the Bolsheviks—the majority—as distinct from the defeated minority, the Mensheviks. In subsequent years, after 1917, the two labels acquired important political meaning. The term "Bolshevik" became part of the official name of the Communist Party, in paretheses;** whereas Menshevik became equivalent to the "enemies of the people."

It was still in 1917 that in another crucial vote, Tamm, who belonged to the Mensheviks, voted with the Bolsheviks. Lenin attributed much importance to Tamm's support, thus, lending Tamm certain political prominence, and this is how Tamm's political past became widely known in the Soviet Union. This could have helped Tamm in Soviet times, but it could just as well have hurt him, because the story was proof that he had been a Menshevik. Tamm was not a dedicated Menshevik, but arguing later that he was not would have sounded as if he were trying to whitewash his political past. He was deeply disturbed

---

** Communist Party of the Soviet Union (Bolsheviks).

by Lenin's Bolshevik dictatorship, and by 1922, he had withdrawn from politics and restricted his activities to theoretical physics.

Tamm spent another academic year, 1927–28, in Western Europe, this time as part of the Soviet efforts to re-establish scientific connections. He especially enjoyed his stay in Leiden, where Paul Ehrenfest was his host. The two men felt mutual respect and friendship, and Ehrenfest's recommendation helped Tamm to visit some of the best research laboratories in Europe. He met famous physicists, including Albert Einstein, Walter Elsasser, Pieter Zeeman, Oskar Klein, Erwin Schrödinger, Niels Bohr, and, most importantly for Tamm, the taciturn Englishman Paul Dirac. Tamm was anxious to prove himself worthy of his colleagues. He classified them according to his estimates of whether or not he could measure up to them. He felt himself on equal footing with Ehrenfest, but found Dirac much above him.

In the period *before* Stalin's terror in the second half of the 1920s, scientists like Tamm could find a lot to be enthusiastic about in the Soviet Union. Tamm tried to convince his Western colleagues about the merits of the Soviet system. Dirac was receptive to Tamm's ideas. They had met when Dirac came to Leiden to visit Ehrenfest. Dirac was as tight-lipped as his reputation predicted he would be, but the extrovert Tamm managed to become friendly with him. For years, they enjoyed unique interactions.[3] Dirac and Tamm met briefly again in 1929 in Moscow when Dirac was in transit from Japan to England.

Paul Dirac, O. N. Trapeznikova, and Igor Tamm in 1928 in Leiden. Photograph by Lev Shubnikov, founder of cryogenic physics in Kharkov and victim of Stalin's terror. Olga Trapeznikova, a low-temperature physicist, was Shubnikov's wife.

At a meeting of theoretical physicists in the early 1930s in the Soviet Union. From left to right, 1-Dmitrii Ivanenko, 6-Niels Bohr, 8-Lev Landa, 10-Yakov Frenkel, 14-Vladimir Fock, 15-Igor Tamm. *Source*: Courtesy of G. A. Sardanashvily, Moscow.

In the summer of 1936, Tamm's love for mountain climbing infected Dirac, and the two went off on a tour of the Caucasus. Kapitza was the only other physicist Dirac was interested in seeing in the Soviet Union. However, the war and the isolationism of Stalin's Soviet Union did not help Tamm and Dirac's friendship. When in 1956 they met again in Moscow, they did not return to their past closeness. That the meeting took place at all was surprising because of Tamm's deep prior involvement in the secret nuclear weapons program, which made it difficult for him to meet with foreigners.

From the mid-1930s, the lives of Soviet physicists took a drastic turn for the worse. Lev Landau (see chapter 5) was arrested, and his life was only miraculously saved. Some of his colleagues were not only arrested and incarcerated but sentenced to death and executed. Among them were great talents and some of Tamm's friends. His favorite pupil, Semyon Shubin, perished and so did another very gifted young physicist, Matvei Bronshtein, for whom Tamm had served as reviewer of his dissertation.

Tragedy struck his immediate family as well. His younger brother Leonid was arrested in 1936. In the factory where he worked, there was a terrible fire compounded by a horrific explosion due to the accumulated flammable gas. The authorities accused Leonid of starting the fire. He was well respected by his associates, a leader, and just as hard-working as he was brave. When the fire started, all others ran away, seeking cover, and Leonid alone tried to suppress the flames with his own hands. He did not succeed completely, but his actions prevented the fire from spreading and saved a large part of the factory. Nonetheless, he was tried and sentenced to ten years of incarceration without the right of having correspondence. His family did not know then that this formulation of the sentence was a euphemism for immediate execution. His wife was exiled.[4]

Igor Tamm was lucky. It happened that at the height of the terror, he and his family went for a summer vacation in a remote place. While they were there, a family member brought Tamm a summons from the secret police to appear at their infamous headquarters, the Lubyanka, in Moscow. Tamm arrived at Lubyanka in the evening of the day indicated in the summons. He was ordered to return the next morning. On that morning, a newspaper article by Stalin about some abuses of power by the authorities appeared. This may be why when Tamm returned to Lubyanka, they merely talked with him and then let him go.[5]

Tamm had graduated from Moscow State University in 1918, and for some time he taught there and at several other institutions of higher education. From 1934, and for the rest of his life, the P. N. Lebedev Physical Institute of the Academy of Sciences (Fizicheskii Institut Akademii Nauk [FIAN]) was his principal affiliation. He was closely associated with the internationally renowned physicist Leonid Mandelshtam from 1920 until Mandelshtam's death in 1944. In 1933 Tamm had been elected a corresponding member of the Soviet Academy of Sciences, and he organized a section of theoretical physics at FIAN.

Tamm was engaged in conventional war-related research during the first period of the war. The situation changed in 1943 when the struggle against Germany took a turn, and the existence of the Soviet Union was no longer threatened. Tamm's section of theoretical physics was re-established at the FIAN. He had not yet been invited to participate in the atomic bomb project, though. Distrust of Tamm stemmed not only from his political past but also from his family background.[6] The Soviet security organs considered him a "*volksdeutsch*."[†]

The paternal side of Tamm's family was of German origin, and the maternal side was Cossack. His father showed heroism under the German occupation during World War II. Tamm's parents and his sister were late in departing from Kirovograd before the German occupiers arrived. Tamm's sister broke her leg just when the war started and lay in a hospital. She was released from the hospital before the fracture healed, and she had difficulty walking. When, after Kiev's liberation, Tamm managed to visit the family, his mother had just been buried and his sister fell ill again. This time she was diagnosed with breast cancer, and it was Tamm's task to find a surgeon to operate on her. The operation had to be performed without anesthetics since there were none available. After the operation Tamm took the amputated breast for laboratory analysis,

[†] The expression "Volksdeutsche" referred to ethnically German people, where Volk is folk in the sense of "race." The term was used by the Nazis for people belonging to the German "race" but living beyond the borders of Germany proper.

and the malignancy of the tumor was confirmed. Tamm wanted the family to move to Moscow, but at this point his father was arrested for allegedly cooperating with the Nazi occupiers. He was saved by the testimony of his colleagues, who attested that he helped others and saved the lives of the persecuted. Finally, Tamm managed to bring his father and sister to Moscow.

In 1946, Tamm received his first assignment on the atomic bomb project, though he still did not have access to its most classified sections. He produced a paper on the width of very intense shock waves, which could not be declassified for the next twenty years. When the development of the hydrogen bomb started—not just feasibility studies, but real work—Tamm was directed to organize a special group at his section of FIAN to deal with its theoretical problems. Andrei Sakharov and Vitaly Ginzburg were among its first members. Within a short time, they came up with crucial ideas for the hydrogen bomb.

Tamm had already started working on the hydrogen bomb in 1948, that is, before the Soviet atomic bomb was tested at the end of August 1949. When in spring 1950, he moved to the secret installation Arzamas-16, he took with him two of his young associates, Sakharov and Yu. A. Romanov. Ginzburg had to be left behind because he lacked security clearance. Tamm stayed at Arzamas-16 only until the completion of the successful test of the first Soviet hydrogen bomb in 1953.

In the early days at Arzamas-16, office space was at premium. Even Tamm did not have a private office but shared one with Sakharov and Romanov. There were no set office hours, and the colleagues seldom talked with each other about anything but the problems that they had to solve. Sakharov patiently suffered the two others' permanent smoking. Tamm lived alone; his wife came only for short visits. The work was compartmentalized to such an extent that the members of Tamm's group and those of the other theory group led by Yakov Zeldovich were not allowed to know about each other's work. Even those in charge of these groups could seldom travel beyond the classified areas of the installation and only then with good reason. Tamm was a good organizer and articulate in expressing his thoughts. He spoke fast, and his colleagues joked that if fast speaking had a unit, it should be called "one tamm."[7]

In spite of the political disadvantages of his family background, which might have justified greater caution, Tamm behaved bravely when his conscience dictated such behavior. As described in greater detail elsewhere (see chapter 2), in January 1951, when the deeply religious mathematician M. M. Agrest was fired from Arzamas-16, Tamm openly expressed solidarity with him.

Tamm and most of the other leading theoretical physicists on the hydrogen bomb project maintained their interest in fundamental research during their years at Arzamas-16. They had the ability to choose the right research problems and to convey them to their young associates. In the early 1950s, the acumen of theoretical physicists working at Arzamas-16 may have exceeded that of those working in various research institutes in Moscow. Hence the comparison in which it was said that Sarov (in whose vicinity Arzamas-16 was located) might be called the New Moscow; and Moscow, the Old Sarov. Interest in challenging physics was among the driving forces for the scientists at Arzamas-16; another was Soviet patriotism.

Tamm served in a most dedicated way in the development of the Soviet nuclear bombs. The question arises, how could Tamm and others with similar negative experiences serve Stalin and the Soviet Union and with such dedication and produce these terrible means of mass destruction. Just as the possibility of nuclear weapons emerged, theoretical physics was no longer merely a *l'art pour l'art* science for the satisfaction of scientists' curiosity or the advancement of their careers. It had become a decisive force in defense, and not just for Stalin and the communist regime. It was indeed *national defense*.

Stalin was a shrewd enough politician to recognize the value of the scientists' patriotism—and everybody else's—and to play on it. During World War II, it helped him that the war of Hitler's Germany against the Soviet Union threatened the existence of the Russian people and the other peoples of the Soviet Union. When the struggle against Nazi Germany was labeled the Great Patriotic War, it was not a mere slogan forced onto the Soviet people—it was reality.

After the German attack on the Soviet Union, on June 22, 1941, the fledgling Soviet efforts in nuclear research were stopped, and even those few scientists who had been engaged in it were redirected to work on traditional weaponry. Although Stalin distrusted the intellectuals—while counting himself to be one—he recognized their value. On September 15, 1941, shortly following the German attack, the State Committee of Defense under Stalin's leadership forbade sending scientific researchers and instructors in higher education to the front and, generally, forbade employing them outside of their areas of expertise. This decision further enhanced the dedication of these specialists to the cause, and strengthened their feeling of responsibility before their fatherland. Stalin's approach to the scientists conspicuously differed from Hitler's. When Max Planck warned Hitler about the consequences of forcing the Jewish scientists out of Germany, Hitler made his famous statement: "Our national policies will not be revoked or modified, even for scientists. If the dismissal of Jewish scientists means the annihilation of contemporary German science, then we shall do without science for a few years."[8]

After World War II, the Soviet citizens were indoctrinated in the dangers of American imperialism. The Soviet physicists took pride in showing that they were capable of performing as well as the Americans. Nonetheless, at the time of the Soviet quest for the first Soviet atomic bomb, the Soviet physicists were not allowed to seek their own solutions. They had to limit themselves to doing exactly what they were instructed to do, which amounted to copying the

American approach (without knowing that it was the American approach). However, in the development of subsequent atomic bombs and the hydrogen bomb, they had the opportunity to seek their own original solutions.

The first Soviet hydrogen bomb was successfully tested in August 1953. It was not yet a bona fide hydrogen bomb, only a boosted atomic bomb, only about twenty times more powerful than the Hiroshima bomb. But it is also true that fusion reactions took place in it (see chapter 3). Upon the success of the test, Tamm and the other scientists were showered with awards, bonuses, and goods generally unavailable to ordinary Soviet citizens, including new apartments, dachas, and cars. The scientists knew that had the test failed, they would have been severely punished. Tamm confided in his daughter that the squadron to execute them had already been formed, and had the scientists not succeeded, its action would have been swift. Of course, with Stalin's death in March 1953 things might have changed, and the test took place a few months after Stalin's death. Still, before the test, Lavrentii Beria had also disappeared—he had been the powerful supervisor of the Soviet nuclear project who in the past erected a protective umbrella for its participants.

Tamm's scientific production was not huge by volume, some seventy papers and two monographs, including nonresearch contributions, such as reviews of the works of other physicists. His research contributions covered four principal areas: (1) macroscopic theory, (2) theory of the atomic nucleus, (3) theory of fundamental particles and their interactions, and (4) applications. The contribution for which he and his colleagues were awarded the Nobel Prize belonged to the first area. The work was done in the period 1937–1944, in part jointly with Ilya Frank. It was the theoretical interpretation of the Vavilov-Cherenkov Effect, as it was known in the Soviet Union, or the Cherenkov Effect, as it was known in the rest of the world.

The Cherenkov Effect was an experimentally discovered phenomenon in whose foundation there was a beautiful mechanism of emitting light by fast-moving particles. The effect was then used to establish techniques for the detection of charged particles, fast moving in air, water, ice, and other media. For the particle itself that radiates, the effect leads to a new mechanism of resistance as a consequence of collisions with many other particles in the same medium.

Another of Tamm's achievements was the discovery of phonons in 1929 and over subsequent years. The word "phonon" comes from the Greek word for sound. It pertains to the electric and thermal conductivity of crystalline solids and ordered liquids. The building units of such condensed phases—atoms, ions, or molecules—may be excited by external interference. The response is collective excitation (and may even be manifested as sound), and this is described as "phonons." When early in the twenty-first century a Russian stamp series depicted the most important scientific discoveries of the twentieth century in Russia, a drawing representing Tamm's phonon discovery was shown next to his portrait. He would have been pleased to see this stamp because he valued this discovery more than the one for which he received the Nobel Prize.

There were other scientific activities by Tamm, not directly related to his theoretical physics, but also significant. One such issue was his concern about how to distinguish between underground nuclear explosions and natural earthquakes—this was related to his participation in the Pugwash movement. The name "Pugwash" came from the first Canadian location of a series of conferences known as the Pugwash Conferences on Science and World Affairs. Although Tamm enjoyed the Pugwash meetings, he had no illusions about the nature of the Pugwash movement. The participants from the West were individual intellectuals dedicated to peaceful coexistence between the two camps, the American-led West and the Soviet-led East. On the Soviet side, the participants were government-controlled scientists who were supposed to voice officially sanctioned opinions. Tamm despised the Western intellectuals who accepted such an imbalanced approach to the discussions. Many years later, in 1988, Tamm's former pupil Andrei Sakharov attended a Pugwash meeting, which he also did not hold in high opinion. He surmised that "Pugwash is worthwhile so long as its efficiency is greater than zero—even if it is not much greater... so let Pugwash do its work. But without me!"[9]

Another of Tamm's concerns was about molecular biology. It pained him to see the growing gap between the progress in the West and the situation in the Soviet Union. Trofim Lysenko had destroyed the science of genetics, and made it impossible to cultivate molecular biology, acting with full support of the Soviet leaders, first Iosif Stalin and later Nikita Khrushchev. Tamm realized that under the circumstances it would be impossible for the Soviet biologists to reverse the trend, but that the nuclear physicists might be able to do it. Tamm convinced Igor Kurchatov, the leader of Soviet nuclear research, about the necessity of taking action. Instead of directly challenging Lysenko, they took measures that could be done within their jurisdiction. In the late 1950s, they organized a special seminar, with Tamm as its chair, for a limited circle of people. Initially, the seminars were held in private rooms of members of the Science Academy, almost like an underground movement. In 1958, they organized a section of radiobiology within the framework of Kurchatov's Institute of Atomic Energy.

Even though Tamm was a theoretical physicist, he gave talks on recent achievements in biology, based on his readings. In 1957, in one of his lectures on the molecular mechanism of heredity, he discussed the genetic code, which at the time was not yet solved. The genetic code describes the mechanism of transfer of heredity information from nucleic acids to proteins. Right after the discovery of the double-helix structure of DNA, the Russian-American physicist George Gamow raised this question of information transfer from nucleic acids to proteins (called eventually the genetic code). In his 1957 lecture at Leningrad University, Tamm mentioned the models suggested up to that point for the genetic code; among them, Edward Teller's unsuccessful attempt at creating a code.[10]

The involvement of nuclear physicists in saving Soviet biology was not merely a matter of altruism. They felt the need to know the consequences of nuclear tests on heredity. Tamm wanted a broad circle of scientists, especially young scientists, to become informed about progress in science. As early as fall 1945 he gave popular-science lectures about atomic energy and wrote popular articles on this topic. Apparently, he felt it incumbent upon himself to inform the broader community about this development.

In 1958, he used the forum afforded by the Nobel Banquet to bring up recent progress in biology as if claiming the right to get involved in it as the quote below shows. At the Nobel Banquet each category is represented by a two-minute speech by the Nobel laureate or by one of them if there is more than one. In 1958, the three physicist laureates decided that Tamm should be the one to speak. In his short speech, he said, "The dividing line between physics and biology is at present a rather sharp one. But a number of impressive recent achievements in biology make one believe that we are perhaps on the eve of an epoch of great discoveries in biology. I venture to express the opinion, that to achieve fundamental success in biology a very close working cooperation of all three sciences, representatives of which are honored by Nobel Prizes, will be indispensable."[11]

After Stalin died, and especially after Khrushchev's originally secret, but soon famous, speech in February 1956 at the Twentieth Congress of the Soviet Communist Party in which he exposed Stalin's crimes, many felt freer and behaved differently than before. Tamm did not change his behavior or the way he reflected on various developments and political issues. He did not need to do either, as he always projected a certain degree of internal freedom.[12] He had high moral principles and was already taking risks in expressing them. However, he preferred taking action when he could expect results to taking action just for the sake of taking actions.

Tamm learned a folk wisdom from his physicist colleague at FIAN, E. L. Feinberg, and liked to repeat it whenever it could be justified: "My God, let me have peace with what I can't change; bravery to fight for what I can change; and wisdom to be able to distinguish between the two."[13] Tamm was already severely ill when in August 1968, following the Prague Spring, Czechoslovakia was invaded by the other Warsaw Pact countries. A group of Soviet scientists, including Andrei Sakharov, prepared a letter of protest against the invasion. It was brought to Tamm for signature and he signed it. Soon afterwards, one of his favorite pupils told him that he should withdraw his signature lest his division of theoretical physics at FIAN suffer from his action, and Tamm withdrew his signature.[14]

Up to 1948, that is, up to his involvement in the nuclear project, the Communist Party did not consider Tamm as a trusted member of Soviet society This was compounded by the leading Soviet ideologue Andrei Zhdanov's personal antipathy toward him. Before the war, Tamm was chairman of the theoretical physics department at Moscow State University. When the University returned from evacuation, Tamm was not given back his chairmanship.

In 1946, Zhdanov eliminated the corresponding member Tamm's name from the roster of the new potential full members of the Soviet Academy of Sciences. On this occasion, an unprecedented episode happened. The outstanding physicist corresponding member M. A. Leontovich, who had been nominated for full membership for the place allocated for a theoretical physicist, turned to the president of the Academy and let it be known that he did not want to become full member if it would mean taking Tamm's deserved place at the Academy.[15] This was a testimonial how much his peers revered Tamm. Leontovich's gesture though did not change Tamm's situation.

Tamm was elected to full membership in 1953, that is, twenty years after his election to corresponding member. His election to full member followed by a few months the successful test of the first Soviet hydrogen bomb. Years later, Sakharov remembered the question Zeldovich had posed about Tamm's exceptionally high value to the project. Zeldovich ascribed Tamm's advantage to Tamm's "high moral level."[16] Even after Tamm had finally been elected full member of the Academy, however, the Soviet leadership continued viewing him with suspicion and did so throughout Tamm's life.

Up to the political changes under Mikhail Gorbachev, it was the general practice that the leadership of the Communist Party exercised veto power over elections to the Academy of Sciences. There were rare exceptions when the academy members did not submissively follow the party instructions; the importance of these rare exceptions appeared magnified in the light of the background of unquestioned dictatorship of the Communist Party.

In 1955, the party leadership wanted to prevent Tamm and Abram Ioffe, the doyen of Soviet physics, from becoming members of the leading body of the mathematics-physics division of the Academy. The party accused Tamm of several negative acts[17]: (1) he did not consider the activities of others with desired objectivity; (2) he made mistakes in compiling the program of the all-union conference on quantum electrodynamics and the theory of elementary particles; and (3) he criticized the ideologically sound activities of the editorial office of the *Journal of Experimental and Theoretical Physics*. To the disappointment of the party organs, however, the elections did not go according to their intentions.

On top of this, the president of the Academy, Aleksandr Nesmeyanov (see chapter 12) supported Tamm's and Ioffe's election to the divisional leadership.

All this happened at the meeting of the physics-mathematics division of the Academy, January 31–February 1, 1955. This meeting went down in the history of the relationship between the Science Academy and the Communist Party as conspicuously atypical for Soviet times. The internal party documents blamed Landau and his circle and Tamm and his circle, and named them the culprits in the opposition to the recommendations of the party. Nesmeyanov's attitude was also noted with very negative remarks.[18] However, from this story it would be a mistake to conclude that the Soviet Academy of Sciences was rebelling against the communist regime or the party leadership in general. It was a rather isolated case, and it was only the physicists who could afford such budding independence due to their privileged position in national defense. But at least it was an early crack in the supposedly monolithic Soviet establishment.

It may seem puzzling that the Communist Party would guard so uncompromisingly its grip over questions that by no means threatened the Soviet regime. The Party and the secret police demonstrated again and again a pathological fear that Western ideas of democracy might penetrate any segment of Soviet society. They were probably right, because any compromise might have become the starting point of a broader demand for democratization in society.

Another telling example occurred later in the same year. A two-day commemorative meeting of the fiftieth anniversary of Einstein's theory of relativity was scheduled for November 30–December 1, 1955. The meeting took place within the framework of the physics-mathematics division of the Science Academy. There was a preparatory commission consisting of Tamm, Landau, Vitaly Ginzburg, and Evgenii Lifshits. Already suspicious, the Communist Party was carefully monitoring the activities of the commission. Party officials criticized the choice of three of the presenters, Landau, Ginzburg, and Lifshits, claiming that they were not sufficiently versed in the theory of relativity, while some whom they considered to be the real experts were not in the program. This is already extraordinary that the Communist Party would be meddling in the details of the scientific program of such a meeting.

The party representatives accused Lifshits in particular of spreading teachings about the expanding universe. The Communist Party considered this notion to be contrary to Marxism-Leninism and a product of imperialistic ideology. Another presenter, Zeldovich, was also accused of idealism (as opposed to materialism); he could not be criticized for not being in the field, but the party did not like his scientific views. Zeldovich indeed was of the opinion that the chemical elements were of billions of years of age, and that did not fit the party line. Again, the party organs criticized the Academy leadership, that is, the Presidium, for its permissiveness. The entire ideological struggle had an ironic character because the Soviet Union at this time was making great strides in the fields of cosmology and relativistic astrophysics, which were demonstrated by the presentations of the very scientists the Party did not want participating in the event.[19]

Grave of Pavel Cherenkov in the Novodevichy Cemetery.

Source: Photograph by and courtesy of the author.

Pavel Cherenkov on Russian postage stamp, 1994.

Igor Tamm, Pavel Cherenkov, and Ilya Frank—the three Nobel laureates in Physics, 1958.
Source: Courtesy of Valentina Berezovskaya, Moscow.

It should be noted, though, that even at the January–February 1955 divisional meeting it was not all defeat by the Communist Party for the candidates for the divisional leadership. Mikhail Lavrentiev was elected academician-secretary—the person in charge—and he was the man the Communist Party wanted for this position. Lavrentiev was a noted mathematician who later rose to the position of vice president of the Academy. His most memorable achievement outside mathematics was the initiation of the gigantic scientific center in Novosibirsk, in Siberia, the Siberian Branch of the Soviet Academy of Sciences.

Soon after Lavrentiev took over the physics-mathematics division, he had to deal with the issue of Nobel Prizes for Soviet scientists. This is of interest to look into not only for general considerations, but also because it greatly concerned Tamm's Nobel Prize in 1958. In the archives of the Academy, there is a Resolution No. 19 from about November 1, 1955, in which it is stated that "the divisional leadership does not find it advisable to nominate Soviet scientists for the Nobel Prize since this prize cannot be considered international as demonstrated by the lack of Nobel awards to outstanding individuals of science and culture of our country (D. I. Mendeleev, L. N. Tolstoy, A. P. Chekhov, M. Gorky)."[20]

In order to understand the origin of this odd resolution, its circumstances need to be recalled. In late 1955, there was disappointment in Moscow that the expected Nobel Prize for Nikolai Semenov did not materialize (it would, though, the next year, in 1956) and the physicists thought they would have nothing to lose by their protest. Unbeknownst to them, by then, the Cherenkov Effect was being considered by the Physics Nobel Committee; the prize for it would be awarded in 1958. The possibility of a Semenov Nobel Prize was still not dismissed, and the chemistry division did not issue a statement similar to the one by the physics-mathematics division. The mathematicians did not need to bother, because there is no Nobel Prize in mathematics. It is noteworthy that the physics-mathematics division discussed the matter as a question of divisional nominations, whereas the Nobel Prize institution excludes collective nominations. Here, however, there is evidence that the Nobel Prize institution was willing to tolerate, even encourage, the involvement of collectives for getting nominations for Soviet scientists—they appeared eager to have Soviet participation in—as they termed it—the "Nobel movement."[21]

The Soviet organizations—the Science Academy and the Communist Party—took it upon themselves to interfere in the nominations originating from the Soviet Union. They opposed nominations for Landau and Tamm. In both cases the reason was the role these two scientists had played in demanding greater freedom for academicians in Science Academy matters.[22] It is well documented that in 1958, Sheldon Glashow, a future Nobel laureate physicist, having been awarded the necessary financial support, wanted to spend some time in Tamm's group. Glashow wrote to Tamm asking to be accepted as a visitor. In turn, Tamm asked the Academy for permission, but he never received a response, neither permission nor a denial. It is impossible to know exactly where his request ran aground, at the Academy, the Interior Ministry, the Communist Party, or the Ministry of Foreign Affairs, because all these institutions could be involved in deciding such questions.

Tamm's possible interaction with foreign scientists was frowned upon. It did not help this particular case that Glashow's parents had immigrated to America from Russia (even if they were escaping the anti-Semitic czarist regime at the beginning of the twentieth century).[23] Tamm was never informed about the reason he could not receive the foreign visitor. For quite some time, Soviet diplomats encouraged Glashow, telling him that he would receive his Soviet visa, though not for a visit to Tamm but to another group in Moscow.[24] However, his Soviet visa never materialized, and Tamm received no explanation either. It was a humiliating situation for Tamm, and it puzzled his colleagues in the West.[25]

There was, though, pressure on the Academy of Sciences to advance the Nobel recognition of the discovery of the Cherenkov radiation. The political considerations favored Cherenkov's sole candidacy, whereas correct scientific reasoning favored also including the two theoreticians, Tamm and his associate Frank. The Nobel Committee for Physics invited Igor Kurchatov to submit nominations of worthy candidates for the 1958 physics prize. Kurchatov asked Vitaly Ginzburg and E. K. Zavoiskii to prepare the scientific justification for the nomination of Cherenkov, Frank, and Tamm. Kurchatov then sent the nomination to the Science Academy. The Academy was supposed to forward the nomination to the Ministry of Foreign Affairs to arrange for the documentation leave the country and get to Sweden. It seems bizarre that Kurchatov had to follow such a complicated procedure rather than simply mail his nomination to Stockholm. To understand how things worked at that time and what it entailed finally for Tamm (and Frank) to become laureates, I narrate further details about the events leading to the 1958 physics prize.

At my request, the long-time former secretary of the Nobel Committee for Physics, Anders Bárány, looked up the 1950s in the archives at the Royal Swedish Academy of Sciences.[26] His findings are summarized as follows. Cherenkov was nominated regularly in the 1950s, but it was not until after the discovery of the antiproton in 1955 at the University of California, Berkeley, that the Nobel Committee for Physics started to seriously consider him for the prize.[27] This was due to the important role the Cherenkov counters played in the antiproton discovery. This was also the starting point for the more widespread use of the technique. Lamek Hulthén, Professor of Mathematical Physics at the Royal Institute of Technology in Stockholm, wrote a nomination for Cherenkov in 1956. It was repeated in 1957, when a nomination for Cherenkov came also from the Soviet Union; it was by V.N. Kondratev. The same year Wolfgang Pauli nominated Cherenkov together with Tamm. Erik Ingelstam, Professor of Optical Physics, also at the Royal Institute of Technology, nominated Cherenkov, "possibly together with Frank or with both Frank and Tamm." Finally, Vladimir Fock in Leningrad nominated Cherenkov together with Sergei Vavilov. Fock must have known that Vavilov had died in 1951, but he apparently did not know that the rules of the Nobel Prize excluded posthumous awards (in the Soviet Union, posthumous awards were common). In 1958, Hulthén again nominated Cherenkov, but this time together with Vavilov—in his case we must suppose that he did not know that Vavilov had passed away. At the end of his nomination, Hulthén even excuses himself for previously not having understood the important part played by Vavilov! Ingelstam repeated the same nomination he made in 1957, that is, Cherenkov, "possibly with Frank or both Frank and Tamm."

Cherenkov's candidacy was reviewed as early as 1952 by Axel Lindh, a member of the Nobel Committee for Physics. He expanded his review in 1955, and in 1957, he wrote a detailed final report. In this he argued that Vavilov had given Cherenkov important impulses to continue his investigations, although Vavilov's explanation of the effect was wrong. Frank and Tamm, he said, had made the correct interpretation, and it was through their work that the Cherenkov Effect became known as a usable detector. Lindh suggested that the Committee seriously discuss a prize in this area.

The Nobel Committee for Physics writes in its reports to the Royal Swedish Academy of Sciences in 1956 and 1957 that they realize the importance of the Cherenkov Effect for detector work. In 1958, they have a discussion about a possible Nobel Prize for the discovery of the antiproton and stress the fact that Cherenkov counters played an important role in it. Following the suggestion in Lindh's 1957 report, they discuss a prize for the discovery and explanation of the Cherenkov Effect. They also mention that Vavilov had died in 1951. The Nobel Committee then proposes a shared Nobel Prize in Physics to Cherenkov, Frank, and Tamm, and the Academy follows this proposal. It is noteworthy that there is a reference in the archival documents to a letter signed by Landau, Nikolai Andreiev, and Abraham Alikhanov stating that Cherenkov, Frank, and Tamm should share the prize (see also chapter 6). The Nobel Committee received this letter some time in 1958, and it was not handled as a formal nomination.

Naturally, Tamm was pleased to be awarded the Nobel Prize, but he maintained that he received it for less significant work than what he considered to be his principal discovery, the phonons. He was keenly aware of some missing names on the roster of Nobel laureates. He considered three omissions especially hurting: two were Leonid Mandelshtam and G. L. Landsberg for the discovery of what the Russian literature calls "spectroscopy of combination scattering" and is known as "Raman spectroscopy" in the rest of the world. The third omission was Zavoiskii for the discovery of electron paramagnetic resonance (EPR). Tamm did not find his own award fully justified and declared: "To some others the prize was given for something accidental, including me."[28] He repeatedly nominated Zavoiskii for the Nobel Prize. Mandelshtam occupied a special place in Tamm's thinking; he revered him as his teacher and said frequently, "I owe him everything; everything."[29]

Let us, however, return to the story of political interference in connection with the 1958 Nobel Prize for the Soviet physicists. After the Nobel award had been announced, the three awardees for a while could not respond to the Swedish queries about whether or not they would be able to accept it, let alone attend the ceremonies. That same year the Swedish Academy had awarded the Nobel Prize in Literature to the great Soviet writer Boris Pasternak.[††] Soviet officialdom found it difficult to fathom seeing the Soviet physicists together with the "heretic" Pasternak at the same ceremony. The physicists could give their affirmative response to the Swedes only after Pasternak had been blackmailed into declining to accept his Nobel Prize (it was hinted that if he accepted the award and went to Stockholm to receive it, he might not be let back into the country).[30] It seems that nothing was left to chance in the Soviet regime. When the three Soviet physicists returned home after the Nobel

---

†† The Swedish Academy is responsible for the Nobel Prize in Literature. It has eighteen members: writers, literary scholars, historians, and a prominent jurist. It is independent of the Royal Swedish Academy of Sciences, which is responsible for the Nobel Prizes in physics and chemistry.

festivities, there was a meeting with the representatives of Moscow society at the auditorium of the Moscow Museum of Technology. Tamm was the principal speaker. After he finished, someone shouted, obviously by prearrangement: "Long live Soviet science, which under the direction of the Communist Party of the Soviet Union is leading the way for the whole humankind to world peace." The audience duly greeted the slogan with applause.[31]

After the Nobel episode, Tamm's brave attempts to democratize the Soviet Academy of Sciences continued. One might have thought that he gained encouragement from his Nobel award, but it is impossible to discern any difference between the way he acted before and after his great distinction. At the end of 1956, for example, all the divisions of the Academy were holding their own meetings in preparation for the next election of the president of the Academy. Following the separate divisional meetings, the General Assembly of the Academy was supposed to elect the president, and no opposition was anticipated, not only to re-electing the current president, but to the procedure either. Based on past experience, it was expected to be a perfunctory exercise. Aleksandr Nesmeyanov was the current president; he was reasonably popular; and the Communist Party did not want to replace him, either. He was the only candidate for the post.

However, at the physical-mathematical divisional meeting, Tamm criticized the way the general assemblies of the Academy were conducted. There was hardly any discussion, and the elections of the Academy leadership were always predetermined by higher party authority. There was no dissent ever in the voting. Tamm wanted the members of the Academy to have the right to a free debate. Of course, it did not escape anybody's attention that what Tamm said about the Academy of Sciences was applicable to the whole Soviet system, so Tamm's views could be taken as criticism of its foundations. Tamm mentioned in particular the re-election of Nesmeyanov whose first five-year term was ending. Since it was the intention of the higher party organs to re-elect him, it was taken for granted that this is what would happen; and, in fact, it did. But Tamm managed to make his point and suggested that the president first make a report to the membership before his re-election was decided. Tamm spoke at the divisional meeting, and the physicists and mathematicians agreed with him, but all the other divisions were for re-electing Nesmeyanov *without* further discussion.

When the voting then came, Nesmeyanov won, but this time it was not unanimous, making this election an unprecedented event. The general assembly decided that the re-elected president should make a report at their next meeting outlining his plans for the future. This was also a first. Incidentally, Tamm and Nesmeyanov were on friendly terms, and Tamm made it clear that his criticism was not directed against Nesmeyanov as a person, but was instead in connection with the presidency. Nesmeyanov did not take Tamm's move personally; in fact, he found it justified, and the two remained on friendly terms for the rest of their lives.[32]

Up until the late 1950s, elections of new corresponding members and full members at the Soviet Academy of Sciences were rare occasions; later they took place once every two years. In the 1958 elections, party interference was still substantial. As before, the intention was that the party organs rather than the Academy membership should decide the outcomes. The Central Committee of the Communist Party had a section dealing with science and higher education to exercise party control over the Science Academy and the institutions of higher learning. Most of the associates of this section had been trained as scientists. There was, then, the so-called communist fraction of the Academy, that is, the group of party members among the academicians who got together before every election and decided whom they would support, following the intentions and recommendations of the party leadership. From this point, it was a clear path to the elections.

This was, however, 1958, and things could no longer be taken for granted as far as the wishes of the party were concerned. The party organs saw the writing on the wall and were taking precautions to prevent election results that from their point of view could be disastrous. The documents revealing the behind-the-scenes machinations of the party remained classified for a long time. It was only in 1994 that the original list of names was published, along with the party's evaluation of the situation in the division of the physicist and mathematician academicians. A document stated, in part: "In the Division of Physico-mathematical Sciences of the Soviet Academy of Sciences, an incorrect situation has formed according to which a group of non-party-member scientists, in particular academicians L. A. Artsimovich, A. I. Alikhanov, L. D. Landau, M. A. Leontovich, and I. E. Tamm, tend to ignore the intentions of party organizations and try to impose their own high scientific authority over party influence, especially in the questions of personnel that *have always been the non-alienable concern of the party* (italics added)."[33]

The document listed the names of those physicists-communists who had been defeated in previous elections as a result of the insufficient vigilance of the Presidium of the Academy. The defeated candidates had been recommended by the party's Central Committee; in their stead, pupils of Landau and a pupil of Tamm had been elected along with other non-party-member physicists. The lamentation goes as far as hinting at that party membership could disadvantage physicists from being elected to membership of the Academy. This party document could be considered to be a testimonial that, at least on this occasion, the Science Academy, and in particular its division of the physical-mathematical sciences, was an island of democracy in an utterly undemocratic environment; and the scientists whom the party document described as "perpetrators" were the unsung heroes of democracy.

The next illustration of Tamm's demeanor comes a few years later; the story also shows the limits of how far the regime let him voice his opinion.[34] On February 6, 1962, Petr Kapitza gave a speech at the general assembly of the

Soviet Academy of Sciences as part of the discussion of the account by E. K. Fedorov, Secretary General of the Academy. Kapitza lamented the damages from the denial of cybernetiscs in Soviet science and the nonunderstanding of Einstein's famous equation, $E = mc^2$ (E energy, m mass, c the speed of light in vacuum). He criticized the Soviet "philosophers" for their errors, though he carefully avoided any criticism of the philosophy of Marxism-Leninism. He mentioned in passing the philosophers' erroneous actions in the case of the uncertainty principle in quantum theory and the theory of resonance in the investigation of chemical reactions. Due to the private initiative of a section editor, the periodical *Ekonomicheskaya Gazeta* (*Economics Journal*) published Kapitza's speech. This was followed by an uproar in party press.

Dmitrii Ivanenko and Igor Tamm (lower right corner) in discussion at a meeting. In between the two, on the left, Rudolf Peierls; Victor Weisskopf is on Tamm's right.

Source: Courtesy of G. A. Sardanashvily, Moscow.

Tamm felt that Kapitza needed support and wrote a strong letter to *Ekonomicheskaya Gazeta*. He went one step further than Kapitza. He ascribed the errors not just to the selected individual philosophers; rather, he accused the whole erroneous philosophy of science in the Soviet Union. He claimed that the Soviet philosophical literature broadly distributed negations of such important scientific achievements as the theory of relativity and quantum mechanics. He noted though that only an insignificant fraction of physicists followed the erroneous positions of the philosophers—those who were afraid of getting accused of idealism. Had it not been so, Tamm stressed, it would have been impossible to apply atomic energy for practical purposes, since nuclear technology is based on quantum mechanics and the theory of relativity.[35]

Just as Tamm had felt the need to give additional support to Kapitza, now the *Gazeta* felt the same was needed for Tamm. They turned to another internationally renowned authority, the physicist Vladimir Fock, who produced a strong supportive statement in which he unambiguously joined Tamm's and Kapitsa's positions. It turned out, however, that Tamm and Fock as well as the *Gazeta* went a little too far, and the party authorities prevented the Tamm and Fock letters from being published. They were duly filed in the archives. *Ekonomicheskaya Gazeta* was also a party publication, and the section editor, I. D. Sobko, who initially published Kapitza's statement, was moved to another job and was never heard from again.

During the late 1960s, as Tamm grew increasingly frail, Andrei Sakharov came to visit him from time to time. The two discussed poems in which at the time Sakharov was deeply interested and talked about politics. One of their topics was the role of individuals in shaping history. They tried to answer questions of "what if..." They tried, for example, to chart the fate of their country if instead of Stalin, Lev Trotsky had become the supreme leader of the Soviet Union. Tamm estimated that there would have been ten times fewer victims, while Sakharov opined that the number of victims would have been a hundred times smaller. Tamm compared Trotsky with Beria as both could be rational and perceptive. Tamm met Trotsky when Trotsky was responsible for national defense in the Soviet government. Trotsky visited the plant where Tamm worked at the time and he easily grasped ideas that must have been rather foreign to him. Concerning Beria, Tamm noted that Beria could completely ignore whatever disadvantage someone had as long as he could be useful for Beria's goals, such as nuclear weapons. For example, when Tamm wanted to involve the excellent physicist Leontovich in the work on the hydrogen bomb, Beria was informed about Leontovich's background, which made him unsuitable in the eyes of the security organs. However, Beria was not shaken and decided on the spot to involve Leontovich.[36]

Another celebrity who came to visit Tamm was the renowned author Alexander Solzhenitsyn. They talked about a great variety of topics. Solzhenitsyn was working on his book *First Circle* and wanted to learn about how Stalin looked in the Mausoleum. The writer had never visited the Mausoleum, and by this time Stalin had been removed from it. When the anti-Solzhenitsyn campaign started in the Soviet Union, Tamm felt solidarity with him and called him "poor Solzhik."[37] When in 1970 Solzhenitsyn was awarded the Nobel Prize, he came to Tamm asking for his advice on what to do.

Tamm considered himself to be a poor organizer, but his section at FIAN, his *Teorotdel* (Theoretical Section) appeared very well organized, as if it had organized itself—which may have been the attribute of a good organizer. Tamm respected good organizers, among them Igor Kurchatov and Abram Ioffe. He respected the ability to build experiments and construct apparatuses. He counted Mandelshtam, Landsberg, and Zavoiskii among those virtuoso scientists who had such traits. Further, he especially appreciated the ability to provide criticism. He considered Wolfgang Pauli, Paul Ehrenfest, and Lev Landau to be excellent critics. He found scientific criticism to be a very fruitful necessity for healthy scientific progress. He stated, "Without them [the critics], our progress would tremendously slow down! It's a pity that not everybody understands this."[38]

Tamm considered sectarian zealotry, pseudoscience, and unprincipled complicity to be the most dangerous enemies of science, and called them the three-headed dragon. He hated Lysenko and his associates, including a man called Nikolai Nuzhdin whom Khrushchev badly wanted to be elected an academician. One can imagine how the world-renowned Soviet nuclear physicists regarded Lysenko when the agronomist told Khrushchev that underground nuclear tests frightened the Earth and as a consequence would stop food production. Further, he declared that sparrows could move around so quickly on their tiny feet, even on snow in wintertime, because their feet were fueled by atomic energy. Khrushchev warned the scientists that if Nuzhdin were not elected, he, Khrushchev, would disperse the Academy. During the next election campaign, Tamm, Sakharov, and the biologist Vladimir Engelhardt all argued against electing Nuzhdin, and the Academy did not elect him. Khrushchev was no longer in the position to do away with the Academy because in October 1964 he had been retired by his Politburo colleagues. It was indicative of the changes when in 1964, in preparation for the national holiday of November 7, among the series of portraits of Soviet heroes along the spacious Lenin Avenue in Moscow, Lysenko's portrait was quietly replaced by that of Tamm.

The unusual modern tombstone over Tamm's grave at the Novodevichy Cemetery is the work of sculptor V. Sidur. Years before Tamm's death, Vitaly Ginzburg took Tamm to visit Sidur's workshop. Tamm had mixed feelings about the sculptor's works; some he liked; others he did not understand. Following Tamm's death, the family asked Sidur to produce a tombstone. The authorities wanted to erect a bust in traditional style, but the family felt this was too static to represent Tamm's mobility. They preferred Sidur's modern statue, which may or may not correspond to Tamm's taste in art, but well expresses his maverick nature, contrasting with the many tombstones around in classical style.[39]

The three-starred (the three golden stars of a Hero of Socialist Labor) Yakov Zeldovich.

Source: Courtesy of Olga Zeldovich, Moscow.

2

Yakov Zeldovich

SOVIET PROMETHEUS

> Yakov B. Zeldovich (1914–1987) was one of the most intelligent participants of the Soviet nuclear program. He went from secondary school directly to postgraduate studies, skipping university education. Most of his career he spent on weapons and explosives, but never abandoned basic science. During the last two decades of his life, he focused on astrophysics and cosmology. His contributions often followed up discoveries by others. He recognized the need for higher mathematics for scientists and engineers and wrote outstanding books to enhance mathematical literacy.
>
> He was patriotic in strengthening the defense of the Soviet Union, and was recognized by three stars of Hero of Socialist Labor, four Stalin Prizes, and other awards. He was not a "dissident," avoided taking stands, and appeared expedient, but was devastated when he realized that the leaders of the country might use its weapons for purposes that he condemned. He thrived on his successes, enjoyed life, and died when he seemed still full of energy and plans. Today, there is no public remembrance in Russia of this most decorated giant of the Soviet era.[1]

Yakov Borisovich Zeldovich* loved to give lectures and was a popular speaker, vigorous and businesslike. On one occasion he arrived in a packed lecture hall a few minutes before the start of his presentation, took off his jacket, placed it on a chair, and went to see some people in the building. The audience was left with his jacket with the three Hero of Socialist Labor stars on it. Some daring youths tried on the jacket with the three stars. We cannot know whether he purposely left his jacket there on display before his presentation, but it is known that he prominently displayed his distinctions when he wanted to impress some bureaucrats with whom he had to deal.

---

* Prometheus seems a fitting comparison to this great scientist. Prometheus was a Titan in Greek mythology who stole fire from Zeus and gave it to mortals.

The three stars were awarded to Zeldovich, one each in 1949, 1953, and 1956, during a seven-year period marked by three important events in the development of the Soviet Union's nuclear weapons capabilities: the first atomic bomb, the first hydrogen bomb, and the perfection of the hydrogen bomb with the application of radiation implosion. The latter was Andrei Sakharov's "third idea," which may have emerged from a discussion between Zeldovich and Sakharov. Zeldovich's career was almost a miracle, but it was yet more miraculous that he could embark on it in spite of his family background. His maternal grandparents and his mother's sister lived in Paris. His father's sister was arrested in 1936. His brother-in-law was executed in 1937 during the Great Purge, and his sister-in-law lived in America. But he was among the most valuable assets of Arzamas-16, the secret project developing the Soviet nuclear weapons, and the project served as his protection.

Many considered Zeldovich to be the most intelligent of all the scientists of the Soviet nuclear elite. He was exceptionally versatile and changed fields almost as others changed topics in a conversation. He maintained that when entering a new field, it took considerable effort to learn 10 percent of the general information on the new subject; but even that 10 percent made it possible for him to begin independent work in it. Then, it was not difficult to reach a 90 percent understanding of the new subject and that level of knowledge carried pleasure and facilitated creativity. The next 9 percent was almost impossible to learn, according to Zeldovich, who added meekly that not everybody had the ability to master it. The last 1 percent he considered hopeless. His advice was, rather, to switch fields again after the first 90 percent.[2]

It appears that Zeldovich did not think it was too hard to move from one field to another, but, of course, it takes a good scientist who is capable of independent thinking in the first place to do this. About such changes, he was in accord with the evaluation by another outstanding theoretical physicist Walter Gilbert. The Nobel laureate stated that the only special trait required to make one capable of such changes was the ability to make correct decisions, that is, to distinguish between right and wrong. If one has this ability, the rest, acquiring knowledge of a new field, is not difficult.[3] Of course, again, this is true for a Gilbert or a Zeldovich, but not necessarily for the majority of people engaged in scientific research.

Zeldovich was one of the most decorated Soviet scientists, so it is surprising that his only public commemoration, in the form of a bust, is in Minsk, the capital of Belarus, one of the independent successor states of the Soviet Union. The explanation for the Minsk sculpture is that he was born in that city. It was Soviet law that the birthplace of two-time heroes of the Soviet Union or of Socialist Labor was obliged to erect a bust in their honor. Russia today is full of the most diverse forms of commemoration of Soviet-era heroes, not only scientists but military leaders and even communist politicians. Commemoration of many of Zeldovich's peers can be found in the names of research institutes, streets, and squares, where busts and statues of them abound. But there is none for Zeldovich in Russia proper, with the exception of his tombstone in the Novodevichy Cemetery.

Zeldovich was born in Minsk on March 8, 1914; however, his family moved to St. Petersburg within half a year. St. Petersburg was renamed Petrograd as World War I broke out because the Russian czar did not want the capital of his empire to carry a German-sounding name. The next name change came under the Soviets, when the city was renamed Leningrad. The name was changed back to St. Petersburg in 1991.

Zeldovich came from an intellectual Jewish family. His father, Boris Naumovich Zeldovich, was a lawyer and his mother, Anna Petrovna Zeldovich, was a graduate of the Sorbonne, the famous university in Paris. She was a translator from French and was a member of the Writers' Union. One of Zeldovich's daughters, Marina Ovchinnikova, characterized her as "small and energetic, sharp and quick-witted, our restless grandmother."[4] Zeldovich acquired a deep love and knowledge of literature at home and often cited poetry in his scientific articles. Boris Pasternak was his favorite poet; it was therefore appropriate that his attitude toward science in general and discoveries in particular were compared to Pasternak's, whom Anna Akhmatova described as follows (in Stanley Kunitz's translation)[5]:

> He has been rewarded by a kind of eternal childhood,
> With the generosity and brilliance of the stars;
> The whole of the earth was his to inherit
> And his to share with every human spirit.

Zeldovich started as a prodigy and learned more via self-education than from formal instruction. He was ten years old when he entered secondary school (high school in American terminology), and sixteen when he graduated. He was too young for college. Already at the age of twelve he was considering his future, and he talked with his father about the different branches of knowledge. Mathematics, physics, and chemistry attracted him, but characteristic of his realistic assessment of his abilities, he felt he would not be exceptional in mathematics. He vacillated between physics and chemistry and, finally, did not have to decide between them. But it is interesting that his school experience gave him the impression that physics was a closed subject, with all the fundamental theories in place. His physics teacher presented it as a classical subject, reciting Newton's laws in both Russian and Latin. In contrast, chemistry did not have a fundamental theory; students learned about concepts such as catalysis and valence, and to Zeldovich, it seemed too empirical. He felt that a bridge was missing between physics and chemistry. He finally discovered the connection in the atomic theory discussed in Ya. I. Frenkel's book *Structure of Matter*.

With this background, it is not surprising that Zeldovich found his intellectual home at Nikolai Semenov's Institute of Chemical Physics. Semenov's goal was to establish the connection between physics and chemistry, exactly what Zeldovich had found wanting. However, Zeldovich started out as a laboratory assistant at the Institute of Mechanical Processing of Mineral Resources, which was concerned with a plethora of different fields in the geosciences, mineralogy, chemistry, and technology. Zeldovich developed great respect for the great scientists in these areas. Even in his later writings, he singled out A. E. Fersman, whose most important contribution was mapping and discovering various metals, minerals, and other useful materials in the extended territories of the Soviet Union. He found radium and uranium ores, among others, and his research included radiation geochemistry and geoenergetics. Years later, the availability of uranium ores would be among the factors determining the success or failure of the Soviet atomic bomb program. The field impressed Zeldovich, but he did not continue his studies in the geosciences.

Although at the Institute of Mechanical Processing of Mineral Resources, he was only one of the technical personnel, his tasks were far from routine, and he felt that there were possibilities for growth. He was genuinely interested in experimental observations and captivated by visible effects. His broad interest in science was expressed by his participation in excursions to other research institutes. In one such excursion, he visited Semenov's department of chemical physics (as it was then) of the Leningrad Institute of Physical Technology headed by Abram Ioffe (today, it is the Ioffe Institute). Zeldovich was so much taken by what he saw that he decided to work there in his free time.

Things progressed quickly, and within a couple of months he became an employee of what had already become the Institute of Chemical Physics, with Semenov as its director. When Zeldovich announced his intention to change fields, his previous work place asked the Institute of Chemical Physics for a vacuum pump in exchange for him, and the pump was duly delivered. In Semenov's Institute, Zeldovich felt as though he was in the midst of scientific progress on a world scale. In 1932, for example, James Chadwick's telegram informed the attendees of an institute seminar about the discovery of neutrons, and the seminar responded with an enthusiastic telegram. At that time the Soviet scientists were still part of the international community. One of Zeldovich's senior colleagues, Yulii Khariton, had been Chadwick's doctoral student in Cambridge, England.

Zeldovich felt that he would need further formal education, He signed up for a correspondence course at the Faculty of Physics and Mathematics of Leningrad University. This allowed him to continue in his job. However, his previous misgivings proved correct; to him, the speed and scope of structured instruction was too confining and inadequate. He ended his studies and never graduated. Instead, he learned from the excellent associates of Semenov's institute, including Simon Roginskii, Yulii Khariton, Matvei Bronshtein, and Semenov himself. Zeldovich participated in experiments, read about theories to interpret the experiments, and attended the seminars. He learned a lifelong lesson about the usefulness of learning about experiments even though he preferred doing theoretical work.

Skipping university graduation, he pursued a PhD equivalent, the Candidate of Science degree, on the basis of his research on adsorption and catalysis. When he defended his dissertation in September 1936 he was twenty-two years old. His road to becoming a scientist seems more unorthodox today than it was at the time, and it was not unique either. At that time, in Germany, for example, to earn a doctorate in physics, one did not need a formal university diploma; rather, students could go directly for their doctorates. In contrast, doctoral studies in chemistry had to be preceded by a university diploma. Zeldovich used to be asked if he felt uncomfortable because of not having a university diploma, and his response was that he often did, but not after his election to the Science Academy.

Zeldovich continued his steep rise in the scientific hierarchy, and it was fortunate for him that his field was the fast-moving area of explosives and detonations. His next step was to earn the Doctor of Science degree, which bears only a superficial resemblance to the German Habilitation, the requirement for being allowed to deliver courses at university level. The Doctor of Science degree was more a certificate about significant achievements in independent research. Zeldovich chose the theory of nitrogen oxidation as the topic of his dissertation, which he defended at the end of 1939. One of the reviewers of his dissertation was the famous electrochemist Aleksandr Frumkin (today, there is a Frumkin Institute of Electrochemistry in Moscow). Zeldovich was only twenty-five years old. It is noteworthy that his colleagues wanted him to be elected corresponding member of the Soviet Academy of Sciences even before he earned his prerequisite higher doctorate. It did not happen, but it was a sure sign of his recognition as a scientist among his peers.

It is to Zeldovich's credit that his success in making jumps in his education and career never made him conceited. Even much later in life, his close colleague and equally exceptional physicist, Andrei Sakharov noticed that "it often seemed to him [Zeldovich] that he was a dilettante and was not sufficiently professional in regard to some questions and he made enormous efforts in order to overcome his deficiencies." Sakharov spoke these words when he delivered a eulogy on December 7, 1987, at a meeting devoted to Zeldovich's memory.[6]

Young Yakov Zeldovich.

Source: Courtesy of Olga Zeldovich, Moscow.

In Semenov's Institute, Zeldovich became a member of Khariton's laboratory, which excelled both in experiment and in theory. As a theoretician, Zeldovich considered himself to be Lev Landau's pupil, although they were not connected organizationally. Landau and his group operated in Petr Kapitza's Institute of Physical Problems (Institut Fizicheskikh Problem [IFP]). But Landau had a nationwide impact on the development of theoretical physics in the Soviet Union. His "teorminimum" exam—a set of comprehensive, tough tests in theoretical physics and mathematics—were famous, and physicists were welcome to take it regardless of their affiliation. Zeldovich did not take the exam, but he was at the level of those who had. Looking back, Zeldovich characterized Landau's talent as "harmonious, and his judgment harsh, but almost always fair."[7]

Even when he was already a much-decorated physicist, around 1960, Zeldovich enjoyed visiting Landau's theoretical physics course at Moscow State University. At the time, Zeldovich was still at Arzamas-16, but there was a moratorium on nuclear testing, and his schedule was more flexible than usual. Also, he was feeling increasingly that he should be moving away from nuclear physics and Arzamas-16, and back to Moscow and fundamental research.

When in 1962, soon after Landau's tragic car accident (see chapter 5), Zeldovich visited Landau, he wished him a fast recovery and that he "become the former Landau again." Landau's response reflected his old self: "I'm not sure whether or not I will become the former Landau, but at least, surely, I will become Zeldovich."[8] Landau meant, of course, that even if he were not able to become his old great self, he was certainly be able to become at least a lesser physicist. It was as witty a response as it was unfair.

In reality, Landau held Zeldovich in high esteem. In 1946, Landau gave Zeldovich a helpful recommendation, contributing to his election to corresponding member of the Soviet Academy of Sciences. Landau did not make such recommendations lightly, and his endorsement carried weight. Zeldovich was thirty-two years old when Landau wrote these words about him:[9]

> Ya. B. Zeldovich is without doubt one of the most gifted theoretical physicists of the USSR. The big set of his works in the field of the theoretical investigation of burning is especially worthy of mention. These studies are the best and most important in this field and not only in the USSR but world-wide.
>
> His works are characterized, along the application of the "usual" techniques of theoretical physics by invoking also hydrodynamics. It is very rare among theoretical physicists that someone might be able to exploit both areas, but this is characteristic and a valuable feature of Zeldovich's activities, enabling him to attack problems that neither the experts of hydrodynamics, nor the "typical" theoretical physicists might do.
>
> It is stressed that Zeldovich's scientific activities are yet far from their zenith. On the contrary, his works demonstrate continuous scientific development.

Zeldovich had a long engagement with nuclear physics, which started years before the Soviet nuclear weapons program had begun. Right after the discovery of nuclear fission at the end of 1938, he and Khariton started working on this topic on their own initiative and outside their official work hours. They were intrigued by the new direction in physics and its possible applications. It was also a natural continuation of studies at the Institute of Chemical Physics that were related to chain reactions and, in particular, to branched chain reactions (see chapter 8). The process generated by the neutron bombardment of uranium was a branched chain reaction. They dealt with a variety of aspects of nuclear fission, including the question of critical mass. They began publishing papers in 1939, but by the time they had immersed themselves in this area, such works had become classified and could no longer be published.

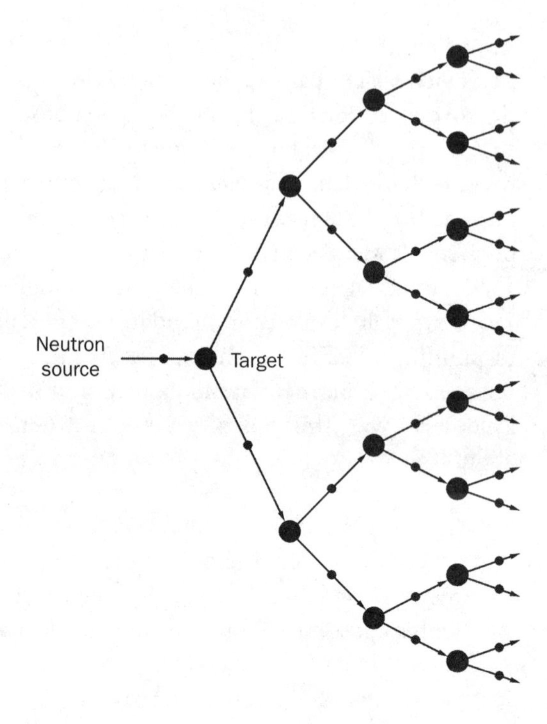

Scheme of a branched nuclear chain reaction: the large spheres represent uranium-235 nuclei and the small ones, neutrons. The products of the fission of uranium are two smaller nuclei that are not shown in the scheme.

During the first two years of the war, 1941–1943, Zeldovich was involved with a broad range of traditional weapons. First, he perfected the charges for antitank weapons to ensure their penetration of the tank's armor. When he reported on his innovations to the artillery officers, they suggested that he work on improving the famous powder-charged rockets, known by their popular name *Katyushas*. Here, Zeldovich and his colleagues, including another outstanding physicist, Ovsei Leipunskii, applied their fundamental physics to making the *Katyushas* the fearsome weapons of the Red Army.[10]

The Institute of Chemical Physics was evacuated, moving from Leningrad to Kazan for the most critical period of the war. When it was decided to move it back, it did not return to Leningrad; rather, Moscow became its seat of operation. This facilitated the concentration of nuclear research under the general direction of Igor Kurchatov, who invited Semenov, Khariton, and Zeldovich when he started recruiting participants. The first Soviet uranium-graphite reactor began operating on December 25, 1946, almost exactly four years after the American one on December 2, 1942, in Chicago.

Soon after the Institute of Chemical Physics re-established itself in Moscow, Zeldovich was appointed as head of its theoretical department, and he served in this position between 1946 and 1948. In this period, he was also a professor at the Moscow Engineering Physics Institute. In 1948, there were big changes in Zeldovich's life. During the preceding years, he had continued publishing papers on nonclassified research, and his publishing activity got him into trouble at this time, although it might be argued that this was merely an excuse to construct accusations against him. It was part of the anti-Semitic campaign, which was not officially spelled out as being against Jews; rather, it was against "cosmopolites." Zeldovich and Leipunskii were accused of disclosing state secrets in their publications. Even their non-Jewish superior at the institute was accused of overlooking those secrets and issuing the permission to publish. Although Zeldovich by then worked mostly in the secret nuclear laboratory, he was fired from the Institute of Chemical Physics, and moved full time to Arzamas-16. Leipunskii who was a lesser name in the atomic bomb project, was ordered by the director of the institute to move to the far-away Semipalatinsk Proving Ground in Eastern Kazakhstan, which became a good hiding place for him.

Both Zeldovich and Leipunskii became members of projects under the top government figure Lavrentii Beria, whose direct supervision served to protect them. Beria's priority was the success of the atomic bomb project; he was not interested in concocted accusations against its scientists. However, even Beria could not have stopped the anti-Semitic measures in which some of his subordinates in the secret police participated with zest. The initiator of the campaign had been Stalin himself.

Zeldovich became the principal theoretical physicist of the Soviet atomic bomb project at Arzamas-16. He also gave a presentation about the possibilities of nuclear fusion reactions on December 17, 1945, at a meeting about the development of the hydrogen bomb. His coauthors were I. I. Gurevich, I. Ya. Pomeranchuk, and Khariton. At the time, the Soviet leadership had not yet reached a decision to go forward with this development. They wanted to focus on the creation of their atomic bomb, which was subsequently successfully tested at the end of August 1949. The Soviet decision to develop the hydrogen bomb was made on February 26, 1950, a mere four weeks following President Truman's announcement on January 31, 1950, about the continuation of research in the United States on all atomic weapons, *including the hydrogen bomb.*

The initial Soviet model was similar to the American "Super," as the Americans often called the hydrogen bomb in the initial stage of their work. In this design, an atomic bomb at one end of the device would start the thermonuclear reaction. Extensive calculations were performed for this model—just as in the American program. The work was conducted under the leadership of Zeldovich, and a large group of physicists and mathematicians assisted him. Like the Americans, the Soviet scientists eventually found that this initial model lacked feasibility.

For the hydrogen bomb program, another theoretical group joined Arzamas-16. It was from the Physical Institute of the Academy of Sciences (Fizicheskii Institut Akademii Nauk, [FIAN]); it was led by Igor Tamm and included Andrei Sakharov. They contributed fruitfully to the development of the Soviet hydrogen bomb (discussed in chapter 3).

Zeldovich was involved both in the development of the first the atomic bomb, and, subsequently, the hydrogen bomb. The principal task was creating ever-more-powerful nuclear charges and making them lighter, smaller, more easily deliverable, and economical. The need to economize was always an important consideration with the especially precious tritium.

Zeldovich remained at Arzamas-16 until October 1965. He dreamed about returning to pure science. Even during the years of the most intense weapons work, he continued fundamental research, traveled to Moscow from Arzamas-16 to visit research institutes, attended seminars, and gave presentations there. He kept up with world literature in his fields, and when he was finally freed from Arzamas-16, he was able to immediately rejoin world-level science. He started asking for release from classified work at the end of the 1950s, and finally, Khariton helped him by reassuring the Soviet leadership that Zeldovich had done what was expected of him. By then, Zeldovich was a full member of the Science Academy.

However, Zeldovich's road to the full membership had not been smooth. In 1953 most leading physicists who had participated in the creation of nuclear weaponry were elected full members. Zeldovich was an exception, and no explanation has been found for this slight. It embarrassed his colleagues, especially Sakharov, who was elected on this occasion to full membership, skipping the corresponding member stage.

The next time, in 1958, the directives of the Communist Party categorically opposed Zeldovich's election, but the Academy membership did not follow them. There was a tremendous difference between 1953 and 1958. In 1958, he was elected unanimously by the Division of Physical-Mathematical Sciences. It was not only the nonparty members Landau, Tamm, Leontevich, and Artsimovich but also the party member Kurchatov who supported Zeldovich and called him a genius.[11] The party representatives criticized Kurchatov for not following the party line, not even trying to mask the fact that they possessed information from the secret police about the scientists, especially those who participated in the nuclear programs. Such information was collected by wiretapping and by informers, and it left no doubt about the negative views of these outstanding scientists concerning the Soviet leadership. Nonetheless, this could not be the sole reason for what happened to Zeldovich; he was always rather careful, certainly more so than, for example, Landau, who was elected full member of the Academy in 1946, skipping corresponding membership.

When in 1958 Zeldovich was elected full member of the Academy, he was still only forty-four years old. Given his versatility, an intriguing question was whether he himself singled out any one among his achievements. The question came up when some of his colleagues considered nominating him for the Nobel Prize. According to Alfred Nobel's will, one discovery or invention had to be singled out, rather than lifetime achievement. The nominators and Zeldovich agreed that the achievement to be singled out should be his work on self-organizing systems in active media, that is, on phenomena including flames, ignition, and detonation. Zeldovich considered most of his other work to be of more transient significance. The nominators thought that understanding the phenomena mentioned had been interesting to people from prehistoric times, and in this respect they likened Zeldovich to Prometheus. The nomination had been complied and submitted, alas, Zeldovich was not selected for the award.

Zeldovich always enjoyed strong support from his family. His first wife, Varvara Pavlovna Konstantinova was a scientist in her own right and worked at the Institute of Crystallography of the Academy. They met in 1932 when he was eighteen, and she twenty-five. They had two daughters, Olga and Marina, and one son, Boris. Their life together lasted thirty-nine years, from 1937 till her death. Their family backgrounds were strikingly different. He came from a well-organized family and she came from a cheerful, seemingly disorganized one with many children and relatives. Zeldovich became especially friendly with one of Varvara's brothers, Boris Konstantinov, a future academician and vice president of the Academy; but this was yet far in the future. In 1937, another of her brothers, also a scientist, fell victim to Stalin's terror. This is why, at the time, the marriage was considered by a family member "an act of courage" on Zeldovich's part.[12]

Varvara Pavlovna formed "a cult" for Zeldovich at home; she guarded his working conditions, and the "magic words, 'Papa is working' stopped any fussing or noise."[13] No wonder the children "were rather afraid and shy" of their father.[14] Daughter Marina noted that he did not have time for his children, whereas he was very relaxed with his colleagues. It was always their mother who went to the parents' meetings at school, which was just as well, because the daughters shivered at the thought that he might go and let the teachers see his "complete orthogonality" to what they did in school.[15] All in all, though, Zeldovich's children remember a "happy childhood."[16] In later years, son Boris and Zeldovich were often seen together immersed in discussions of physics.[17] Boris was a student of the Faculty of Physics of Moscow State University, as his sisters had been, and became a successful scientist and a corresponding member of the Academy. After the collapse of the Soviet Union, he moved to Florida and became a professor at University of Central Florida.

Zeldovich's letters to the family from Arzamas-16 showed "happiness, warmth, and certainty in his love" for them.[18] But the long separation, even if broken by frequent visits to Moscow, took its toll on Zeldovich's private life. He had his youngest daughter, Annushka, from a relationship in Sarov. Annushka's mother, O. K. Shiryaeva, an architect, was a political prisoner. She was sent to the infamous Kolyma in Eastern Siberia, where their daughter was born in 1951. Zeldovich supported Shiryaeva and eventually achieved her and her daughter's liberation.[19] Zeldovich had a total of six children, all of whom he recognized as his children. All of them were trained as physicists.[20]

After the death of Zeldovich's first wife, he married Anzhelika Yakovlevna Vasileva. She guarded her husband's working conditions by keeping away everybody whom she considered a disturbance to him. She did not live long, and Zeldovich married for the third time; his third wife was Inna Yurevna Chernyakhovskaya.

Throughout his life, Zeldovich was never a director of a big institute; never had a pretentious office with an army of secretaries protecting their boss; he thrived on constant interactions with his colleagues. In the period 1941–1943 numerous institutes of science and education were evacuated to Kazan, and most evacuees from the Moscow and Leningrad academic institutions occupied a modest three-storey building. There were great names among them, such as Kapitza, Sergei Vavilov, Landau, Tamm, Grigory Landsberg, Semenov, Frumkin, and many others, including Zeldovich.

He always had close interactions with other outstanding physicists, especially Khariton and Sakharov. They were friends who respected each other and were united not only in their interest in physics but also in belonging to the elite of Soviet society. The friendship of Khariton and Zeldovich had its roots in their youth. When the two appeared together at a festive celebration, as, for example, the 1982 fiftieth anniversary of the Institute of Chemical Physics, each wearing his three stars of Hero of Socialist Labor, it was quite a scene.

Zeldovich was a sportsman, yet his physical attributes were not very impressive. From a distance he did not strike anybody as extraordinary. He was short, stocky, and round-faced; he wore round glasses, was not handsome, and did not appear distinguished even when wearing his three-starred jacket. However, as soon as people interacted with him, their impression of him changed. He captivated his partners in conversation. I met him only once; it was in April 1987, at a conference at the Institute of Crystallography in Moscow. During a break I was having a conversation with the director of the institute, Boris Vainshtein. Zeldovich appeared unannounced; he needed to discuss something with Vainshtein. It was obvious that Vainshtein was impressed that Zeldovich had approached him. Both were full members of the Academy, and Vainshtein was also an institute director, but Zeldovich was *Zeldovich*. I do not remember a word of our conversation, but

I remember the man as if I had met him yesterday. I find it difficult to express in words what was so memorable about him; the encounter made a lasting impression on me.

He was an enthusiastic lecturer who had a natural ability to convey his passion for science and captivate his audience. "Zeldovich appeared in the auditorium in simple clothing, a plaid shirt with unbuttoned collar. Without any board, he began to speak very simply and intelligibly. His emotions caught up the audience, and the contact with the listeners was complete. Before us was a man all caught up, lively, like a little bouncing ball."[21] He has been described as vivid, charming, with a good sense of humor and shining eyes, often with some mischievousness in them. He was a devoted friend and a caring colleague but could appear intolerant because of his inability to compromise, and sometimes this led to the cooling off of his friendships.[22]

Considering the plethora of his scientific achievements and the large number of different areas of physics in which he was involved, some foreign scientists supposed that "Zeldovich" was a collective label for a whole group of researchers, just as the name N. Bourbaki was. This perception was facilitated by the fact that for a long time Zeldovich could not travel and did not appear at meetings. For Zeldovich, foreign travel gained mystical power, perhaps because it proved unattainable until the last years of his life. In this he was not alone in the Soviet Union and Eastern Europe. And he was not alone among Soviet big-name scientists, who meticulously marked on a world map all the locations from where they had received invitations, but were never allowed to visit.[23]

There have been conflicting observations about the extent of Zeldovich's unwillingness to enter into conflict with the authorities. But the notion is quite generally held that he avoided politics, even in his conversations. He preferred following instructions to disobeying them. For example, it was forbidden to carry letters away from Arzamas-16. The purpose of this rule was to keep everything under the control of censorship; whereas the purpose of breaking it was to speed up interactions with friends and family. Zeldovich observed the rule, and many interpreted this as unfriendly cowardice. But to break it would have indeed represented a risk for him, especially since he traveled to Moscow more often than most, and his carefulness was probably justified. It was known that the secret police tried to trap scientists preparing for such travel, by asking them to take a letter or a package with them, only to "unmask" them as violating the rules of the installation.

Zeldovich also preferred to let his articles be censored at the manuscript stage rather than fight for leaving them intact. It was characteristic of his general demeanor when he reasoned that "it's not worth getting involved in!"[24] In 1962, there was a conference in Tartu about modern physics and astronomy, for which Zeldovich prepared with great enthusiasm. During the meeting, there was a campaign to collect signatures protesting atmospheric nuclear testing by the Soviet Union. Zeldovich disappeared before he could be asked for his signature. He later told a colleague that he could not reconcile signing such a protest while working on the very tests the signatories were protesting against.[25]

Just as he avoided political discussions, he taught his son how to evade topics that made him uncomfortable. Zeldovich's advice was to always have a few simple jokes or anecdotes to tell that would divert attention from the undesired topics. But in debates on scientific issues he was not at all conformist; he was one of the few who could stand up to Landau. When in 1952 Semenov was being attacked for ostensibly falling for idealism in his work, Zeldovich came to the meeting and made a strong statement in Semenov's defense. On his jacket, he conspicuously displayed his star of Hero of Socialist Labor (he had only one at the time). The audience was duly impressed by his distinction, which at that time was quite rare. The fact that Zeldovich's award had not been announced in the media, although such things as a rule were, only enhanced the impact, because the secrecy implied that he had received it for something extraordinary. In fact, the awardees of classified programs were discouraged from wearing their stars in public lest it invite curiosity about their activities.

He liked wearing his stars, especially when they were supposed to help a cause, as when he came to Semenov's defence, or, for example, when he appeared at the trial of the father of one of his colleagues. He hoped to impress the judge, not so much to gain favor, but to steer the procedure in the direction of fairness. This happened in 1957, and the "crime" was criticism directed at N. S. Khrushchev. Another occasion on which he displayed his stars—by then, he had three—was when Zeldovich (supposedly) went to visit Mosfilm, the Soviet Hollywood, and wanted to impress the actresses. Alas, he was barred from entering because the guards were sure that having three stars could not be for real.[26]

Zeldovich, as most of the other Soviet physicists, considered it his patriotic duty to participate in creating nuclear weapons for the Soviet Union. They did not question whether or not the development of these horrific weapons was justified. However, when the first Soviet nuclear explosion was executed at the end of August 1949, among the physicists observing the test, Zeldovich and also Kurchatov stayed quiet. Others expressed their joy; some shouted "long live Comrade Stalin!"[27] Zeldovich felt grave responsibility for his seminal contribution to the creation of Soviet nuclear weaponry, which most probably strengthened Stalin's recklessness in international affairs. In 1950, soon after the first Soviet atomic bomb was exploded, North Korea invaded South Korea. Zeldovich was tormented by the thought that the Korean War might have not happened had the scientists not placed atomic bombs in Stalin's hands.[28]

Marina Ovchinnikova supposed that his feelings of responsibility hastened his departure from Arzamas-16. She makes a strong case for Zeldovich's psychological troubles when she described his visit to a psychiatrist, ostensibly suffering because bodyguards—euphemistically labeled "secretaries"—were shadowing him. The psychiatrist must have had a hard time making a politically correct diagnosis.[29] On the other hand, Zeldovich was known to play tricks on his bodyguards. For example, he swam far into the Irtysh River, in the vicinity of the Semipalatinsk proving ground, knowing that his frightened bodyguard could not swim (a surprising deficiency for a bodyguard).

It was an expression of courage that Zeldovich never wavered in unambiguously identifying himself with Jewish causes. The overt anti-Semitism in the Soviet Union during the last years of Stalin's reign began in 1947–1948. The famous actor of the Moscow Jewish Theater and well-known leader of the Jewish Anti-Fascist Committee, Solomon Mikhoels, was assassinated in January 1948—his death was attributed to a road accident.[30] It was the deed of Stalin and the secret police, but nobody could openly accuse them of the crime. When Zeldovich learned about the tragic event, he and a friend, as a token of solidarity, went to the Jewish Theater to attend the farewell for Mikhoels.

Zeldovich's departure from the secret project did not reduce his feelings of responsibility for his participation in the creation of the atomic bomb and later the hydrogen bomb. In October 1973, at the time of the Yom Kippur War between Israel and its Arab neighbors, he heard gossip at the Ministry of Medium Machine Building (the authority for nuclear projects) about the possible deployment of a Soviet atomic bomb against Israel. Zeldovich decided that he would commit suicide if this were to happen. He composed a letter to leave behind, but knew that the letter would disappear if it fell into the hands of the authorities. He wanted to be sure that the reason for his suicide became known, and so he left a copy of his suicide note with his good friend, the physicist academician Ilya Lifshits. As is known, no atomic bomb was used against Israel in the Yom Kippur War, and after the war ended, Zeldovich collected the copy of his suicide note.[31]

Zeldovich's demeanor may have also changed with time. Back in the early 1950s, there was a controversial story, which has become known as the Agrest Affair. It has been interpreted as an example of negative behavior on the part of Zeldovich, when in reality it was the regime and the atmosphere in which the scientists all existed that was reprehensible. Mattes Agrest was the head of the mathematics section in the theoretical division at Arzamas-16. He was a religious Jew with a large family. When the anti-Semitic upheaval reached the atomic project, he became a thorn in the eyes of the security organs. Just being a Jew did not suffice as a reason for dismissal because at that time it meant that the leading echelons would have been to a large extent emptied. As Agrest's past was scrutinized, it came out that he had studied Judaism and had graduated as a rabbi. He may also have had relatives in Israel, and this finally served as official pretext for his dismissal. On January 13, 1951, he was told that he had twenty-four hours to leave the installation.

Having relatives in Israel at that time seemed a valid reason for dismissal even to Agrest's most sympathetic colleagues. However, they found the twenty-four-hour time limit cruel. Tamm and others protested, and the time limit was extended to one week. On the day of the Agrests' departure, Tamm left work early, announcing that he was off to help the Agrests to pack their belongings. Sakharov—he was far yet from becoming a "dissident"—made two gestures. He shook Agrest's hand, which was a demonstration of solidarity, and he offered Agrest and his family to stay in his unoccupied Moscow apartment.[32] Two of Agrest's Jewish colleagues, however, appeared indifferent to his plight: his immediate superior, Zeldovich, and the scientific director of Arzamas-16, Khariton. Zeldovich's and Khariton's reactions to Agrest's plight was in sad contrast with the solidarity expressed overtly by Tamm and Sakharov. It may be argued, though, that it was easier for the non-Jewish Tamm and Sakharov to express solidarity with Agrest than for the Jewish Khariton and Zeldovich. Agrest subsequently found employment doing other classified work.

Zeldovich had nothing to do with Agrest's dismissal, but Sakharov "heard that Zeldovich (although Zeldovich denied this) had upset Agrest greatly by making him work on the Sabbath."[33] Apparently, Agrest tried to reconcile observing the Sabbath with his work schedule at Arzamas-16 in such a way that though he came to work on Saturdays, he restricted his activities to *discussions*; he would not *write* anything. In 1981, that is, three decades after the events, Zeldovich wrote a letter to Agrest concerning their encounter on a Saturday.[34] Zeldovich was nonreligious and for him Saturday was like any other day; he never paid attention to whether any day was a holiday or a workday. On that particular occasion, the two were having a prolonged discussion during which—resisting Zeldovich's prodding—Agrest could not bring himself to write anything on the board with a piece of chalk.

The episode kept tormenting Zeldovich. Eventually, Agrest expressed regret that he did not inform Zeldovich that he did not write on the Sabbath.[35] We can sympathize with Agrest for his plight, but the humiliation Khariton and Zeldovich must have felt also elicits sympathy. They felt unable to stand up for a Jew, because they themselves were Jews. Both Khariton and Zeldovich understood the prevailing anti-Semitic sentiments and actions during the last years of Stalin's reign. Many years later, Zeldovich told the mathematician A. D. Myshkis "about a group of young people who were trying to revive Jewish customs, right down to their clothing, keeping the Sabbath, and so forth. [Myshkis] expressed doubts about the reasonableness of this, but [Zeldovich] talked with certainty about people's right to behave this way."[36]

Yakov Zeldovich and Valentine Telegdi in 1960, during the High Energy Physics Conference in Kiev.

Source: Courtesy of Olga Zeldovich, Moscow.

In the mid-1970s, the already three-starred academician Zeldovich suffered anti-Semitic attacks in the guise of criticism of his immensely popular mathematics book, *Higher Mathematics for Beginners*.[37] The book had appeared in several editions, both at home and internationally, and had grown out of Zeldovich's unhappiness about how mathematics was being taught. He found it unsatisfactory that the subject came too late in the curriculum after other subjects that needed mathematics were already being taught. He likened the situation to serving lunch, but making the salt and pepper for it available only at the time of the afternoon tea.[38] In the book, Zeldovich did not follow all the rigorous approaches of pure mathematicians, but it was a highly didactic volume that brought life to a subject that many outside the field dreaded. Encouraged by the book's success, Zeldovich cooperated with Myshkis in producing two further volumes, *Elements of Applied Mathematics* and *Elements of Mathematical Physics*.

Zeldovich had an excellent testing ground for pedagogy in his own family, where virtually everybody was a physicist, and they held family seminars. When his daughters went to school, he hired S. S. Gershtein, a future academician, to tutor them. The girls were excellent students who could still learn from Gershtein, but it was not lost on the young physicist that this engagement came in handy for him during a difficult period of unemployment.

Zeldovich has been characterized on occasion as accommodating toward the communist regime. However, he never joined the Communist Party, and he never signed any letter condemning Sakharov for his "dissident" activities. When Sakharov was in exile, it was forbidden to make references to his work—it was common practice under the Soviets to make any undesirable person into a nonperson. When Steven Weinberg's book *The First Three Minutes* was to appear in Russian translation, Zeldovich was asked to prepare a supplement. In it, he made a reference to Sakharov, which the censors wanted to remove. Zeldovich's response was that if the reference went, his supplement would go with it. On another occasion, the censor threw out a reference in the manuscript of a popular article prepared for the large-circulation magazine *Priroda* (*Nature*) on the grounds that the editors should not show Sakharov in a positive light in a paper read by so many people. A compromise was reached: instead of being included in the more conspicuous list of references, it was mentioned in a footnote, making it less likely that the readers would notice it; but the reference to Sakharov stayed in the piece.

Nonetheless, as Sakharov noted, "hurt feelings and mutual coldness crept in" between the former close colleagues, Zeldovich and Sakharov, especially in the period of Sakharov's exile to Gorky.[39] Sakharov perceived that Zeldovich "strongly disapproved" of Sakharov's social activism and was puzzled why Zeldovich did not support him. Their close relationship was never re-established during the short period of time after Sakharov's return from exile and before Zeldovich's death.

Yakov Zeldovich is a shining example of the level of physicists who participated in the Soviet nuclear program. It was characteristic of him that during his period at Arzamas-16, he stayed versed in the progress of basic science. Not only did he keep up with the literature, but he also contributed to it. Had it not been known that his principal engagement between 1947 and the early 1960s was the nuclear weapons program, it would have been difficult to discern that he was absent from ordinary academia for such a long time. As one of his colleagues remarked, Zeldovich's "lifespan encompassed several scientific biographies."[40] It was not only that Zeldovich benefited from staying abreast of the happenings in modern physics; he helped others to do the same. He transmitted knowledge and information in a creative way. This is why Sakharov noted that "in the area of fundamental physics, much of my research arose from my contacts with him, under the influence of his work and ideas."[41] Both Sakharov and Zeldovich were exceptional physicists, but Sakharov's name has gained additional eminence because of his human rights activism, which has been recognized by a Nobel Peace Prize. There are some who believe that Sakharov's glory may have pushed Zeldovich back into the shadows.[42]

Zeldovich was fifty years old when he finally returned to pure science in full force, becoming an associate of the Institute of Applied Mathematics of the Soviet Academy of Sciences. Mystislav Keldysh was the director of the institute, who in 1961 had followed Aleksandr Nesmeyanov as the president of the Science Academy. As a mathematician Keldysh was involved in the nuclear weapons program as well as in the space program. He greatly supported Zeldovich, who created a division of astrophysics in the institute, and took up cosmology, which had always been among his interests.[43] By doing so, he reached back to a Russian scientist in the early twentieth century, Alexandr Friedman, who around 1920 had created a theory of the expanding universe that contradicted Einstein's views of cosmology. Einstein originally rejected Friedman's theory but later accepted it. Friedman died prematurely in 1925, without realizing that his two brief papers would considerably impact the development of modern cosmology.

Zeldovich's associates contributed significantly to the question about the origin of the universe and were quoted in the Nobel lectures about the discovery of remnant heat in outer space. The existence of this remnant heat served as unambiguous evidence of the correctness of Gamow and his associates' Big Bang model. Nuclear physics and astrophysics are not so far from each other as they seem at first glance. Many of the processes theoretical physicists investigated on paper and later by computer had been going on in space from the start of the existence of the universe. The joint application of astrophysics and nuclear physics made it possible to understand how the universe began and how the chemical elements had formed.

A closer look at this topic may be instructive to learn a little about Zeldovich's approach to fundamental problems in science. Regarding the beginning of the universe, Zeldovich was fond of the model that supposed a cold start and subsequent development. This was one alternative. Another was the scheme originated by Gamow and his associates, which has become known as the Big Bang model, a name originally coined to ridicule it. The Big Bang model started with an initial explosion producing enormously high temperatures. Zeldovich correctly understood that gaining evidence about the presence or absence of a relic (background) radiation in outer space would be decisive in demonstrating the correctness of the cold model or the hot model. He formed a small research group consisting of three people: himself, A. G. Doroshkevich, and I. D. Novikov. Once the question about the relic radiation had been resolved, their pioneering approach received recognition in Arno Penzias's Nobel lecture. Referring to the contribution of the Soviet scientists, Penzias said that it was "the first published recognition of the relict radiation as a detectable microwave phenomenon."[44]

Penzias and Robert Wilson discovered this relic radiation by serendipity and and only their subsequent literature search uncovered Doroshkevich and Novikov's relevant paper.[45] In contrast, Doroshkevich and Novikov were so thoroughly familiar with the literature that not only did they point to the possibility of detecting relic radiation—if it existed in the first place—but suggested using a specific instrument used at Bell Laboratories for this purpose. Unfortunately though, Doroshkevich and Novikov misinterpreted the then-available measurements published by the associates of Bell Labs, and in 1964 they deduced that there was *no* relic radiation.[46] The conclusion from this misinterpretation was the easier than the Big Bang model to accept because it was consistent with Zeldovich's cold universe model.

Yakov Zeldovich and Pope John Paul II in 1980 at the Vatican.

Source: Courtesy of Olga Zeldovich, Moscow.

Yakov Zeldovich with one of his most famous associates, the academician astrophysicist R.A.Sunyaev, in the 1980s.

Source: Courtesy of Olga Zeldovich, Moscow.

It came as a great disappointment to Zeldovich when he learned about Penzias and Wilson's observation of the remnant heat, as it signified the failure of the cold model. But it was a manifestation of Zeldovich's devotion to scientific truth that he immediately acknowledged not only the importance of Penzias and Wilson's observations, but also the validity of Gamow and his colleagues' model. Incidentally, this recognition also required political courage, because ever since Gamow had defected from the Soviet Union in the early 1930s, he had become a nonperson there.

In a rare glimpse into his weapons work, Zeldovich revealed to one of his associates that his not very well-founded choice of the cold model was a reflection of the working habits in the nuclear project. There, they were often forced to make instant decisions and make a choice among possible solutions when further deliberations might have been called for.[47] The discovery of the existence of relic radiation reached Zeldovich toward the end of 1965. He regrouped at once, and started giving enthusiastic lectures about the relic radiation and the hot universe—the topic seemed to fit his temperament eminently, and wherever he went, packed auditoriums greeted his presentations.

In spite of his generally cheerful demeanor, Zeldovich also experienced dark moments. For example, not long before he died, he recounted to a Western visitor, K. S. Thorne of the California Institute of Technology, the discoveries he could have made but did not—and others made them, in the West. The visitor was surprised because Zeldovich made a great many important contributions to physics and astrophysics. In fact, his work led other scientists in both the Soviet Union and the West to make significant contributions to the field.[48] P. J. E. Peebles of Princeton University, one of the major players in cosmology, noted: "Through all my career in cosmology, I could be sure that if Zeldovich was not hard on my heels it was because he was racing far ahead."[49] However, Zeldovich's dark moments came not only toward the end of his life. In 1967, when conversing with Sakharov about the works each liked best in their own oeuvre, Zeldovich noted, "My works too often sink without a trace."[50]

Zeldovich and Sakharov were two great physicists in the Soviet nuclear program and beyond. They lived and worked for years side by side at Arzamas-16, and they had excellent interactions. Many of their results in the project must have come out of their discussions, but no priority controversy ever surfaced about their collaboration. According to Yurii Smirnov, a close observer, Zeldovich's talent as a physicist matched Sakharov's, but, as noted above, Sakharov has outshined Zeldovich due to his human rights activities.[51] Smirnov emphasized Zeldovich's enormous contributions to atomic defense, physical foundations for the internal ballistics of solid-fuel rockets, scientific schools in chemical physics, hydrodynamics, the theory of combustion, nuclear physics, the physics of elementary particles, and astrophysics.

Smirnov noted that Zeldovich and Sakharov complemented each other. When Zeldovich entered a new field, he preferred to define and solve a problem and move toward generalization. He found it useful to review the new field he was entering, and from this, instructive review articles were born. Sakharov in a similar situation, when entering a new field, appeared to perceive the physical laws and the connections between various phenomena, and he could move directly toward technical innovation.[52] Their ways of reasoning were very different. Sakharov was more abstract and formal; Zeldovich preferred analyzing concrete problems.[53]

Both Sakharov and Zeldovich were characterized by "their democratic attitude, devoid of any hint of grandeur or overbearing, and their openness and accessibility; they destroyed the traditional picture of academicians as objects of universal reverence and worship."[54] The two were very different in character and temperament, and their working habits were also very different. Zeldovich got up very early in the morning, exercised vigorously, and worked till the evening, taking only a few short breaks. He had his rest in the evening. When Sakharov worked on a problem, it grabbed him and stayed with him without a moment of interruption.

Zeldovich was sometimes compared with Enrico Fermi for his versatility and because his science was his life. In the first period of his career, Sakharov also showed complete devotion to solving the problems of physics for defense. Eventually, however, physics gradually moved to the background, yielding to his political activities. Ultimately, curing the illnesses of society became Sakharov's principal concern, as though he felt that the physics could be accomplished by others but that his human rights and other political activities were truly unique. In this respect, Leo Szilard comes to mind. Fermi and Szilard together faced the danger that totalitarian states might acquire the most deadly weapons. Fermi contributed enormously to the physics of the American projects; for him, science was his life. For Szilard, science was a means to solve life's most pressing problems of human society—the survival of democracies.[55] In this sense, Sakharov was closer to Szilard than to Fermi, and Zeldovich closer to Fermi than to Szilard.

Generosity was characteristic of Zeldovich. He was generous with Sakharov, for example, in sharing all he knew about the latest achievements of modern physics. Another example of his generosity was that he welcomed Tamm and his colleagues at the weapons installation rather than being upset about the arrival of another theory group, which he might have considered a threat or at least a rival. Yet another example of his generosity was discovered by Boris Gorobets among the archival materials of the late Evgenii Lifshits, shortly after Zeldovich's unexpected death of heart failure. Gorobets found an unpublished evaluation of Lifshits's works in cosmology written by Zeldovich. Gorobets (the son of Lifshits's second wife) was so moved that he wrote a poem honoring Zeldovich's memory.[56] It is quoted here in the free-style translation of chemistry Nobel laureate and poet Roald Hoffmann[57]:

In memory of Ya. B. Zeldovich

One hears a string orchestra.
The earth sad
and the pillared hall too.
A genius leaves, as a wave,
into a tunnel bereft of light or shade,
into a world without phonons.
While in science superstrings
are young and ripple,
canons crack.
The cosmic foam
boils up. Chopin's March…
And the icon, his face.

Andrei Sakharov's statue on Sakharov Square in St. Petersburg with university buildings in the background.

Source: Photograph by and courtesy of the author.

3

Andrei Sakharov

SOVIET CONSCIENCE

Andrei D. Sakharov (1921–1989), nuclear physicist and former pupil of Igor Tamm, was not well known before he became the most famous Soviet "dissident" and a fearless fighter for human rights during the last two decades of the existence of the Soviet Union. His other distinction was "father of the Soviet hydrogen bomb." While he was with the Soviet hydrogen bomb project, he earned three stars of a Hero of Socialist Labor and jumped from being a Candidate of Science to full member of the Soviet Academy of Sciences at the age of thirty-two—an unprecedented achievement.

Eventually, he started questioning the justification of nuclear tests, and ultimately, he was eased out of the Soviet nuclear weapons program. He developed an interest in Soviet human rights violations and became a leading figure in fighting for the persecuted and prosecuted. He was awarded the Nobel Peace Prize in 1975, well before he entered the most probing period of his struggles. He risked his freedom and his health and was exiled from Moscow and subjected to continuous harassment, but nothing deterred him from his dedication to science and his human rights activities. He never abandoned his interest in fundamental physics, and performed pioneering research, including controlled thermonuclear reactions and cosmomicrophysics.

He was one of the most influential scientists in his contribution to making the Soviet Union a superpower. He was also one of the most influential individuals who forced Mikhail Gorbachev and the Soviet leadership to retreat, contributing to the collapse of the Soviet empire.

Few other scientists did as much to achieve the superpower status for the Soviet Union as Andrei Sakharov, and few fought its human rights violations with as much dedication and self-sacrifice. His gradual transformation was a most remarkable change. A third component of his life's work was his outstanding achievements in fundamental physics.

Andrei Dmitrievich Sakharov was born on May 21, 1921, in Moscow.[1] There were two strong influences at home. He grew up religious, and although at the age of thirteen he stopped believing in God, he retained a lifelong respect for those who did. The other influence was his father's physics. Dmitrii Sakharov taught physics and was a prolific author of physics books, both textbooks for undergraduates and popular science. Sakharov's mother taught gymnastics in school. Sakharov was a sensitive child and characterized himself by an "awkwardness in dealing with people" that lasted throughout his whole life.[2] Despite this "awkwardness," however, he never avoided the company of others.

Once he reached school age, he received instruction at home for the first five years, together with a few other children. When he finally went to school, he did not enjoy it; he did not make friends, and was not very good with his hands. His parents decided to keep him at home for a few more years before letting him go back to school for the final years of general schooling. At the time, attending school was not obligatory in the Soviet Union, and many families of the Soviet intelligentsia preferred to start instruction at home rather than send their children into the chaotic school system. Over time, the Soviet school system became highly disciplined and rigorously structured.

Young Andrei was observant and deep thinking. Relying on his readings, he discovered that most natural phenomena can be reduced to interactions between atoms and expressed by mathematical formulas. He participated in competitions in mathematics and physics, but racing against time had a paralyzing effect on him, so he stopped such activities. At the age of fourteen, he started to do physics experiments at home. Reading books about physics was even more enticing for him, and he extended his reading toward other areas of science. He was impressed by Paul de Kruif's *Microbe Hunters*. This book had impacted many other youngsters of his generation on their way to becoming outstanding scientists, including several future Nobel laureates.

In the fall of 1938, Sakharov became a freshman at the Faculty of Physics, Moscow State University. World War II started on September 1, 1939, but it caught up with the Soviet Union only on June 22, 1941. In Sakharov's senior year the university was evacuated to Ashkhabad, Turkmenistan, near the Iranian border. In spite of the war conditions, he received a good education. It was unusual in his generation, but Sakharov never joined the Komsomol, the communist youth organization. This was not a stand for or against anything; it just happened that way. It is true, though, that he found the ideology of Marxism-Leninism—one of the mandatory subjects in his studies—hard to digest, but he did not have any objection to what it represented.

After graduation, and throughout the war, Sakharov worked as an engineer in a munitions plant in Ulyanovsk, about 500 miles east of Moscow. He showed himself to be skilled and excelled with technological innovations. In addition to his daytime work, he continued his self-education and worked on theoretical problems. He came across Yakov Frenkel's book on quantum mechanics and relativity, which was a revelation for him. He met his future wife, Klavdia "Klava" Vikhrieva at work. She had completed four years of study of chemical technology at the Leningrad Institute of Technology before the war, and worked in the laboratory of chemical analysis. They married on July 10, 1943, and had three children, Tatyana "Tanya" in 1945, Lyubov "Lyuba" in 1949, and Dmitrii "Mitya" in 1957.

Sakharov's father had known the renowned theoretical physicist Igor Tamm (see chapter 1) who worked in the famous Physical Institute of the Academy of Sciences (Fizicheskii Institut Akademii Nauk [FIAN]). Sakharov sent Tamm his first manuscripts of a theoretical nature, but received no response. Sakharov did not become discouraged, and in December 1944, he turned to Tamm when he wanted to begin his doctoral studies. This time, he and his father did not leave anything to chance. Sakharov's father notified Tamm about his son's intentions. Tamm received Sakharov cordially, and though he noted substantial holes in Sakharov's knowledge, he was upset only by Sakharov's lack of English. He advised the young man to learn it right away. His other suggestion was to start reading the classics on relativity and quantum mechanics to which Frenkel's book proved to be a good introduction.

FIAN and Tamm's group in it had a friendly atmosphere, which helped Sakharov adapt to his new environment. He greatly benefited from Tamm's seminars. There was the intimate Friday seminar in Tamm's private office, where only his own small group discussed current research problems. There was then the Tuesday seminar in theoretical physics in a big conference hall, which was open to all physicists. It was of the colloquium type; the participants were assigned papers in recent literature to report on at the next seminar. It was in Sakharov's nature that he did not take anything for granted and was grateful to have the privilege of being part of Tamm's seminars.

In addition to other requirements, graduate students were supposed to fulfill some teaching obligations. Sakharov taught three semesters of physics at the Moscow Energetics Institute, where he encountered innovative physics. He also encountered social issues of which, for the time being, he remained a passive observer. Some of the physicists at the Institute were engaged in research that, Sakharov would say in retrospect, could have led to the design of what today is known as the laser. Alas, the researchers involved happened to be Jewish; during these postwar years their situation was becoming increasingly precarious, and they were not allowed to develop their research potential.[3]

Sakharov had noted anti-Semitism already during the war, and more so during the ensuing years. He was touched by this, because he sensed certain "Jewish" qualities that attracted him. He described one of his youthful friends in the following way: "his inner purity, his contemplative nature, and a melancholy empathy that seems to be an innate Jewish characteristic."[4] In the postwar Soviet Union anti-Semitic incidents were increasing, with official sanction. The euphemism "cosmopolitanism" did not hide the reality of anti-Semitism.[5] Sakharov felt the injustice of persecution deeply.

Still, Sakharov tried to keep a distance from everything not related to his studies. In fall 1947, he completed his dissertation for the Candidate of Science degree (PhD equivalent). It dealt with interactions and transformations of elementary particles. Today, they are called "fundamental particles," and many more are known than in Sakharov's time. Two of Lev Landau's (see chapter 5) pupils, both future academicians, Isaak Pomeranchuk and Arkadii Migdal, judged his dissertation.

Sakharov proved to be a gifted researcher with original ideas. He was hard-working; his work was his hobby. He produced many results, and he published them in due course, though publishing was not a trivial matter. There was considerable red tape before a manuscript could be submitted for publication. It had to be certified that the work did not contain classified information, such as military applications. When Sakharov later wrote articles that had political and societal considerations, he learned that many other kinds of information were also classified in the Soviet Union, including data about crimes, accidents, natural disasters, alcoholism and other ills of society, work safety, damage to the environment, industrial and agricultural production, and so on. It was a long list, and it made publishing extremely difficult.

Sakharov's excellence was noticed by the people who were setting up the Soviet nuclear weapons program. Igor Kurchatov offered him a prestigious position at his Institute of Atomic Energy, which other outstanding scientists had already joined. Sakharov resisted attempts to entice him away from FIAN, preferring fundamental research in Tamm's group. When he finally shifted his research from fundamental questions to the nuclear area, he was not asked to do it; he was merely told to do so. Yet this did not make him feel uncomfortable because Tamm had been charged with organizing a small group of theoretical physicists at FIAN to assist the work on thermonuclear explosives—the hydrogen bomb.

Sakharov became a member of this highly classified project in June 1948, and his status was elevated to the position of Senior Scientist. Soon, he got a raise in salary, and his family was given a room of 150 square feet in a communal apartment.[6] "Communal" meant many rooms, with one family living in each room and all the families sharing a common hall, toilet, and kitchen. The room was small and the arrangement uncomfortable, but it was a definite improvement. Before, the Sakharovs had lived in rented spaces and were at the mercy of landlords. This change prompted Yakov Zeldovich to quip that this "was the first use of thermonuclear energy for peaceful purposes."[7]

The thermonuclear or hydrogen bomb is also called the "fusion bomb," because in it an enormous amount of energy is liberated in the fusion reaction between light atoms, such as, for example, the reaction between one deuterium and one tritium (both are heavier isotopes of hydrogen) yielding one helium, one neutron, and energy:

$$D + T \rightarrow He + n + energy$$

The adjective “thermonuclear” indicates that very hot conditions are necessary for the reaction to take place and that the nuclei of the atoms are involved in it. Energy liberated in this reaction derives from the mass difference: the mass of one helium atom and one neutron is smaller than the mass of the deuterium and tritium taken together. The mass-to-energy conversion is described by Einstein’s famous equation, $E = mc^2$, where E is energy, m is mass, and c is the speed of light.

The energy produced in stars comes from thermonuclear reactions at a temperature of many millions of degrees, which is still lower than the temperature necessary for such a reaction under terrestrial conditions. Matter is considerably compressed in the stars, which facilitates the reaction and thus allows it to happen at “lower” temperatures. In 1932, George Gamow gave a lecture in Leningrad on the energy production in stars, which Nikolai Bukharin, a leading Soviet politician, attended. Bukharin was impressed by the presentation and in an ensuing conversation with Gamov, the idea of setting up a thermonuclear reaction under terrestrial conditions came up. Bukharin offered to put the electric power of the entire Moscow district at Gamow’s disposal for the experiment, but Gamow declined.

The concept of the thermonuclear bomb figured in a conversation between Enrico Fermi and Edward Teller at the early stage of the Manhattan Project, but its development was shelved as they concentrated on the atomic bomb during World War II. After the war, both the United States and the Soviet Union gradually embarked on efforts to produce thermonuclear bombs, whose power was orders of magnitude higher than that of the atomic bombs. Sakharov spent twenty years on developing and perfecting the Soviet hydrogen bomb.

In the thermonuclear bomb, the fusion reaction leads to explosion, which is the purpose in its military application; but an explosion would not be possible to utilize for peaceful energy production. Almost from the beginning of his involvement with the project, Sakharov was fascinated by the concept of *controlled* thermonuclear reactions, which would solve the world’s energy problems. He gained Tamm’s enthusiastic cooperation, and they together developed the idea of the magnetic thermonuclear reactor, or MTR.[8] Eventually, this research was continued at the Leningrad Institute of Physics. It did not promise immediate practical application; hence, it was decided not to keep it classified. It made waves when in 1955, Kurchatov accompanied Nikita Khrushchev and Nikolai Bulganin on a state visit to Great Britain, and gave a public lecture on the topic. For decades, research on *tokamak*, the Russian name of the device, has been going on in many places.[9] In 1952, in related studies, Sakharov suggested new experiments for the development of devices to utilize chemical explosions and nuclear explosions to produce extremely strong magnetic fields.[10]

Andrei Sakharov as freshman of Moscow State University.

Source: Courtesy of Alexander Vernyi and Lyubov Vernaya-Sakharova, Moscow.

Andrei Sakharov in 1953, following his election to the Soviet Academy of Sciences.

Source: Courtesy of Alexander Vernyi and Lyubov Vernaya-Sakharova, Moscow.

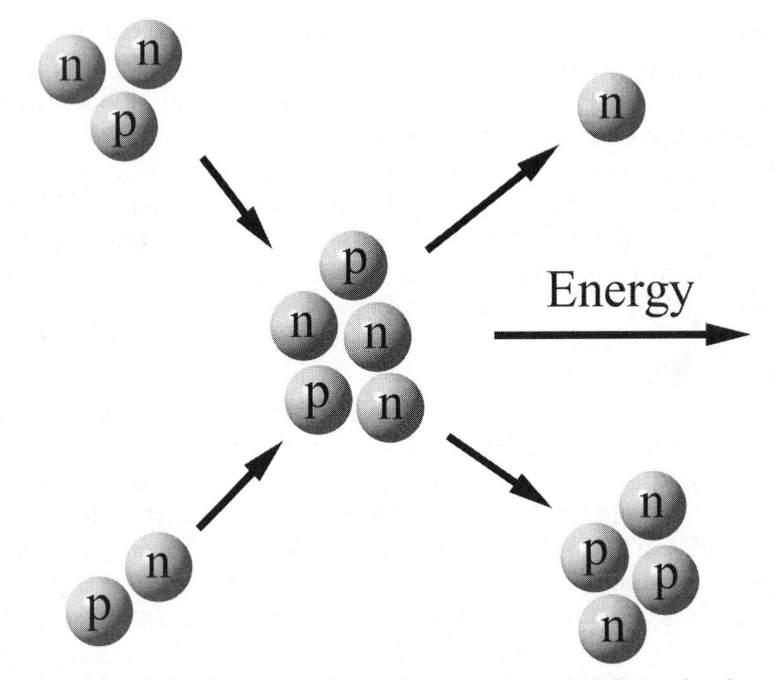

Schematic representation of a thermonuclear—or fusion—reaction: a deuterium nucleus (one proton plus one neutron) and a tritium nucleus (one proton plus two neutrons) are fused, yielding a helium nucleus (two protons plus two neutrons), a neutron, and energy.

Initially, the primary task for Tamm's group was "to verify and refine the calculations produced by Yakov Zeldovich's group" at the secret installation Arzamas-16.[11] While Tamm's group was operating at FIAN, Vitaly Ginzburg (see chapter 6) was also a member. When Tamm moved to Arzamas-16, Sakharov and another young physicist went with him, but not Ginzburg. The task was exciting physics, and it was even more enticing that it was deemed essential. Sakharov considered himself a soldier in working on this project, and the participants "were possessed by a true war psychology."[12] This was soon after the Great Patriotic War—the Russian name for the war against Nazi Germany—in which a foreign power tried to subjugate the proud Russian nation and the Soviet Union. Now the country was making extraordinary efforts and sacrifices to create the necessary defense against a yet-more-powerful adversary.

None of the Soviet scientists protested working on the hydrogen bomb. The situation was very different from that in the United States where there was a highly charged debate whether or not the United States should develop it. At the end of October 1949, the General Advisory Committee (GAC) of the Atomic Energy Commission concluded that the United States should *not* develop the hydrogen bomb. There were only nuanced differences: the majority opinion signed by J. Robert Oppenheimer and a number of other members stated that not developing such a bomb would be "a unique opportunity of providing *by example* some limitations on the totality of war."[13] The minority opinion by Enrico Fermi and Isidor Rabi, both Nobel laureates, considered it wrong on ethical principles to develop such weapons and suggested, instead, "to invite the nations of the world to join us in a *solemn pledge* not to proceed in the development or construction of weapons of this category" (italics added).[14] The Soviet government did not seek the opinions of its scientists, and they did not volunteer their opinions, either. Making decisions was the prerogative of the leaders of the Soviet Union, ultimately, Stalin.

In the 1980s, in writing his *Memoirs*, Sakharov commented on the merits of Edward Teller's stand in the American debate about the hydrogen bomb. Teller was not a member of the GAC, but he recognized the bomb's potential and could not imagine that Stalin would be talked out of acquiring it. He understood that showing an *example* or making a *solemn pledge* made no difference. On the contrary, Stalin would have considered such behavior a weakness or deceit on the part of the Americans. Sakharov noted that "in the 1940s and 1950s my position was much closer to Teller's, practically a mirror image (one had only to substitute 'USSR' for 'USA,' 'peace and national security' for 'defense against the communist menace,' etc.)... Unlike Teller, I did not have to go against the current in those years, nor was I threatened with ostracism by my colleagues."[15]

There was an additional source of skepticism about the necessity of the hydrogen bomb in the United States; many in responsible positions did not believe that the Soviet scientists would be up to the task of creating it. Teller had a different opinion. He had met some of the Soviet physicists in Western Europe in the late 1920s and early 1930s and held them in the highest esteem. Besides, he understood the potential of a totalitarian regime to focus its resources on selected targets and reach them despite the generally poor conditions of its economy and infrastructure.[16]

The inclusion of Tamm's group into the thermonuclear weapons project very soon brought great advancement. Three new concepts emerged in quick succession. Sakharov called them the "first," "second," and "third" ideas in his *Memoirs*, because he was careful not to give away classified information in his reminiscences. Today, we can identify them in that order as the so-called layered-cake design (*sloika* in Russian) of arranging the fission and fusion fuels; the application of the solid fuel lithium(6) deuteride (*lidochka*); and the radiation implosion approach. The first and third ideas originated from Sakharov; the second, from Ginzburg. Sakharov himself emphasized the collective nature of their work, that the third idea was not his alone; he was merely one of its chief authors.

In this connection, the question arises whether intelligence from the American program might have contributed to the development of the Soviet hydrogen bomb. The idea of radiation implosion—compression of the fusion fuel by radiation generated from an atomic bomb—already figured in the April 1946 meeting at Los Alamos about the thermonuclear bomb. Klaus Fuchs attended this meeting. He was one of the originators of the idea, together with John von Neumann, and he passed the information on to his Soviet contacts.[17] This does not necessarily mean that the idea became part of the Soviet program—apparently it did not. On the American side, it did not either, and only in spring 1951 did the Teller-Ulam design based on radiation implosion come about.[18] Stanislaw Ulam was a Polish-American mathematician who was involved in the feasibility studies of the hydrogen bomb at Los Alamos.

In the Soviet project, the third idea, radiation implosion, was (re)discovered in early 1954. It is remarkable how an idea can be around yet needs to be reinvented when all the conditions are ripe for its utilization. It is not rare in the history of scientific discoveries that an idea is suddenly found revolutionary even though it had been floating around without many experts recognizing its utility.[19] This is why Teller (along with Ulam) and Sakharov (along with his colleagues) in their respective programs deserve much credit for hitting on the idea of radiation implosion, and doing so at the right moment. Nonetheless, it cannot be ruled out that both earlier and later intelligence played a role in the successful development of the Soviet hydrogen bomb.[20] This may—or may not—be another reason why Sakharov always disclaimed being the sole author of the third idea.

Almost immediately after Sakharov had suggested the first idea, he was invited to join the Communist Party. He declined, expressing his disagreement with some of the party's past actions. He felt apprehensive that becoming a party member would mean more rigorous subordination and the probability of an administrative position, which Sakharov did not want.

The first Soviet atomic device was tested successfully on August 29, 1949. The Soviets did not announce the test, even after it had been executed, but the Americans determined from their analysis of air samples that it had happened. The event shook the American experts and helped those in the United States who wanted to develop the hydrogen bomb. The successful test of the Soviet atomic bomb also gave a boost to Stalin's determination to develop the hydrogen bomb.

The hydrogen bomb project meant a greater challenge than the atomic bomb for the Soviet scientists. First of all, there was no American blueprint to follow. There may have been some fruits of espionage, but they were certainly not on the scale of those for the atomic bomb. The Soviets' ace in atomic bomb intelligence, Klaus Fuchs, was no longer a participant in the American project: he had returned to Britain and would soon be arrested for spying. The Americans themselves did not yet know how to produce the hydrogen bomb. For the time being, they were busy examining whether or not the so-called classical Super would work.

The growing importance of Sakharov's role in the project was signified by his transfer to Arzamas-16 in 1950 and, further, by the assignment of bodyguards to him in 1954. He had this protection until 1957, but he was never comfortable with the arrangement. Sakharov remained at Arzamas-16 for eighteen years. He and his colleagues had to live within the confinement of the installation, bordered by barbed wire, with various restrictions. There was also a large labor camp, and the presence of the prisoners was a constant reminder of an alternative in case the scientists failed in their task. Sakharov never questioned the importance of or justification for the work he was doing.

There were changes in Sakharov's family life. Soon after he had moved to Arzamas-16, his family followed him, and they lived at the installation for the next six years. From about 1956, they divided their time between Moscow and the installation, and the children went to school in both places. His children grew up; each of his two daughters married a physicist. The eldest, Tanya, gave birth to Sakharov's first grandchild, Marina. Eventually, Tanya and Marina ended up living in London. Lyuba and her husband lived in Moscow.

From the time of Sakharov's arrival at the installation, he continued his work in Tamm's group. Tamm stayed at Arzamas-16 until August 1953, and left immediately after the first success of the hydrogen bomb project. Tamm impacted Sakharov's interest in physics, but even more, his outlook on everything else. Tamm was a keen observer of the events around him and his consistent decency served as example for Sakharov. For example, Sakharov was in full agreement with Tamm, who declared in connection with anti-Semitism that there was "one foolproof way of telling if someone belongs to the Russian intelligentsia. A true Russian intelligent is never an anti-Semite. If he's infected with that virus, then he's something else, something terrible and dangerous."[21] In this quote, the word "intelligent" must have meant member of the intelligentsia. Tamm regularly listened to the BBC, both in Russian and in English, and told his young associates about the news at breakfast. Sakharov learned about important events in the world outside the Soviet Union from Tamm.

Yakov Zeldovich was another person who played a beneficial role in Sakharov's life at the secret installation. Zeldovich had come to Arzamas-16 before the Tamm group and was the leading theoretician for the atomic bomb project. He had a keen interest in keeping up with the development of physics and was generous in sharing his knowledge with Sakharov. Zeldovich had a certain degree of social conscience and noticed when his colleagues needed assistance. He was not quite willing to expose himself in such situations, but instead urged Sakharov to get involved. He assessed that Sakharov's word might carry more weight than his.

When there was a commission from Moscow investigating the scientists' views in connection with the controversy between Lysenko and Mendelian genetics, Sakharov did not hide his acceptance of the Mendelian theory. The commission left him alone, but not everyone was treated with such leniency.

Lev Altshuler, the head of one of the experimental groups was threatened with losing his job for his views. Zeldovich suggested that Sakharov turn to the director of the installation to protect Altshuler. Sakharov's interference was successful, but in retrospect, he wondered why Zeldovich had prompted him to act rather than doing something himself. Then he realized that Zeldovich was in the precarious situation of being Altshuler's friend, and both of them were Jewish. On the other hand, Zeldovich held a high and important position at Arzamas; yet he seemed to refrain from taking risks that might jeopardize his situation.[22]

Sakharov's excellence as a researcher and his reserved demeanor made his judgment sought after beyond narrow scientific problems. He was often in the group of physicists who were invited for discussions with Lavrentii Beria, the supreme head of the nuclear project. These meetings had their special choreography; those who were summoned arrived in Moscow and had to wait for the actual appointment. The wait sometimes lasted a whole week—it was not a subtle expression of who was important and who was not. Then finally the group was collected and taken to the meeting with Beria.

On one occasion Beria received Sakharov alone. At the end of the encounter Beria invited Sakharov to pose a question. Sakharov was taken aback for a moment, but then blurted out what he had on his mind. He asked Beria, "Why do we always lag behind the USA and other countries, why are we losing the technology race?" Beria's response was very general about the lack of research and development and of a broad manufacturing base. Beria must have realized that whereas the Soviet State had the capability to focus on and solve a selected problem even if it was a huge challenge, this capability was a limited one. On Sakharov's part, it was a genuine question; at the time he did not realize that it reflected on the long-term endurance of the Soviet regime. Twenty years later, he would arrive at his own answer when he understood that "insufficiently democratic institutions and a lack of intellectual freedom and free exchange of information were to blame."[23]

In 1953, when Stalin died, Sakharov commented to his wife: "I am under the influence of a great man's death. I am thinking of his humanity."[24] He felt that he and Stalin toiled toward the same goal, making the country strong. At this time, Sakharov still considered the Soviet regime to be the way of the future that other countries would one day emulate. There were, however, eye-opening changes in Soviet society following Stalin's death. One of them was the freeing of Jewish doctors, who had been accused of plans to assassinate Stalin and other Soviet leaders. Now the accusations were declared baseless, and the accusers were arrested.

Beria's fall followed Stalin's death by a few months, but preparations for the forthcoming test went ahead at full speed; it appeared that nothing was being overlooked. A new director who would replace Yulii Khariton, the scientific leader of Arzamas-16, was even designated in case of failure. The designated replacement was the mathematician academician Mikhail Lavrentiev. He traveled with the leading Arzamas-16 figures to the test site near Semipalatinsk in faraway Kazakhstan. As it happened, after the successful test, Lavrentiev returned to Moscow.

Not long before the target date for the test, it was discovered that the fallout problem had not been considered. Frenetic activities ensued in which the Soviet scientists thoroughly examined the American manual on the effects of nuclear explosion, the so-called Black Book.[25] Finally, evacuation of the civilian population was hastily arranged from the regions that were deemed susceptible to fallout. In retrospect, the evacuation proved to be a prudent decision.

The tension among the creators of the hydrogen bomb and among all others involved in the test was mounting. It was further enhanced when one week before the test, Georgii Malenkov declared at the session of the Supreme Soviet that the Soviet Union possessed the hydrogen bomb. Malenkov was the chairman of the Council of Ministers, and to make this announcement prior to the test was uncharacteristic of the secretive Soviet leadership. It showed their eagerness to boast of this achievement, which would testify to their ability to carry on without Stalin and Beria.

The test took place on August 12, 1953. On its eve, Sakharov followed Zeldovich's advice and took a sleeping pill to make sure that he would get a good night's sleep and be in full command of his capacities on the great day. As soon as the blast happened, it was obvious that its power was unprecedented. The minister of Medium Machine Building (the nuclear authority by a camouflaged name), Vyacheslav Malyshev, told Sakharov about Malenkov's congratulatory telephone call and kissed the physicist. Clearly, Sakharov was the hero of the day. Malyshev and Sakharov immediately walked over ground zero. This may have been too hasty an action. Many years later, Sakharov suspected that this experience may have been the cause of health problems for Malyshev and himself.

Soviet nuclear history counts the August 12, 1953, test as the first deliverable hydrogen bomb not just in the Soviet Union but worldwide. No doubt, it was deliverable; it was dropped from an airplane. But it is questionable whether it was a genuine hydrogen bomb. It used a mixture of fission fuel, uranium-235, and fusion fuel, lithium(6) deuteride. Its yield was about 400 kiloton (0.4 megaton) TNT-equivalent, very low as far as hydrogen bombs go. The American nondeliverable thermonuclear device of enormous weight and size, "Mike," had been tested on November 1, 1952. Its yield was 10,400 kiloton (10.4 megaton) TNT-equivalent. The ultimate hydrogen bomb in the Soviet program, incorporating the third idea, the radiation implosion approach, would be tested only in 1955.

Following the successful test of August 12, 1953, Sakharov and his colleagues were taken on special trips and visits. He was especially impressed by the ballistic missile plant, an operation on an even larger scale than Arzamas-16. Sergei Korolev was the chief constructor of missiles, and his name was as classified as the names of the creators of the nuclear weapons. Hardly anybody knew anything about Korolev's past. He was arrested in the 1930s, and it was by chance that the aircraft designer Andrei Tupolev, who was also incarcerated, saved him and involved him in building his aircrafts. At the time they both worked in a so-called *sharashka*—a labor camp where the prisoners were highly skilled scientists and engineers. Following their first encounter, Sakharov and Korolev developed a good relationship; their activities overlapped in that Korolev's missiles were to deliver Sakharov's warheads.

Sakharov was promoted to Tamm's position as head of their theory group at Arzamas-16 when in 1953 Tamm left the installation and returned to Moscow. Sakharov was still very young, and he still had the lowest scientific degree, Candidate of Science (PhD equivalent). However, he was so much favored by the Soviet leadership that they talked about promoting him to membership of the Science Academy. He did not yet have the prerequisite Doctor of Science degree. Only a Doctor of Science may be elected to corresponding member and a corresponding member may then be elected to full member, the pinnacle for a Soviet scientist (as it is for a Russian scientist today). Before the October 1953 elections to the Academy of Sciences, the last elections had been held in 1946, so there was a heightened level of excitement during the preparations for the coming elections. The timing of the test of the hydrogen bomb was most fortunate for the nuclear scientists expecting promotion.

Even before the test, Kurchatov had already announced his plans to nominate Sakharov for corresponding member. For this, Sakharov had to obtain the Doctor of Science degree, which was hastily arranged. Arzamas-16 had such a rich population of high-level scientists that it had the right and the proper organizational structure to approve scientific degrees, although it could not confer them. This was the prerogative of the State Commission of Accreditation, but in such cases it was a mere formality. Once Sakharov had received this promotion, Kurchatov changed his mind and declared that Sakharov should be elected full member of the Academy at once. Not only was his nomination approved by the division of physicists and mathematicians, the vote was unanimous. Then, on October 23, 1953, Sakharov was elected academician—that is, full member—by the General Assembly of the Academy. He was thirty-two years old at the time of his election, and he had moved from Candidate of Science to academician virtually overnight.

Of the other heroes in this book, Tamm was elected full member on this occasion; he was fifty-eight years old; and Yulii Khariton, the scientific head of Arzamas-16, was also elected full member at this time; at forty-nine-years of age. Both had been corresponding members of the Academy for years. The thirty-nine-year-old corresponding member Zeldovich had been nominated, but not elected to full membership; this was embarrassing, not the least for

Sakharov. Both Tamm and Sakharov received their first Hero of Socialist Labor distinction, while Khariton, Zeldovich, and a few others received it for the second time. Sakharov was among those who were awarded the Stalin prize with its huge monetary reward and an expensive dacha in Zhukovka, a Moscow suburb.

In the meantime, the Soviets decided to launch a second weapons laboratory, which became known as Chelyabinsk-70. It came about amid discussions of issues similar to those in the United States when the question of organizing the Livermore laboratory in addition to Los Alamos arose. Those who favored the establishment of a second installation cited the benefits of competition; opponents were afraid that a second installation might have a negative influence on the strong concentration of high level scientific personnel of the first one. Comparison of the two eventually showed a marked difference. Arzamas-16 had a large number of Jewish scientists in leadership positions; Chelyabinsk had hardly any. Sakharov writes that although he was not Jewish; he was usually counted together with the Jewish lot. Some officials in the Ministry of Medium Machine Building referred to Chelyabinsk-70 as "Egypt" and to the dining room of Arzamas-16 as "the Synagogue."[26]

Andrei Sakharov and Igor Kurchatov in 1958 at the Atomic Energy Institute in Moscow.

Source: Courtesy of Alexander Vernyi and Lyubov Vernaya-Sakharova, Moscow.

The next step in the quest for the hydrogen bomb was the application of radiation implosion for compressing the thermonuclear fuel, which was tested on November 22, 1955. The test was successful, but was accompanied by personal tragedies. A little girl died in a bunker where her mother had placed her for safety. The bunker collapsed under the impact of the shock wave generated by the explosion. A young soldier died in the trenches where his platoon was supposed to be sheltered. There were also other injuries. Sakharov had mixed feelings: "We were stirred up, but not just with the exhilaration that comes with the job well done. For my part, I experienced a range of contradictory sentiments, perhaps chief among them a fear that this newly released force could slip out of control and lead to unimaginable disasters."[27] Marshal Mitrofan Nedelin, who represented the Soviet leadership at the test, invited senior personnel from all the organizations involved with the project for a celebration. He asked Sakharov to make the first toast on this occasion. The toast and the marshal's response have become a memorable exchange.

Sakharov was no orator, but he wanted to say something more than the usual banalities. He expressed his desire in the following way: "May our devices explode as successfully as today's but always over test sites and never over cities."[28] To me, this should have been considered nothing more than an expression of peaceful desire by a scientist. Apparently, to those present, it meant something more. Sakharov noted that there followed a silence as if he "had said something indecent." To Nedelin, the statement must have appeared out of the ordinary because he felt compelled to tell his audience a lewd parable whose message was to leave politics to the politicians: "An old man wearing only a shirt was praying before an icon. 'Guide me, harden me. Guide me, harden me.' His wife, who was lying on the stove, said: 'Just pray to be hard, old man; I can guide it in myself.' Let's drink to getting hard."[29]

The marshal's rude response brought home how Sakharov's words had been interpreted. Even if Sakharov's words could have been dismissed as an expression of pacifist sentiment, the response put him in his place. It was a humiliating experience for the scientist-hero in front of a large and sensitive audience. Sakharov understood where his place was; or, rather, where people like Nedelin thought the scientists' place was. They were expected to make discoveries and produce bombs, but there should not even be a pretense that they might have a word in decision making, not even at the level of expressing desires. Sakharov learned a life lesson. A few years later he got a second one from a higher authority, and yet more explicit.

In July 1961, Nikita Khrushchev convened a meeting of political leaders and the atomic scientists. By then, in addition to his preeminent party position, he was also Chairman of the Council of Ministers; thus he was, in one person, the supreme authority of the country. This gathering took place just a few months before the test of the so-called Czar Bomb on October 31, 1961, the largest nuclear blast in the world.

Soviet nuclear might was at its peak, and Khrushchev was taking advantage of its potential. There was international tension and the Soviet Union was to announce the end of a moratorium on testing. Just a few weeks before this meeting, Khrushchev had met with the new American president, John F. Kennedy. The Berlin wall had just gone up. Khrushchev was eager to let the scientists demonstrate their support and preparedness for strengthening their Soviet fatherland. He opened the meeting, and then leading representatives of those present talked about their work. The decision to resume testing was political; the scientists were not invited to discuss the decision.

When it was Sakharov's turn, in addition to reporting about the work, he voiced his opinion that he did not expect much benefit from the resumption of testing. However, he did not feel quite satisfied that his message got across and after the speech wrote a note for Khrushchev in which he warned that the new tests would jeopardize the test-ban negotiations and disarmament. Khrushchev received the note, read it, and pocketed it. When all the speeches were over, he invited the participants for dinner, following a break. I mention this because even if Sakharov's note had upset Khrushchev, he had time to cool down; but this was not the case.

When the gathering reconvened, Khrushchev started with a speech before any food had been served. First, he raised his glass but then lowered it back to the table, signaling that this was not going to be a toast. The tension in the hall was mounting. Khrushchev devoted his entire speech to Sakharov's note, which he did not read to his audience. Among other things, he said that[30]

> [Sakharov] has moved beyond science into politics. Here he is poking his nose where it does not belong. You can be a good scientist without understanding a thing about politics. Politics is like the old joke about the two Jews traveling on a train. One asks the other: "So, where are you going?" "I am going to Zhitomir." "What a sly fox," thinks the first Jew. "I know he's really going to Zhitomir, but he told me Zhitomir so I'll think he is going to Zhmerinka."
>
> Leave politics to us—we're the specialists. You make your bombs and test them. And we won't interfere with you; we'll help you. But remember, we have to conduct our policies from a position of strength. We don't advertise it, but that's how it is! There can't be any other policy. Our opponents don't understand any other language.... Sakharov, don't try to tell us what to do or how to behave. We understand politics. I'd be a jellyfish and not Chairman of the Council of Ministers if I listened to people like Sakharov!

There were about sixty guests present that evening, and none looked in Sakharov's direction. Only one of the sixty, one of his physicist colleagues, Yurii Zysin, came over when the meeting ended and told Sakharov that he supported his position. Khrushchev was not the first and would not be the last to warn the physicist not to meddle in politics. In February 1987, when Sakharov was back in Moscow following a seven-year exile in Gorky, he participated in the Forum in Moscow for a Nuclear-Free World and the Survival of Mankind.

When he criticized the policies of Gorbachev's government, Evgenii Velikhov, Vice President of the Science Academy, countered by saying that "scientists should not interfere in politics."[31]

In 1987, Sakharov simply ignored Velikhov's remark; but in 1961, Sakharov's humiliation by Khrushchev was complete, and so was his alienation. Nonetheless, he continued working at Arzamas-16 for years and left the project only in 1968. Sakharov was not the type whose emotions determined his actions. His humiliation was of secondary importance; the primary importance was that he found additional testing unnecessary and harmful. He had come to this conclusion as a result of research, which was characteristic of his approach to crucial concepts.

He had started worrying about the biological consequences of testing as early as the mid-1950s. When he expressed his reservations about testing, a high-level official, Nikolai Pavlov, told him, "If our work and our testing are giving us strength for that battle [against the forces of imperialism]—and they certainly are—then the victims of that testing, or any other victims, don't matter."[32] Pavlov was competent and well informed; he respected Sakharov and called him "our gold reserve."[33] His reckless evaluation of the collateral damage from testing did not differ much from Edward Teller, who from the late 1950s did everything he could to prevent cessation of testing.[34] Teller knew that fallout affected reproductive cells, leading to mutations and abnormalities in future generations. He posed and answered the question: "Are abnormalities harmful? Because abnormalities deviate from the norm, they may be offensive at first sight. But without such abnormal births and such mutations, the human race would not have evolved and we would not be here. Deploring the mutations that may be caused by fallout is something like adopting the policies of the Daughters of the American Revolution, who approve of a past revolution but condemn future reforms."[35]

In his youth, Sakharov was interested in genetics, and this interest was rekindled by the fact that the long-term biological consequences of nuclear explosions are nonthreshold events. Nonthreshold events are those for which there is no minimum amount of radioactivity below which no damage might be possible—any small amount might induce damaging genetic changes. Sakharov felt increasing responsibility to learn more about the biological effects of the tests and to disseminate this knowledge. His curiosity encouraged him to go in this direction. An added reason for his inquiry was the deplorable state of biology in the Soviet Union. The biologists were under the tight grip of Trofim Lysenko, and any unbiased investigation of heredity was considered heresy. Only the physicists were in the position to carry out investigations without interference.

Fortunately, Sakharov was not alone in this undertaking. Tamm had also been worried about the state of biology in the Soviet Union and had helped Sakharov to become versed in its problems. Kurchatov, being the director of a big institute concerned with nuclear energy and a powerful science administrator himself, also decided to take action. He created a section in his institute for biologists who had lost hope for employment in bona fide biological research institutes. The possible biological effects of nuclear events, such as testing, provided a legitimate justification for such research in a nuclear physics institute.

The physicists were counting on the assistance of the president of the Soviet Academy of Sciences, Aleksandr Nesmeyanov, who was rather hesitant but at least did not oppose taking steps meant to salvage biology. In time, he himself facilitated some measures in this direction (see chapter 12). The main culprit in allowing Lysenko to continue to reign a decade after Stalin's death was Khrushchev. Lysenko succeeded in convincing the new Soviet leader that his approach, and only his approach, to agriculture could save it. Lysenko denied that genetics was a science, while, ironically, he was the director of the Institute of Genetics. He believed that acquired traits were hereditary; and he fought his opponents with the harshest means. Sakharov's study eventually wandered into an area of science that had long been taboo in the Soviet Union. He stated that "the simplest non-threshold effect is the influence on heredity.... A single ionization event is sufficient to cause irreversible change—a mutation—in a gene." The struggle against Lysenkoism is a separate and important chapter in the history of science in the Soviet Union; however, our interest here is limited to Sakharov's contribution to uncovering the biological consequences of nuclear testing.

Sakharov started seriously dealing with this issue at the time when in the United States the idea of the so-called clean bomb was advanced by some scientists and welcomed by the government. The suggestion was that a clean hydrogen bomb would cause no radioactive contamination of the environment. While it was true that such contamination could be reduced, it was impossible to eliminate it completely because the trigger of the hydrogen bomb was a fission bomb and the thermonuclear explosion itself also added to the radioactive contamination. About one-third of the damage originated from the production of strontium-90 and cesium-137 isotopes during the explosions, and they had immediate harmful effects. About two-thirds of new cancer cases or genetic disorders arose from carbon-14. The neutrons, freed as a consequence of thermonuclear explosion, interact with the nitrogen in the atmosphere, and such interactions produce carbon-14 isotopes of carbon. It is a long-lasting radioactive isotope with a half-life of five thousand years and would harm living organisms, including humans, for thousands of years. The production of carbon-14 accompanies all thermonuclear explosions whether or not they come from "clean" bombs. Sakharov's overall estimate was that *for every one-megaton-TNT-equivalent detonation of nuclear bombs there would be ten thousand human victims.*[36]

The relationship he established between the amount of testing and the anticipated number of victims was frightening. By 1957, nearly fifty-megaton (fifty-thousand kiloton) TNT-equivalent explosions had taken place. Sakharov estimated that such events would eventually result in a half million victims. For the time being, Sakharov's working on the problem of the hazards of the clean bomb seemed to serve Khrushchev's political aims, because it discredited American claims about the "humane" character of the clean bomb. Khrushchev's government had just declared a moratorium on testing, and all the blame could be shifted to the Americans. When Sakharov communicated his estimates, he indicated a lower limit and a higher limit, the way scientists usually provide estimates. To strengthen Sakharov's points, the Soviet editors—and his paper appeared in various versions and in various venues—deleted the lower limits in Sakharov's estimates, and quoted only his upper limits. This made his numbers appear yet more frightening than he had intended, and reduced their credibility before expert readers. Khrushchev's involvement went beyond just sanctioning the publication of Sakharov's data; he introduced editorial changes in the articles. Khrushchev did not discuss the changes with Sakharov; it was Kurchatov rather than Sakharov who discussed Sakharov's manuscripts with the Soviet leader.

During the late 1950s, the battle of testing and moratoriums continued. Sakharov found it unacceptable that Khrushchev was using the testing of nuclear weapons to further his political goals, ignoring the opinions of the scientists. The largest-ever nuclear bomb was detonated on October 30, 1961, on the Novaya Zemlya Archipelago. It was a fifty-megaton explosion, but the bomb could easily have been enhanced to be a 100-megaton version. This giant bomb was called alternatively the "Czar Bomb" or the "Big Bomb." The test was timed to coincide with the Twenty-Second Congress of the Communist Party of the Soviet Union. It was the surest sign of Khrushchev's megalomania. It had no strategic significance; and no such bomb has been detonated since.

By the time of the Big Bomb, Sakharov had already drifted away from the official Soviet aims, and he was appalled that the health hazards had no restraining effect on the Soviet leadership. Seen in this light, it is most puzzling that Sakharov at about this time came up with a proposal for the efficient deployment of the Big Bomb in its 100-megaton version. Also, he used the plural "100-megaton charges" rather than singular "charge." In his words[37]:

> After the test of the Big Bomb, I was concerned that the military couldn't use it without an effective carrier (a bomber would be too easy to shoot down). I dreamed up the idea of a giant torpedo, launched from a submarine and fitted with an atomic-powered jet engine that would convert water to steam. The targets would be enemy ports several hundred miles away. Naval experts assured us that the war at sea would be won if we could destroy the enemy's harbors. The torpedoes' bodies would be made sturdy enough to withstand exploding mines and to pierce anti-torpedo nets. When they reached their targets, the 100-megaton charges would explode both underwater and in the air, causing heavy casualties.

When Sakharov discussed his proposal with the high-ranking naval officer P. F. Fomin, the admiral, according to Sakharov, "was shocked and disgusted by the idea of merciless mass slaughter."[38] This was the darkest point in Sakharov's career, and had it not been described in his own *Memoirs*, it might be hard to believe the authenticity of the story. Yet there were other projects in which Sakharov anticipated military orders before they materialized. His overzealous behavior can only be explained by his getting blinded by the almost limitless possibilities of using physics for destruction. Yet these works and ideas happened at the time when he was already getting disillusioned with the arms race and was increasingly opposing the tests. But in spite of his clashes with Khrushchev, outwardly everything continued as business as usual, and for the successes of 1961, in spring 1962, he received his third Hero of Socialist Labor star along with other distinctions. Khrushchev may have humiliated the physicist, but he recognized the value of his contribution.

Sakharov's award and his description of his activities suggest consistency, but there are those who believe that this is too simplistic a perception of what Sakharov represented at that time. His suggestion to the Soviet Navy described above tormented him by the time he was writing his *Memoirs* during his exile in Gorky. His physicist son-in-law, Alexander Vernyi, however, questions the sincerity of the admiral's reaction to Sakharov's proposal. Admiral Fomin was not merely a naval officer; he was also a ministry official at the time, actively engaged in the application of nuclear weaponry for the Soviet Navy. This included the possibility of inducing a gigantic tsunami by underwater nuclear explosion. According to Vernyi, an admiral feeling shame for what he is engaged in would represent a drama of Shakespearian proportions. It is more realistic to group Admiral Fomin together with Marshal Nedelin than with a tragic Shakespearean hero.[39] One of Sakharov's biographers, Gennady Gorelik, simply states, "Sakharov's tale tells us more about his conscience than about historical reality."[40] We may also suppose that even an admiral's insincerity must have a reason. Could it be that Fomin camouflaged the Navy's engagement in the creation of tsunamis even before the most distinguished Soviet physicist, whose explosives were to be used to execute the plans?

The negotiations for a test ban led to a resounding success when on August 5, 1963, an agreement was signed in Moscow, to take effect on October 10 of that same year, forbidding tests in the atmosphere, in outer space, and underwater. Sakharov was among those who felt the satisfaction of having contributed to this most important step to limit biological damage in future generations. At this time Sakharov was already thinking of his life after nuclear weapons; he started contemplating his return to basic science but stayed for five more years at the installation. At the same time he was more actively participating in the elections to the Science Academy. and contributed to the defeat of a Lysenko protégé in the 1964 elections.

Zeldovich played an important role in Sakharov's return to basic science. Zeldovich never completely tore himself away from science, and tried to stay informed about major developments, especially in cosmology and astrophysics. Zeldovich and Sakharov were two very different individuals, but the differences in their lifestyles did not hinder their splendid symbiosis in discussing science. When in 1950 Tamm and his young associates appeared at Arzamas-16, forming a second theory group where Zeldovich and his group had been alone, friction, or at least some rivalry, might have developed between them. Instead, truly creative interaction ensued between the two groups and the amicable and fruitful cooperation between Zeldovich and Sakharov was its most conspicuous manifestation. Zeldovich benefited from having such a genius partner as Sakharov for discussions, and for Sakharov, Zeldovich was his window to basic physics during the years when Sakharov immersed himself in the development of the hydrogen bomb. This is not to say that Sakharov did not do any fundamental physics during his immersion in weapons work. We have already mentioned his work in the early 1950s on controlled thermonuclear reactions. There were further problems to be solved related to the physics under extreme conditions, and there he found sufficient challenges. In the 1960s, he was captivated by the magnificence of unsolved problems in cosmology. In 1964 he produced his first paper about the puzzles of the expanding universe.

One of his best-known works, from the mid-1960s, was related to the evolution of the universe. It is usually referred to as "baryogenesis"—the origin of baryon asymmetry of the universe, or BAU. Baryons are protons and neutrons, and they make up most of the mass of the universe. The issue of BAU deals with the fundamental question of why our world is made of matter and not antimatter, or why is there matter at all in the universe? When the Big Bang happened and the universe was formed, it was extremely hot, representing enormous energy, and it produced both particles and antiparticles. With gradually decreasing temperature, the particles and antiparticles annihilated each other in pairs. Had they been around in equal numbers, the universe would have become empty. It turns out that there must have been some excess of matter over antimatter, even if a very tiny excess, in the early universe—this is "baryon asymmetry"—and from this, it followed that the universe now consists of matter. The big question is, why was there an excess of matter over antimatter? Physicists have been looking for an answer, and Sakharov facilitated these attempts. He did not provide the answer, but in 1967, he set up three conditions that must be fulfilled by any explanation of baryon asymmetry.

The first condition is that processes must exist that are able to change the baryon number. The second is that there must be a bias in the laws of nature favoring matter over antimatter. The third is that the processes that can change the baryon number must happen in a state not characterized by thermal equilibrium. This latter condition seems difficult to grasp at first, but it seems quite natural if we take into account that from the moment of the Big Bang the universe has existed in the state of thermal transition—continuous cooling. The concept of BAU is a most attractive manifestation of the union of the physics of fundamental particles and cosmology. Sakharov's contribution to the investigation of BAU played a role in the emergence of cosmomicrophysics.[41] The quest to understand BAU has continued.

During Sakharov's last years at the installation, the most intriguing task he was engaged in was the investigation of the antiballistic missile (ABM) systems. He summarized his views on the ABM systems as follows: (1) a foolproof ABM defense was not possible; it was very expensive, and it could be neutralized at much lower expenditures; and (2) the deployment of an ABM system would be a destabilizing factor because it would upset the strategic balance between the two superpowers.[42] The arms race would continue, but at a higher cost and sophistication. This study was Sakharov's last significant contribution to the work at Arzamas-16. As both superpowers realized these problems, the result was the 1972 Treaty on the Limitation of Antiballistic Missile Systems. By this time, however, Sakharov had already been dismissed from Arzamas-16.

When Sakharov returned to Moscow, he was already immersed in what is usually referred to as the "dissident movement." Even prior to his dismissal, steps were taken to punish him, such as when he was removed from his position as department head and his salary was reduced. This was not the first and would not be the last decrease in his earnings, but Sakharov lived frugally and spent much less than he earned. His family did not spend much money either, and he had a considerable bank account. No such governmental action would stop Sakharov from his political activities.

Sakharov's involvement in social questions started with his taking stands on general political issues. For example, he signed a collective letter in opposition to any attempt to rehabilitate Stalin (meaning "exonerate," but the Russian language uses the word "rehabilitate"). He joined the signatories of a protest against a change in the criminal code that was meant to make it easier to prosecute dissidents, and prosecute them severely. He participated in a silent demonstration of protest on December 5, 1967, the Day of the Soviet Constitution. So far, these were general actions, but soon he became involved in interventions on behalf of individuals. He was interested in a wide variety of issues wherever he found injustice. He became dedicated to the efforts to save Lake Baikal from pollution. He was a great asset for every movement he decided to join. He never wanted to take over the leadership of any movement, and he never wanted to impose his views. He had excellent connections; at this time he could still contact even the leaders of the country. Sakharov was a unique phenomenon; a conspicuous member of the Soviet Establishment; one who did not care about his perks but whose status as an academician, past role in creating Soviet military might, and three Hero of Socialist Labor stars radiated authority and induced respect.

In 1968, Sakharov published his essay "Reflections on Progress, Peaceful Coexistence, and Intellectual Freedom." He warned the human race of the dangers of "thermonuclear extinction, ecological catastrophe, famine, and uncontrolled population explosion, alienation, and dogmatic distortion of our conception of reality."[43] He argued for *convergence* between the socialist and capitalist systems at all possible levels: economic, social, and ideological. His arguments were benevolently naïve, but the world appeared hungry for such an approach. In the first two years of its publication, almost twenty million copies of the essay were printed. To characterize its popularity, one of the best-known Soviet dissidents called it "handy as a spoon at dinnertime."[44] Sakharov survived the publication of the essay unscathed, but the authorities persecuted those who helped to disseminate it. Its appearance tragically coincided with the Prague Spring and its ruthless suppression. Sakharov had become one of the best-known Soviet dissidents. It was inevitable that his path would cross with that of another great dissident, the famous writer Aleksandr Solzhenitsyn. In 1970, Solzhenitsyn was awarded the Nobel Prize in Literature and was exiled from the Soviet Union. His nationalistic and religious views were in conflict with Sakharov's liberalism, but the two had mutual respect for each other.

Sakharov's first wife, Klava, gave him much freedom and never complained about his political activities. She did not even voice apprehension over activities that must have seemed to put their very comfortable life at risk. She became ill, and when she was properly examined the diagnosis was incurable cancer. She died in early 1969, and her ashes were buried at the Vostryakovsky Cemetery in Moscow. Apparently, Sakharov made no attempt to secure a plot at Novodevichy Cemetery to which he undoubtedly would have been entitled.

Upon his departure from Arzamas-16, Sakharov was transferred back to FIAN in Moscow as a senior scientist in Tamm's department, continuing where he had started twenty years before. When Sakharov was departing from Arzamas-16, he closed out his bank account and gave away his substantial savings to the Red Cross and other organizations. The next decade was a story of courage and dedication. Sakharov did not have to be talked into waging protests; he was on the lookout for causes, participated in setting up watchdog organizations for human rights abuses, and fought against the use of psychiatric incarceration. The inhuman nature of the Soviet regime appeared before him in its naked reality. He attended trials, fought for religious freedom and against anti-Semitic discrimination, and for the right to emigrate. The number of causes he took up was limitless.

In 1970, Sakharov met Elena (Lusia) Bonner, his future second wife and future partner in his heightened human rights activities. Bonner's mother came from a Siberian Jewish family, and her father was an Armenian communist leader, Gevork Alikhanov, who perished during Stalin's 1937 campaign of terror. Elena Bonner was a war hero in the medical service; later she trained as a pediatrician. She was a highly cultured woman and had been a fearless human rights activist long before she met Sakharov. She and Sakharov married in 1972. There are those who ascribe Bonner's damaging influence over Sakharov, as if she had radicalized him and sacrificed him for purposes of her family, for example, to secure permissions for some family members to emigrate, rather than only for universal human rights. However, Sakharov's *Memoirs* reveal unambiguously that he received enormous strength and purpose from his association with Bonner.

When Sakharov was awarded the Nobel Peace Prize in 1975, and was not allowed to travel to Oslo to accept it in person, Bonner substituted for him. There was a spontaneous torchlight demonstration in Oslo on the evening of the Nobel Prize Award Ceremony. I happened to be visiting at Oslo University. The demonstration impressed me. It was very different from the May 1 demonstrations in Hungary in which I had taken part, which were never spontaneous, and in which participation was never voluntary.

During the days of the Nobel festivities Sakharov was in Vilnius, Lithuania, to attend the trial of one of his closest friends, another human rights activist. Alas, Sakharov was not allowed to enter the courtroom. In Oslo, his acceptance speech and his lecture were both read by Bonner. It was particularly moving when Sakharov's Nobel lecture mentioned by name many human rights activists who were in jail. He asked the participants of the ceremony to consider these activists as his personal official guests at the ceremony.

The way the Soviet leadership reacted to Sakharov's actions multiplied their impact. They fought against Sakharov with ferocity, but without much success. They tried everything they could to silence Sakharov, but the effect was often the opposite. One of their most despicable acts was organizing a letter by Sakharov's fellow academicians criticizing his activities. Forty academicians signed the letter. There were various stories about how the signatures had been collected. Kapitza declined to sign it. Zeldovich was not asked or could not be located. When Anatoly Aleksandrov, a future president of the Academy was called, his wife answered the phone and told the caller that her husband was drunk and could not come to the phone.[45] The letter was part of a big media campaign against Sakharov in which the most diverse segments of the population represented themselves with published letters condemning Sakharov's activities.

The most powerful means of protest for Sakharov was a hunger strike, and he exercised it for the first time in June 1974. He wanted to turn attention to the plight of political prisoners and timed his action to coincide with President Nixon's visit to Moscow. He continued his research in physics, but he could devote only a limited amount of time to it in view of all his other activities and the emotional toll they took. He met with world-renowned physicists when they visited Moscow, such as Wolfgang Panofsky, Sidney Drell, and Victor Weisskopf. During the 1970s, the Sakharovs had a rough life, but he brought a lot of attention to the human rights movements due to his fame. There is no doubt about his personal courage, but there were some privileges that his position and his awards ensured that distinguished him from his comrades in these movements. Among his perks, for example, was that he could at any time order a car with a driver from the car pool of the Science Academy to take him and his wife wherever he wanted to go. He did not experience financial hardship due to his considerable allotment as a full member of the Academy in addition to his salary at FIAN.

In January 1980, it was announced that the Presidium of the Supreme Soviet decided to strip Sakharov of his state awards. Although Sakharov refused to hand over his stars and ribbons, officially he was no longer accorded the privileges of (three-time) Hero of Socialist Labor. This did not change much in Sakharov's appearance because he never wore his distinctions on his jacket anyway, but it was a warning sign on the part of the authorities that they were ready to take harsher steps against him. The next one would be more extreme.

Sakharov's actions were damaging for the Soviet leadership, because he showed the real nature of the Soviet regime. When he criticized the Soviet intervention in Afghanistan, this was yet another critical point that Soviet officialdom found hard to tolerate. The Supreme Soviet banished Sakharov from Moscow to a place where he would not be able to maintain contact with foreigners. Gorky (today, as in the past, Nizhnii Novgorod) was such a city, and he was exiled there. Sakharov never received official notification about this decision as he had about the revoking of his awards. He spent the next seven years in Gorky, and Bonner went with him voluntarily.

It was a most trying time. The authorities did not spare themselves in coming up with new ways of making the Sakharovs' life unbearable. It seemed that a whole army of secret police was assigned to the case. Sakharov spent a lot of time writing his memoirs, and his manuscripts were repeatedly stolen from him. He demonstrated exceptional willpower in that he never gave up and started anew when necessary. As a result, he had a complete manuscript to publish when his exile was over. There was one step that the authorities did not dare take; they did not make the Science Academy exclude Sakharov from its membership. Moreover, Sakharov remained a member of the Igor E. Tamm Department of Theoretical Physics of FIAN (as it became to be called following Tamm's death) for the entire duration of his exile.

It was a delicate relationship. After Tamm's death in 1971, Vitaly Ginzburg became the head of the department. Ginzburg and Sakharov had worked out a tacit agreement about their relationship when Sakharov's human rights activities and his troubles with the authorities began.[46] Accordingly, Sakharov would not try to involve his colleagues at the department in his various protests, would not invite them to demonstrate or urge them to sign documents of protest. In turn, the members of the department would stay loyal to Sakharov, would help him in his scientific endeavors, and—as his exile would make necessary—would not abandon him in even greater troubles. Sakharov kept his word and adhered to these guidelines, which were never spelled out, and Ginzburg and most of the members of the department never signed any protest against Sakharov's activities. A rare exception happened when a junior associate in the Tamm department was threatened by the institute management who told him that if he did not sign the letter, which was just being circulated, they would not allow him to defend his dissertation. Other junior members in other departments of FIAN were easier targets for blackmail. If they denied signing an anti-Sakharov letter, they were fired. Boris Zeldovich (Yakov Zeldovich's son) was at the time a junior member of FIAN; he refused to sign and was a rare exception in not getting fired; supposedly, his name protected him.[47]

His colleagues in the department visited Sakharov in exile annually, sometimes more frequently—Ginzburg visited him twice—to help him stay informed about developments in physics. Sakharov's dedication to science was evident in that he did not stop doing research under the harsh conditions of his life in Gorky. He worked on the cosmological theories of the pulsating universe. He described the fluctuations of vacuum in the first moments of the birth of the universe, called "Sakharov oscillations." His papers appeared in the early 1980s in the Soviet scientific literature. Sakharov never lost his interest in and dedication to science, but the world around him forced him onto a route where science had to take a secondary role.

Andrei Sakharov with granddaughter Marina (daughter of Tatyana Sakharova and Mikhail Liberman) in 1983 during his exile in Gorky.

Source: Courtesy of Mikhail Liberman, Uppsala, Sweden.

Sakharov's ordeal in Gorky continued for seven long years, and it did not become any easier with time. The KGB devoted limitless resources to irritating him, and did everything possible to minimize his outside contacts. They spied on him all the time and listened to his conversations. Sakharov was aware that his apartment was being tapped. This notion gave rise to a story about Sakharov and a visiting colleague who were about to discuss a physics problem that included classified information. Both Sakharov and his visitor possessed the necessary security clearance. Yet Sakharov suddenly stopped the conversation, saying that the KGB officers listening in might not possess the necessary security clearance. Nobody hearing the story could be sure whether he was being serious or sarcastic, probably both.[48]

There is no doubt that Sakharov had rigorously observed the secrecy of the Soviet weapons program, and had vowed to adhere to it to the end. And he never hesitated in his belief that the development of the Soviet nuclear weapons was necessary to maintaining peace in a world in which two superpowers were facing each other. This was part of his comprehensive approach to the global political situation.[49] He recognized the need for deterrence, for which parity in nuclear weaponry was a precondition, and he believed in the benefits of negotiations based on trust stemming from cooperation and openness. The contribution by Sakharov and the other Soviet scientists to deterrence and parity can be appreciated in light of the fact that the world lived for decades without direct war between the two superpowers. This was achieved by both sides possessing the most terrible bombs imaginable.

During his Gorky exile, Sakharov had to resort to hunger strikes on several occasions when this seemed to him the only remaining way to wage a protest. They posed considerable hazards to Sakharov's health, but there was no sign that he ever had second thoughts about the course he had taken. The Soviet regime did not seem to budge, nor was there any change in the approach of the Soviet leaders toward him in particular and toward human rights in general. Leonid Brezhnev was followed by Yurii Andropov, and Andropov by Konstantin Chernenko; but they did not differ in policy.

Against this background of changing Soviet leaders, Sakharov stayed in place. The war in Afghanistan continued raging. Sakharov continued protesting that, too. It was during the initial stages of this war that the 1980 Moscow Olympics were boycotted by many democratic countries as a protest against the Soviet war in Afghanistan. The boycott generated a lot of criticism, citing its ineffectiveness in changing Soviet policy. From where I was observing these events—in Eastern Europe—there could have not been a more efficient way of hurting the Soviets than this boycott. At that point, hardly anything could have impacted Soviet foreign policy, but the boycott made the Moscow Olympics incomplete and thus accomplished something unique: it hurt Soviet pride.

Sakharov never for a moment gave up his struggle. Fortunately, the world did not forget him either. He kept writing, and in spite of all the overt and covert police actions against him, his words continued reaching his audience. In April 1983, he was awarded the Leo Szilard Lectureship Award administered by the American Physical Society. Its purpose is "to recognize outstanding accomplishments by physicists in promoting the use of physics for the benefit of society in such areas as the environment, arms control, and science policy. The lecture format is intended to increase the visibility of those who have promoted the use of physics for the benefit of society."[50] Sakharov's talk was smuggled out to America; in it he spoke about international terrorism, the Soviet invasion of Afghanistan, and the need for arms reduction. In conclusion, he positioned himself in the roster of his role models: "I would like once more to remind you of the profound alarm felt by our great predecessors—Einstein, Bohr, Russell, Szilard—for the fate of mankind, and of the ideas they left us. These ideas—about peace, about the danger that threatens mankind, about the importance of mutual understanding and tolerance, about the openness of society, the respect for human rights, the convergence of states with different political systems, the responsibility of scientists—are as important today as when they were expressed for the first time."[51]

Although the Soviet leadership did not dare to go so far as to revoke Sakharov's Academy membership, there was a point during Sakharov's exile when he himself raised this very question. This was obviously a step he felt had to be considered, when nothing else remained in his hands to force the authorities to retreat. In fall 1984, the issue was the government's refusal to let his wife travel abroad. She wanted to see her mother and her children in the United States, but what made the trip vital was that she needed medical attention that she was certain the Soviet doctors under KGB supervision could not have provided for her.

Sakharov composed a long letter to the president of the Soviet Academy of Sciences, Anatoly Aleksandrov, and when the next batch of FIAN visitors came, he asked them to transmit the letter to Vitaly Ginzburg. Sakharov wrote the letter in longhand and also wrote a copy longhand for Ginzburg. The letter has been reproduced in Sakharov's *Memoirs* and in Bonner's book *Alone Together*.[52] Ginzburg also reproduced it in his book about physics and astrophysics.[53]

In his letter to Aleksandrov, Sakharov calls his situation the "most tragic moment" of his life.[54] He describes his predicaments in great detail. The style of the letter is so reserved and unhurried that under different circumstances it could be characterized as pedagogical. His final conclusion is sober and rigorous: "If you and the Presidium of the Academy of Sciences would not find it possible to support my request in this for my most important tragic cause, the travel of my wife, or should your approach and other efforts fail resolving the problem by March 1, 1985, *I request to consider the present letter to be the announcement of my departure from the Academy of Sciences of the USSR* (italics in the original).[55] As if the meaning of his raising the issue of his resignation from the Academy was not clear, Sakharov explained in his private letter to Ginzburg that he was fully aware of what the consequences would be. One would have been a hardening of his financial situation. He had dispersed his considerable savings long before his exile, and his royalties from his publishing activities in the West did not reach him. Fortunately, this most drastic step did not have to be taken; Bonner was allowed to leave for her trip.

From March 1985, there was a new supreme leader of the Soviet Union, the fifty-four-year-old Mikhail Gorbachev. It is now a question of contention whether he wanted to dismantle the Soviet regime with its rigid and ruthless one-party system or was simply being forced to agree to one change after another. He did bring in a new atmosphere, but he was a product of the Soviet regime. An anecdote about him seems to be quite characteristic. Ostensibly, he declared that the Soviet Union must become a democratic country and that to those who didn't understand democracy, it should be explained patiently what democracy was about. If they still did not understand democracy, it should be explained to them again, and again, and again. "But what about those who would still not understand what democracy was?" one of his lieutenants asked. "At that point, they should be liquidated," Gorbachev responded.

In any case, there were still political prisoners in the Soviet Union when Gorbachev had already assured the world that there were none.[56] And Sakharov was still being banished from Moscow and isolated in Gorky when Gorbachev was already in power. One might say that he could have not been aware of every single case of injustice amid so many. This was not quite so, though. Sakharov had written to Gorbachev and the new Soviet leader did receive his letter. He was well aware of the situation, and there were discussions of the terms of allowing Sakharov back to Moscow. This continued to be the situation for twenty(!) long months following Gorbachev's take over of the Soviet leadership. Finally, one day toward the end of December 1986, in a well-publicized move, Gorbachev ordered a telephone to be installed in Sakharov's apartment in Gorky (he had not been allowed a telephone during the years of his exile). The following day Gorbachev graciously "invited" the Sakharovs back to Moscow.

The next day following Sakharov's return to Moscow, his first business was to attend a FIAN seminar. This was symbolic, because back on January 22, 1980, Sakharov had been hijacked on the way to a FIAN seminar and forced into exile. Beyond symbolism, Sakharov was eager to resume doing physics and to see his colleagues. Alas, he did not have much chance for either during the almost three years that remained of his life. Still, he accomplished a lot during this short period. He became a politician in much demand, and now Gorbachev was also finding him useful as a living example of the new Soviet democracy. It was during this time that Sakharov traveled for the first time abroad.

He embarked on his first trip on November 6, 1988; it was to the United States, where he was greeted as a hero and was received by President Reagan in the White House. This was a symbolic meeting and it happened during the last months of Reagan's presidency. Sakharov tried to discuss his reservations about the Strategic Defense Initiative (SDI) with the American president, but he did not get through to him; Reagan kept repeating his usual statements about the protection of world peace by SDI. A more substantive meeting took place in Washington, DC, between Edward Teller and Sakharov, the two respective fathers of the hydrogen bomb. They may have not seen eye to eye in political matters, but there were issues where they were in agreement without ever having discussed them in any great detail. They considered the utilization of nuclear power inevitable in solving the energy problem of humankind, and they both recognized the safety advantage of placing the nuclear reactors underground. Sakharov urged "that people concerned about the potential harmful consequences of the peaceful use of nuclear energy should concentrate their efforts not on attempts to ban nuclear power, but instead on demands to assure its complete safety."[57] Edward Teller could have issued this warning.

There was a banquet honoring Teller for his contributions to the human rights policy of the United Nations, organized by the Ethics and Public Policy Center, a conservative institution, and about 750 people attended the event. The get-together of the two physicists was brief; Sakharov had a busy schedule and had to catch the last plane that evening from Washington to Boston. Upon his arrival to the banquet, Sakharov and Teller had a private chat; then Sakharov addressed the gathering. He was very critical of SDI; he called it a great mistake and pointed to its enormous costs. Sakharov warned that it could lead to nuclear war rather than help avoid it, and that it would hinder arms control. After Sakharov spoke he had to leave, and Teller could respond to Sakharov's remarks only later in the evening. He dismissed Sakharov's criticism and referred to the fact that the Soviet guest had been cut off from proper communications for years and could not have been properly informed about SDI.

Andrei Sakharov and Ronald Reagan in 1988 at the White House.

Source: Courtesy of the Ronald Reagan Library, Simi Valley, California.

Teller could easily brush off other people's opinions if they were different from his own, and in this case suggesting that Sakharov spoke without being informed was especially misleading. Teller should have not supposed that Sakharov would speak out publicly on a vitally important topic without getting to the bottom of it. Teller did not know that Sakharov had investigated the antiballistic missile systems during the last years of his tenure at Arzamas-16, and I have mentioned his arguments opposing their introduction. Furthermore, Sakharov did his best to become informed about the pros and cons of SDI because he anticipated that he would be asked about it. It was especially exemplary that in addition to his opinion he wanted to be versed in the arguments of the SDI supporters.

Upon his arrival in the United States, he was staying in Newton, Massachusetts, with Bonner's children. Sakharov had made arrangements to see Arno Penzias, a Nobel laureate physicist, immediately after a visit to his physician.[58] Penzias responded to Sakharov's summon with a sense of duty; he flew right away from California to Massachusetts and spent a whole afternoon with Sakharov.[59] The Russian physicist wanted "to hear the arguments being presented in favor of SDI from someone he [could] trust, in order to prepare for them."[60] Penzias was not an expert of SDI; he only had his instinctive doubts about its feasibility, but he was familiar with the evaluation of the experts of the American Physical Society (APS), and he trusted them. According to the APS experts, the X-ray laser would not be able to do what it was expected to do. On the other hand, the promoters of SDI—most of all Reagan and Teller—did an excellent job of enhancing the public image of the program. Sakharov found Penzias's presentation very helpful, because he learned about the pro-SDI position with which neither of them agreed, but with which Sakharov felt obliged to be familiar.

Penzias emphasized to Sakharov that an important pro-SDI selling point was that, ostensibly, it offered a world without nuclear weapons;[61] that SDI was more about research than about deploying new weapons; that outstanding scientists were working on the program; and that most useful technological spin-off represented additional benefits from SDI.[62] For his part, Sakharov shared his thoughts about SDI with Penzias. He stressed that the relative costs of SDI for the American economy were much lower than its implementation costs would be for the Soviet Union, considering its economy. The Soviet leadership was apprehensive of SDI, not least because it did not find it possible to create a comparable system for their country. Nonetheless, he found it feasible that the Soviet leaders would "announce a program to calm their people and then negotiate."

A few years later, Penzias told me additional details about his conversation with Sakharov on that November 1988 afternoon. Penzias told Sakharov that he considered Reagan's SDI immoral, because it might induce the Soviets also to change the rules of the game. Penzias added that "the Russians at some point might launch their nuclear weapons. Sakharov said, no, they wouldn't; he said, they would back down, and Reagan would win. Even though he [Sakharov] was fighting against SDI, he thought that the odds were very high that Reagan would win. I [Penzias] still felt that it was immoral. It was a 90 percent bet and a bet with the future of the planet is immoral. There were other means of bringing down the Soviet Union. A Russian scientist I knew called the USSR a low-temperature society with an infinitesimally small specific heat. A slight perturbation would destroy it. A few people in Russia went to the authorities and declared their intention to emigrate and the whole system went nuts."[63] The physical analogy quoted by Penzias sounded attractive, but the slight perturbation would have cost lives and the prolonged suffering of many. Sakharov's vision of what would happen proved to be correct.

Andrei Sakharov and Elena Bonner's grave in the Vostryakovsky Cemetery, Moscow.

Source: Photograph by and courtesy of Alexander Vernyi and Lyubov Vernaya-Sakharova, Moscow.

Andrei Sakharov died on December 14, 1989. He was buried in the Vostryakovsky Cemetery, in Moscow, the same cemetery where his first wife was buried. But he was alone in his grave until 2011 when Elena Bonner died and was buried in the same grave. Sakharov died at the pinnacle of his popularity. A reviewer of his *Memoirs* wrote that his life was "probably the life of the age; he lived the triumph of physics and the catastrophe of Marxism."[64]

A leading Soviet academician, Roald Sagdeev, commented: "We have lost our moral compass."[65] In a poll taken shortly after Sakharov's death, he was found to be the most revered figure in Soviet history. The political component of his life was substantial during the last two decades of his life, and was overwhelming during the last three years. When asked "why he had abandoned his resolve to concentrate on science," his response was that it was because "he could make a greater contribution to politics than to physics."[66] This was reminiscent of Leo Szilard, who felt it incumbent upon him to dedicate himself to easing the plight of refugees and to fight Nazism rather than advance his scientific career.[67]

Sakharov became a symbol. The editor of *Science*, Daniel E. Koshland Jr., wrote "Andrei Sakharov set a standard for the modern hero. Few will achieve his level, but many will fight injustice more fiercely because of his example."[68] We can only speculate about how Sakharov would have viewed the collapse of the Soviet Union, the emergence of Russia and the former Soviet republics as independent states, the evolving Russian society and its precarious road toward democracy, and the fate of Russian science during the past decades. How would he consider all those weighty problems for whose solution he was willing to sacrifice his health and personal freedom and whose solutions do not appear to be much closer today than they were at the time of his death?

PART II

Low-Temperature Physicists

Petr Kapitza in Cambridge.

Source: Photograph by and courtesy of the late David Shoenberg.

Petr Kapitza on Russian postage stamp, 1994.

4

Petr Kapitza

RESPECTED CENTAUR

Petr L. Kapitza (1894–1984) was one of the most brilliant representatives of the great generation of Soviet physicists. He made experimental discoveries in low-temperature physics, and in 1978 was awarded the Nobel Prize. He was also willing to risk his career and life when he stood up to one of the bloodiest dictators of word history. He dedicated his existence to physics, but was also a Soviet patriot.

Kapitza had proved his scientific acumen in Cambridge, England, and when in 1934 he was confined to the Soviet Union, he developed an internationally renowned research center. He took personal responsibility for persecuted and prosecuted scientists, such as the mathematician Nikolai Luzin and the physicists Vladimir Fock and Lev Landau, and probably saved their lives. He acted creatively under a variety of circumstances. He was autocratic but popular, respected and feared, yet has remained a role model in science.

When I became acquainted with Petr Kapitza's name in the first half of the 1960s, he was already a legend.* I never met him in person, but many years later my conversations with one of his pupils, David Shoenberg, brought Kapitza into human perspective for me.[1]

Petr Leonidovich Kapitza was born on July 9, 1894, at the Russian naval base Kronstadt, near St. Petersburg. The Russian Navy was known for its high level of discipline and dedication. Kapitza's father was a military engineer; and his mother (née Stebnitskaya), a teacher, specializing in children's literature. Kapitza had an enriched upbringing; early on he was taken for trips, visiting Switzerland, Italy, Greece, Germany, Scotland, and extended regions of northern Russia. As a child, he was good in arithmetic, but less so in languages. He started his schooling in Kronstadt. First, he attended a "classical gymnasium," ** but this school soon advised him to change to another one with an emphasis on

* Kapitza's associates used this nickname for him, from Greek mythology: Centaur—part human, part beast.

** The classical gymnasium was a high school especially strong in the humanities.

"real" subjects. Today, his bust graces a spot near his first school, which did not find that he met their standards of scholarship.

Petr Kapitza's bust in Kronstadt.

Source: Photograph by and courtesy of the author.

After secondary school, he became a student at the Faculty of Electrical Mechanics of the St. Petersburg Institute of Technology where Abram Ioffe attracted the young Kapitza to do research in his department. In 1914, World War I broke out, and in 1915, Kapitza volunteered to do service at the front as an ambulance driver and served for a few months. In 1916, he traveled to Shanghai, China, where his fiancée, Nadezhda Chernosvitova, was at the time. Together they visited Japan, and then returned to Russia and got married. The revolutionary events and the civil war had great impact on Kapitza's family. His father-in-law was an official of a rightist party, and the revolutionaries shot him. Kapitza and Nadezhda had a son and a daughter, but more tragedy struck. Their son died of scarlet fever. Nadezhda and their newly born baby daughter died of the Spanish flu, as did Kapitza's father. It took tremendous fortitude for Kapitza to hold himself together and continue his studies, which helped him survive this terrible period.

Kapitza's first research papers on the physics of electrons appeared in the Russian-language *Journal of the Russian Physical-Chemical Society*. In 1917, he conducted his summer exercise under the renowned physicist L. I. Mandelshtam in the radio-telegram division of a Siemens plant. When Kapitza graduated as an electrical engineer, he was already a research associate of the Institute of Physical Technology.

During these years, Kapitza and Nikolai Semenov, another aspiring and gifted scientist, forged a friendship for life (see chapter 8). They suggested a new method for determining the magnetic moment of atoms, and coauthored a paper about it. They stopped at the theoretical considerations and never carried through the experimental test of their method. Kapitza's characteristic approach to research had already manifested itself: he could be involved totally in a project, but could then be easily attracted to the next one, just as easily abandoning the previous one. Kapitza did not like to follow a trodden path, not even his own.

In 1921, Kapitza traveled to England as member of a Soviet science delegation whose mission was to renew scientific interactions with the West. In Cambridge, he was greatly impressed by what he saw there, and asked Ernest Rutherford to accept him to his Cavendish Laboratory. Rutherford declined saying that the Laboratory was full, and the matter might have rested there. However, Kapitza pointed out that adding one member would be within the experimental error of 3 percent that usually characterized Rutherford's work in physics. Rutherford liked Kapitza's unusual reasoning and impressed by the young man's determination, he permitted him to stay. It was the beginning of a special relationship between the two physicists lasting a dozen years, until the end of Rutherford's life. Kapitza, in the words of the British author, C. P. Snow, "flattered Rutherford outrageously, and Rutherford loved it. Kapitsa could be as impertinent as a Dostoevskian comedian: but he had great daring and scientific insight."[2]

Indeed, the unusual Kapitza carried out remarkable deeds in Cambridge. He founded an informal seminar—a weekly gathering in his apartment—to report the latest results and ongoing work. It acquired the name Kapitza Club, which continued for a long time after Kapitza had left Cambridge. The venue was small, and not many people attended these seminars, but everyone important in Cambridge physics, including many visitors, gave a talk at the Club. The first session took place on October 17, 1922, and Kapitza talked about magnetism. Soon, they started a logbook, and each time, the presenter signed it next to the record of the topic. This logbook shows how up-to-date the topics of the meetings were. The terse comments hint also at occasional disagreements, which was natural since cutting-edge science was being discussed.[3]

The last session of the Kapitza Club before Kapitza was detained in the Soviet Union, was the 377th meeting, on August 21, 1934. The refugee scientist and future Nobel laureate Otto Stern spoke about the magnetic momentum of the deuteron. Then, in Kapitza's absence, others took over the Club. World War II interrupted the activities, but the participants were eager to resume them after the war. For some time, Kapitza's former British student, David Shoenberg, organized the gatherings.[4] During the second half of the 1950s, the meetings became less frequent, and in 1958 they stopped. Officially, the closing meeting took place in 1966, on the occasion of Kapitza's long-awaited visit to

Cambridge as he came back for the first time following his detention in the Soviet Union. The presenters returned to a topic they could not fully master thirty years before: Kapitza and Dirac discussed the scattering of electrons by standing light waves.

But we get ahead of ourselves with the Kapitza Club story, and now we return to Kapitza's early Cambridge days. In 1923, Kapitza defended his doctor of philosophy dissertation. He investigated extremely strong magnetic fields, which was a new direction at the Cavendish Laboratory. Rutherford made Kapitza the assistant director for research on magnetism of the Cavendish Laboratory. Kapitza's next recognition came when he was elected member of the most prestigious Trinity College of Cambridge University.

Rutherford supported Kapitza's ambitious project; he helped it by encouraging funding agencies to finance the construction and development of new and costly equipment. Rutherford's support was all the more remarkable because Kapitza's research was a sharp departure from the Cavendish tradition of using simple means and building rudimentary experiments, the "sealing wax and string" approach to physics. Kapitza demanded expensive equipment, showed his familiarity with workshop design, and was capable of producing proper drawings for construction. He was both a physicist and an engineer. To facilitate his project, a Russian technician joined him from home. A huge generator was built for Kapitza's experiments by the company Metropolitan-Vickers of Manchester. The new magnetic laboratory of the Cavendish lab was opened in 1926 in the presence of the chancellor of Cambridge University, the former prime minister Arthur Balfour.

In the same year, Kapitza made his first visit back to the Soviet Union since his departure in 1921. Those visits then became an annual feature in his life. He kept up his interactions with his colleagues back home, discussed their research problems with them, and gave lectures. He traveled in response to official invitations issued by high-level Soviet politicians, such as Lev Trotsky and Lev Kamenev. Both Trotsky and Kamenev would eventually become victims of Stalin's terror in his power struggle for the supreme leadership of the Soviet Union.[5] In addition to the formal invitations, Kapitza always requested and was given an official letter guaranteeing him that he would be permitted return to England at the close of his visit.

Kapitza spent some time in Paris during his Cambridge years, and there he utilized his skills in playing chess. There were many cafés in Paris at the time where chess was played for stakes. When Kapitza first appeared at a particular café, he pretended to be a poor amateur. In the end he would usually win.[6] In Paris, Kapitza met academician A. N. Krylov, a former shipbuilding engineer and government official in czarist Russia. He was now co-heading the academic delegation whose members included Kapitza. Krylov's family had immigrated to Paris from Russia after the revolution. Kapitza was smitten by Krylov's daughter, Anna Alekseevna, and in 1927 they married. They had two sons; Sergei, born in 1928, became a physicist, and Andrei, born in 1931, became a geographer (in spite of his father's wish), and was immensely successful.

For the entire duration of his stay in England Kapitza remained a Soviet citizen, and his wife received Soviet citizenship upon their marriage. In 1929, Kapitza was elected both Fellow of the Royal Society (London) as a Soviet citizen and also a corresponding member of the Soviet Academy of Sciences. He was looking forward to a bright career based on his very special status as a Soviet citizen and a leader in British science. He made quiet inquiries whether a foreign national might become an English lord. The thought was bizarre, but when in 1931 Rutherford became Lord Rutherford, it no longer seemed so far-fetched.

The wealthy chemical industrialist Ludwig Mond left a large sum of money to the Royal Society. Fifteen thousand pounds sterling of the bequest went to subsidize a new magnetic laboratory at the Cavendish Laboratory. It was named Mond Laboratory. Kapitza was appointed its director. The Mond Laboratory was opened on February 3, 1933. Again, the chancellor of Cambridge University, Stanley Baldwin, another former prime minister and still the leader of the Conservative Party, opened the new laboratory. Kapitza chatted effortlessly with the esteemed guest and at one point said to the politician Baldwin, "You can believe me. I'm not a politician."[7]

The dignitaries present included not only politicians and the director of the Cavendish, Ernest Rutherford, but also Rutherford's former mentor J. J. Thomson. But for Kapitza, Rutherford was by far the most important of all those present. He coined a nickname for Rutherford, "Crocodile." According to Kapitza, "In Russia the crocodile is the symbol for the father of the family and is also regarded with awe and admiration because it has a stiff neck and cannot turn back. It goes straight forward with gaping jaws—like science, like Rutherford."[8] The exterior of the Mond Laboratory showed a big crocodile in relief, and the key to the new laboratory had a handle shaped like a crocodile. Both were created by Kapitza's friend, the sculptor Eric Gill. Rutherford knew his nickname and did not mind it.

Years later, Kapitza himself also acquired a nickname, "Centaur," but it is not clear whether he liked it. It was coined by the outstanding physicist Aleksandr Shalnikov, one of the charter members of Kapitza's Institute of Physical Problems in Moscow. Once Shalnikov was introducing a visitor to Kapitza who was shocked by Kapitza's rudeness toward his subordinates. The visitor asked Shalnikov whether the director was a human or a beast. Shalnikov's spontaneous response was that Kapitza was a centaur, and the name stuck.[9]

The apparatus in the Mond Laboratory was large and required special physical arrangements. Kapitza built a flywheel, which stored an enormous amount of mechanical energy. This energy was then released into a small wire coil, which exploded. This happened on a time scale of a fraction of a second. The explosion caused a strong mechanical shock that would disturb the sensitive recording instruments monitoring the experiment. This is why the instruments were placed twenty meters away from the experiment and the seismic wave reached the instruments only with a delay after the explosion had happened. For this arrangement, the magnet hall had to be long; it became a convenient gathering place for all who worked at the laboratory. The small rooms for the researchers opened from the magnet hall on either side of the hall.

David Shoenberg was among the students of the Mond Laboratory. He was to be Kapitza's last student in Cambridge. Shoenberg's roots were in Russia, so he had sympathy for Kapitza's heavily accented English. It was not only the Russian accent and sentence construction that at times made Kapitza difficult to understand. His stories and jokes reflected his Russian background, and not everyone grasped easily what he wanted to say. But he had a appealing personality and when his jokes were not understood, his infectious laughter still made his listeners laugh. He was unorthodox in his approach in his lectures. His principle was that letting the students leave the lectures with contradictory ideas would stimulate their thinking. In his popularizing talks, which he gave frequently to scientific societies, he aimed to have 95 percent of his audience understand 5 percent of what he said; and 5 percent understanding 95 percent.[10]

J. J. Thomson and Ernest Rutherford after the inauguration of the Mond Laboratory in Cambridge.
Source: Photograph by and courtesy of the late David Shoenberg.

Crocodile carved into the façade of the Mond Laboratory.

Source: Photograph by and courtesy of the author.

Kapitza gave Shoenberg a research project in which he had to measure the changes in the size of a bismuth crystal in a strong magnetic field.[11] The phenomenon is called the "magneto-striction of bismuth." Kapitza himself had measured the change in the length of a bismuth crystal, but could only determine the change along the direction of the magnetic field. Shoenberg's task was to measure the change in the direction perpendicular to the magnetic field. The changes in the crystal lengths are very small, and Shoenberg had to develop very delicate methods for their determination. Several times during the project Shoenberg was ready to give up, but Kapitza always came up with new suggestions about how to continue. He was full of ideas; many proved useless, but there were invariably some that moved the project along. There are scientists like Kapitza—and Edward Teller and George Gamow were in this category—who are full of ideas. They don't mind pronouncing them even though most of them are worthless, because the few that are valid may be revolutionary. Other scientists—Enrico Fermi and Linus Pauling come to mind—were also full of ideas, but they did not like to make their suggestions out loud before sifting through them and discarding the worthless ones themselves.

Shoenberg learned a lot of physics from his interactions with Kapitza, and he got to know better his mentor's human qualities. He got a taste of Kapitza's temper in the course of this work. When Shoenberg collected all the experimental data, he tried to fit them to the scheme Kapitza had worked out. However Shoenberg tried, the data did not appear consistent with the scheme, but when he slightly modified it, everything fit in perfectly. He told Kapitza about his results, pointing out that Kapitza had made a mistake. His mentor's reaction was "surprise and dismay, he [Kapitza] flew into a rage and said, 'How dare you say such a thing to your teacher? Go and talk to [Paul] Dirac. He checked my calculations and Dirac never makes a mistake.'"[12] Shoenberg did talk to Dirac. It turned out that Kapitza had given him erroneous information about the symmetry of the bismuth crystal, and this had caused the problem. With the correct symmetry, the scheme worked fine. This experience notwithstanding, Shoenberg was so attached to Kapitza that he spent some time with him in Moscow "when Kapitza was building his new institute. He behaved somewhat autocratically in Moscow too and, though his staff respected him, they were a bit afraid of him."[13]

In spring 1934, still in Cambridge, Kapitza succeeded in the liquefaction of helium using an original apparatus of his own creation. He did a lot of work together with John Cockcroft, a pioneer nuclear physicist and future Nobel laureate, who was one of Kapitza's few doctoral students. Kapitza's revolutionary helium liquefier served as the foundation for a mass-produced helium liquefier to be developed in and marketed from the United States. But in England, Kapitza's was not the first liquid helium; Oxford University had liquefied it first, relying on a small machine from Germany built by Francis Simon. When the Nazis came to power in Germany, the Jewish Simon found refuge in Oxford.

On September 1, 1934, Kapitza arrived in the Soviet Union for his usual visit. This time, he had failed to secure the written Soviet permission to leave the country upon the completion of the visit. At this point, he felt it was embarrassing to continue the practice although George Gamow had urged him to get the guarantee. Kapitza told Gamow that a high Soviet official had explained to him: "You must see by now that nobody wants to hold you here by force. And it is somewhat beneath our dignity to give such assurances to the stuffy British Lord [Rutherford], who does not understand these matters."[14] Gamow had experienced great frustration when he was repeatedly refused permission for foreign travel. When in 1933 he was finally permitted to attend the Solvay meeting of physicists in Brussels, he stayed out of the Soviet Union for good.

During Kapitza's 1934 visit, he was told that he would not be permitted to return to England. This was a rude resolution of the relationship between Kapitza and the Soviet state. In 1929, when he corresponded with Kamenev, then the member of the top Soviet leadership responsible for science, they seemed to be equal negotiating partners. Kamenev would have liked Kapitza to return to the Soviet Union full time, and if not full time, at least to spend considerable lengths of time annually in his homeland. Kapitza presented certain conditions and excluded the possibility of predetermined lengths of time. But he was willing to help Soviet science: "I have been following with great interest and satisfaction the tremendous development of scientific research work in Russia and am very glad that I now have the opportunity of helping."[15]

It was Stalin who decided to retain Kapitza.[16] On September 21, 1934, the dictator wrote: "Kapitza may not be arrested officially, but he must be retained in the Soviet Union and not let return to England on the basis of the law about non-returning persons. This will be a sort of house arrest. After that, we will see."[17] A few days later, a deputy prime minister, Valery Mezhlauk, informed Kapitza that he could not leave, that an institute would be built for him in Moscow at the location of his choice, that he could select his co-workers, that his experimental apparatus would be purchased for him from abroad, and that he would be given a spacious home in downtown Moscow.

Gamow's defection in 1933 had been a blow to the Soviet leadership; yet I am not suggesting that Kapitza's retention was its direct consequence, although its impact cannot be excluded. In November 1933, Kapitza wrote to Niels Bohr, "It is better for every man to work in the country and under the conditions which he likes most, and this is why I think it would be much better for Gamow to work abroad."[18] Kapitza senses the possible consequences of Gamow's action on others in the Soviet Union when he mentions "that [Gamow's defection] will make it extremely difficult for young Russian physicists who wish to study abroad to get permission."[19]

The bitter irony of the world situation did not escape Kapitza. Kapitza was a prisoner in his own country, communist Russia, whereas great scientists of similar caliber were being forced out of Nazi Germany. When the Jewish German physicist Max Born was looking for a refuge, he shared his dilemma with Kapitza. In response, the Soviet physicist wrote to him: "You are an unlucky man, everybody wants you and the choice is so great that you cannot make up your mind. Maybe I am somewhat luckier than you are, as I have no choice."[20] But then Kapitza expresses an unexpected but logical idea; he urges Born to join him in Moscow and build his new scientific home there. Born and his wife had planned a brief exploratory visit to Moscow when he received an attractive offer from Edinburgh, which he accepted at once., Born noted in his memoirs that in light of the further developments in the Soviet Union, "accepting Kapitza's offer would have been a catastrophe."[21]

Petr Kapitza's group in the mid-1930s in the courtyard of the Institute for Physical Problems in Moscow. Kapitza is in the middle, David Shoenberg is to his left. Aleksandr Shalnikov is at the extreme left.

Source: Courtesy of the late David Shoenberg.

It is easy to imagine Kapitza's state of mind when he understood that he would not be let out from the Soviet Union, would not be able to continue his life in Cambridge, and would not even be let to go there for a short visit to finish his ongoing projects. But his desperation did not prevent him starting to plan for his science in Moscow. At the end of 1934 it was decided to build his institute in one of the choicest locations in Moscow, on a hill overlooking the Moscow River and the city. Its name was to be Institute of Physical Problems (Institut Fizicheskikh Problem [IFP]) and Kapitza was appointed its director. Today, it is the P. L. Kapitza Institute of Physical Problems of the Russian Academy of Sciences. The naming of the institute had its own peculiar story. The president of the Soviet Academy of Sciences, Sergei Vavilov, died in 1951. At that time Kapitza was in exile from his institute. Vavilov was a physicist, and it was decided to name the IFP after him. For years, there was a tablet at the entrance of the institute, reading something like this:

Awarded the Order of Red Banner of Labor
S. V. Vavilov
Institute of Physical Problems

The Order of Red Banner of Labor indicated there was for the Institute rather than for an individual. When Kapitza was reinstated as director in 1954, he ordered the tablet changed. He diminished the letters referring to Vavilov and enlarged the letters referring to the institute to look something like this:

Awarded the Order of Red Banner of Labor
S. V. Vavilov
INSTITUTE OF PHYSICAL PROBLEMS

It was after Kapitza had died that the Presidium of the Soviet Academy of Sciences changed the name to P. L. Kapitza Institute of Physical Problems. They found another institute to name after Sergei Vavilov.[22]

There has been a lot of speculation about why Stalin forced Kapitza to stay in the Soviet Union. Most probably, it was a confluence of several reasons. We single out one of them that might also bring us closer to understanding Kapitza's personality. Rutherford did everything possible to secure freedom for his favorite pupil, but he thought that the explanation for Kapitza's retention may have been due his "love of the limelight."[23] On one occasion, Kapitza told a group of Soviet engineers that "he himself would be able to alter the whole face of electrical engineering in his lifetime."[24] The Soviet leadership—Stalin himself—may have well been intrigued by this statement and seen in it tremendous strategic importance both for the economy and defense. Lenin used to stress the supreme importance of electrification for Soviet success; his slogan was: "Communism is Soviet power plus the electrification of the entire country."[25]

In 1935, at Rutherford's suggestion, Cambridge University decided to sell the scientific equipment of the Mond Laboratory to the Soviet Union. This gesture greatly facilitated Kapitza's research efforts, but Kapitza would not be allowed to travel to England for the next thirty-two years. His wife, however, could return to England to collect their two sons and belongings and in January 1936 she and their two sons came back to the Soviet Union. The world of science was learning about Kapitza's detention only many months after it happened. Kapitza knew that if he was to have any hope of reversing the decision of the Soviet leadership, any actions of his supporters should be taken quietly rather than in the limelight. However, Stalin was determined to keep him.

Progress was very slow in developing the right conditions in Moscow to enable Kapitza to continue his scientific activities. He understood that the Science Academy was powerless in this matter, and he started corresponding with top Soviet leaders. Starting from 1934, he wrote well over two hundred letters to Stalin, Molotov, and Malenkov. He bombarded the subsequent Soviet leaders with his letters, including Khrushchev, Bulganin, Brezhnev, and Andropov among them. Most of his letters were not just brief notes, but substantial treatises. It was under Stalin that he wrote his most important letters. Some were heroic attempts to gain the freedom of arrested scientists. Others were about the state of Soviet science, the situation of scientists, science education, and other topics on which Kapitza felt his advice might be useful. Sadly, however, there were also letters in which he denounced other scientists and technologists when he had disagreements with them. This was part of his controversial character.[26]

In his long letter of December 1, 1935, to Stalin, he complains about how sordidly he was treated not only by being forced to stay in the Soviet Union but afterward.[27] Once it was decided that he could not leave, and once he committed himself to continue his work and build up his research in Moscow, he should have been assisted in his efforts. Instead, he was being treated as a suspicious stranger, at best. He found it unreasonable that he would not be given close to ideal conditions for his work if keeping him was considered important. There is sad irony in the fact that the only person Kapitza singled out in his letter as a positive example was Karl Bauman, the party leader responsible for science, who was arrested in 1937 and perished in prison.

By the summer of 1936, the building of Kapitza's new institute was completed at about the same time most of the equipment from the Mond Laboratory arrived. Two technicians of the Mond were each given a one year leave to help Kapitza's associates assemble and learn how to operate the apparatus. The Soviet government paid for the equipment, but, of course, the Cavendish scientists working on the transfer, including Rutherford, were not compensated for their lost research activities. They did not complain, once again demonstrating how much they felt that Kapitza was one of them and were willing to help him in whatever way they could.

From summer 1936, Kapitza settled into his Soviet period, which showed continuity for the rest of his life (in spite of the interruption of his internal exile between 1946 and 1954). One of his characteristic traits was his fearlessness in facing the authorities. He waged protests when he became aware of injustices suffered by fellow scientists. Alas, such injustices were so numerous that he could raise his voice only in selected cases. It needs to be stressed, however, that Kapitza was a Russian patriot and a Soviet patriot, and appreciated what the Soviet regime was trying to achieve in science. He understood the characteristics of Stalin's autocratic system, perhaps because he was also inclined to build an autocratic order around himself. Stalin may have felt some respect for Kapitza's strong personality. Kapitza wrote numerous letters to Stalin, whereas the dictator sent him a total of two brief letters. In one of the two, dated April 4, 1946, Stalin referred to Kapitza's letters: "There is much that is instructive in them and I should like to meet you at some time to have a chat."[28] However, they never met.

Kapitza had hardly started his Soviet life when he already felt obliged to wage a protest in defense of a fellow scientist. In 1936, attacks against a noted mathematician, Nikolai Luzin, appeared in a series of anonymous articles in the Communist Party's central newspaper, *Pravda*. These included accusations that Luzin followed ideologies alien to the Soviet order, stole his students' results, published poor papers in Soviet journals, and published his best results in foreign periodicals. Luzin was a full member of the Soviet Academy of Sciences, and an Academy commission ominously endorsed the accusations. It was expected that he would be expelled from the Academy, arrested, and would perish in incarceration. Kapitza learned about the *Pravda* articles while on vacation in the countryside, and immediately wrote a long letter of protest in defense of Luzin. He addressed his letter to the chairman of the Council of People's Commissars (Council of Ministers), Vyacheslav Molotov.[29]

Kapitza did not deal with the accusations against Luzin because he was not familiar with the details of the story. He explained that preventing the publication of poor articles was the responsibility of the journal editors, and that it was in the interests of both Luzin and the reputation of all Soviet scientists to publish their best articles in international journals. His main point was that Luzin's value should be judged by the level of his work. He added that the Soviet Union had to rely on many such scientists who were not necessarily friendly toward the Soviet order. Kapitza noted: "We have not so far succeeded in producing new scientists from among our young people. I explain this by a *very wrong attitude on your part towards science—much too narrowly utilitarian and insufficiently supportive*" (italics added).[30] Further, Kapitza writes, "People like Luzin, who differ from us ideologically..."[31] This is very curious, not in that he separates himself from Luzin but in that he identifies himself with Molotov. The outcome of the case was that Luzin was never arrested and was not even expelled from the Academy of Sciences. The attacks against him stopped.

Soon after Kapitza's institute had begun its operations, it started yielding seminal results some of which could be applied directly for practical purposes. He and his associates developed a technology to liquefy air, produce liquid oxygen and nitrogen, and extract inert gases from air. Simultaneously, fundamental science was cultivated, and the free flow of scientific ideas and experience encouraged. Kapitza's seminars gained a new life at his institute. His huge office provided plenty of space for these meetings. The war-time evacuation, and then Kapitza's absence from the Institute between 1946 and 1954, disrupted the organization of the seminar, but in the second half of the 1950s, it outgrew his office. In addition to the associates of the institute, even more attended from the outside. On February 8, 1956, a milestone event took place in the Kapitza seminars. The biologist N. V. Timofeev-Resovskii and the physicist Igor Tamm talked about modern genetics in open defiance of "official" Soviet biology as determined by the unscientific czar of Soviet biology, Trofim Lysenko. This was a brave and significant step in turning around the tragic history of this most important science in the Soviet Union.

Once again, we get ahead of ourselves in presenting the Kapitza seminars. So now we return to the second half of the 1930s. A brilliant and internationally renowned theoretical physicist, Vladimir Fock, was arrested on February 11, 1937, in Leningrad. Fock had been a professor at the university since 1932. He was very young when he generalized Schrödinger's equation for magnetic fields and for the relativistic case involving charged particles moving in magnetic fields. His methodology of calculating the structure and energy of polyatomic molecules, combined with the American Douglas Hartree's approach, generally known as the Hartree-Fock method, is still being used in quantum chemical calculations.

As soon as Kapitza learned about Fock's arrest, he wrote a letter to Stalin. He lamented the rough treatment of scientists, which would interfere with their ability to work, just as rough treatment of a machine would unfavorably impact the quality of its product. He compared Fock's fate to how Einstein was treated by Germany.[32] He referred to Fock as the most outstanding theoretical physicist in the Soviet Union. He maintained that his arrest must have been a mistake and that the accusations—that Fock had engaged in sabotage by knowingly providing the wrong formula for the evaluation of electrical exploration in geological work—were impossible. Within days, Fock was freed from captivity.[33]

The year 1937 brought sad news for Kapitza: on October 19, Rutherford died. Kapitza wrote to Niels Bohr, "All these years I lived with the hope that I shall see him again and now this hope is gone. . . . I loved Rutherford. . . . I learned a great deal from Rutherford—not physics but how to do physics."[34]

Kapitza found a lot of joy in his work. He prepared his report for *Nature* and for the Russian journal *Doklady Akademii Nauk* (Proceedings of the Academy of Sciences) about the discovery of superfluidity of liquid helium. He was also the one who introduced the term "superfluid."[35] The discoveries on the transport properties of liquid helium would bring two Nobel Prizes in Physics for the Institute for Physical Problems. The first was to Landau in 1962 (see chapter 5) for his theoretical interpretation of Kapitza's experimental discovery. The second was for Kapitza in 1978. The motivation for Kapitza's award was quite general, referring to his basic inventions and discoveries in low-temperature physics. Kapitza referred to the slowness of his award in his Nobel lecture, pointing out that he had left the field in which he received the Nobel Prize thirty years earlier. He delivered his Nobel address on plasma and controlled thermonuclear reactions, a project that recently occupied his mind.[36]

The key elements of Kapitza's discovery, superfluidity and helium, were missing from the citation of Kapitza's Nobel Prize. This was because Kapitza's discovery of the superfluidity of liquid helium was not unique. At about the time of Kapitza's involvement, two physicists in Cambridge obtained similar results independently. There was a slight priority in Kapitza's favor according to the dates of submission of the respective manuscripts, Kapitza's on December 3, and Allen and Misener's on December 22, both in 1937. The two papers appeared back to back in the same issue of *Nature* in early 1938.[37] Soon Kapitza made another experimental discovery. He noticed that when heat flew from a solid surface into helium, there was a temperature jump at the surface. It was later called the Kapitza Jump, and it took a long time for the theoreticians to explain it.

The unprecedented terror of 1937–1938 produced an incident that struck especially close to home. The exceptionally gifted Lev Landau was arrested by the secret police, Narodnii Komissariat Vnutrennikh Del (People's Commissariat of Internal Affairs [NKVD]), on April 28, 1938. Landau was in charge of theoretical physics at Kapitza's institute.† Again, Kapitza took immediate action, and he turned to Stalin. Landau was freed exactly one year later, in 1939, under Kapitza's supervision. The story of Landau's incarceration has been told and retold repeatedly. There have been various versions and its details have been marred by gossip and the lack of hard evidence. Boris Gorobets, a scientist and author of a Landau trilogy, has made a thorough analysis of all available information and separated facts from myths; here we follow his recent account.[38]

Landau had participated in the production of an anti-Stalin leaflet, which was highly critical of the Soviet dictator, labeling him and his regime fascist. This is a fact. Landau's arrest was most likely provoked by his involvement with this leaflet. Not that valid reasons were needed for an arrest in 1937–1938, but in this case, there was one. By this time Landau had developed strong anti-Stalin sentiments. When, years later, Landau was asked about the accusations against him, he kept quiet about the anti-Stalin leaflet and talked about the accusation that he was a German spy. This would not have been a very sophisticated accusation, but NKVD actions were hardly ever based on a sophisticated rationale. There might have been yet another "justification" for the authorities to take action against Landau. When he was still in Kharkov, and the question had arisen whether or not the Ukrainian Institute of Physical Technology (Ukrainskii Fiziko-tekhnicheskii Institut [UFTI]) should shift its activities from basic research to defense-related work, Landau was against the change.

Landau's genius was badly needed in Soviet theoretical physics, in general, and to find the interpretation for Kapitza's discovery of the superfluidity of helium, in particular. When Kapitza, following Landau's arrest, turned to Stalin, he stated that if Landau committed a crime—and he could have not been sure that he had not—he should be punished, but he doubted it very much. There was no response from Stalin. Almost a year later, in a letter to Molotov, Kapitza even brought up the possibility of sending Landau to a special camp for incarcerated scientists and engineers; he believed it to be of the utmost importance to spare Landau's life. He argued that Landau's brain should be utilized for the benefit of the socialist fatherland.

---

† The secret police between 1934 and 1954 was the arm of the interior ministry, the NKVD.

Landau's arrest happened toward the concluding phase of the Terror of 1937–1938; in fall 1938, the Terror was subsiding. The new interior minister, Lavrentii Beria, who replaced Nikolai Ezhov, approached the problem in a more rational way than his predecessor. For example, he wanted the camps to become economically viable; he knew that starving prisoners could not produce. About three hundred thousand people were freed, one-third of all prisoners arrested in the years 1937 and 1938.[39]

For Landau, things were developing painfully slowly. However, after Kapitza wrote to Molotov in April 1939, Kapitza was first invited to see Molotov and then for a visit to the infamous headquarters of the NKVD—Lubyanka. He was received by two high-positioned officials of the Ministry. Again, it is a fact that Kapitza spent hours in the company of these officials. As to what they talked about, various versions have floated around. However, a few days following Kapitza's visit, Landau was freed. Landau learned only many years later that he was freed on the condition that he would remain under Kapitza's supervision.

Kapitza must have learned about the anti-Stalin leaflet during his visit to the NKVD. If he did, it must have shocked him, but he must have instantaneously regrouped and continued his mission. He understood that the leaflet story was different from the false accusations, and that regardless of the easing up of the terror, Landau's life was still in great danger. Kapitza must have found the winning arguments to convince the NKVD officials or, rather, their bosses Beria and, ultimately, Stalin. He must have argued that Landau was irreplaceable and badly needed for Soviet science. Beria could ignore political considerations if economic or defense benefits were at stake, and it seems very probable that Kapitza's arguments had an impact on him. Still, at this time, Landau's value was not yet that high in Beria's eyes—years later it would grow tremendously within the framework of the nuclear weapons program. Mere usefulness would not have sufficed at this point; a number of other people were being incarcerated who might have been no less useful to the economy and defense than Landau. I mention two of the brightest names, who later entered the top tier of the Soviet scientific-technological elite: Sergei Korolev of rocket technology and the airplane designer Andrei Tupolev. The big difference between their situation and Landau's seems to have been the existence of Landau's protector, Kapitza. Stalin respected him and Kapitza had not yet alienated Beria at this point.

It would have been difficult even for Beria to deal with an accusation of a crime directed against Stalin personally, shutting Stalin out of it. The utterly implausible accusation of Landau's being a German spy would fit all parties although for different reasons. For Kapitza and Landau, an implausible accusation was preferable to one that revealed his opposition to the beloved leader. As for Stalin, he must have found it attractive to have an accused spy promising good behavior under Kapitza's tutelage. The alternative would have demonstrated that Landau, a world-renowned scientist and the most gifted person of his generation in science, considered him a fascist and was willing to die fighting against him.

This is also only a probable scenario, but none of the other versions circulating in various memoirs have documentation to support them either. This version is a reasonable explanation for why the accusation that Landau spied for Germany has survived for so long even though it is a myth. It is not a myth, though, that Kapitza took a strong stand in Landau's defense under circumstances in which such bravery was almost unthinkable.

On June 22, 1941, Germany attacked the Soviet Union and made quick initial advances, soon threatening Moscow. Kapitza's institute was evacuated to Kazan along other scientific institutions. For the duration of the war, Kapitza served as member of the Scientific-Technological Council at the State Defense Committee (headed by Stalin). During the war, a portion of liquid oxygen important for the defense industry was produced according to Kapitza's methods (most of it, though, continued to be produced by the classical German technology). Kapitza was appointed to be in charge of the main oxygen authority.

There were other recognitions. In 1939, Kapitza was elected full member of the Soviet Academy of Sciences. He received two Stalin Prizes, one each in 1941 and 1943. In 1945, Kapitza was awarded his first Hero of Socialist Labor star for his contribution to the production of liquid oxygen. In 1974, he would receive a second one on the occasion of his eightieth birthday. He cherished these two golden stars the most and would wear them on December 10, 1978, on his formal dress in Stockholm, at the Nobel Prize award ceremony.

In August 1945, the State Defense Committee created a special subcommittee for the development of the atomic bomb. Igor Kurchatov and Kapitza were its only scientist members. For some time, things developed nicely for Kapitza; he had his own institute, achieved successes both in fundamental and applied science, and was a revered member of Soviet society. This idyllic situation, however, was not to last. He soon clashed with the all-powerful Beria.

From the start of Kapitza's brief involvement in the Soviet atomic project, he felt out of place. He did not like being merely one of the crowd in a nonleadership position. Neither did he like to participate in solving routine tasks where he could not let his imagination fly. The Soviet nuclear weapons project was like that during the creation of the first Soviet atomic bomb, which was built as a faithful copy of the American one. Nuclear physics was not Kapitza's field; there were many scientists in the project whose expertise was more germane in this undertaking. Beria, who was in charge of the project, looked down on scientists. He appreciated them only as long as he needed them.

Kapitza had no knowledge about the role of intelligence in building the first Soviet atomic bombs. Only two scientists, Kurchatov and Yulii Khariton, knew about the information arriving from the United States through espionage. They could not give away the source of their ideas, and it was made to appear merely a lucky coincidence that Kurchatov always came up with the solution for the next step that would fit the project of emulating the Americans. The main characteristics of the Manhattan Project were not classified, and to the surprise of the Soviet authorities, the Americans had published a whole volume filled with information, the so-called Smyth Report.[40] However, the technical details still had to come from espionage, and they proved to be absolutely correct.

Kapitza noted in his long letter to Stalin of November 25, 1945, that the person in charge of the project, Beria, was like a conductor of an orchestra. He had the conductor's baton in his hand, and he waved it, but he did not understand the score.[41] Kapitza could not have known that Beria did possess the perfect (American) score, which he did not have to understand; he only had to make sure that the Soviet scientists faithfully followed it. Kapitza's constant nagging that the scientists should come up with original approaches must have annoyed Beria, creating a situation in which Kapitza no longer represented value for the project Beria cared about. On top of this, Kapitza kept complaining about Beria to Stalin. Kapitza was taking a tremendous risk, because for all he knew, his own life might have been in danger. Kapitza acted, though, as a shrewd politician when he added a post script to his letter of November 25 asking Stalin to show his letter to Beria, because Kapitza considered the letter "not a denunciation but a useful criticism."[42] Kapitza wanted to be excused from the special committee, and by Stalin's direct order of December 21, 1945, his request was granted. Finally, by another direct order of August 17, 1946, Stalin let Beria remove Kapitza from all his positions by direct order, but he did not touch his life.

Kapitza's removal from the Oxygen Authority, responsible for the production of oxygen for industry, was the final step in a long-standing bitter struggle between him and those technologists who favored the classical methods of producing liquid oxygen. The way Kapitza fought this battle was not befitting a great scientist in that he resorted to accusations and could not rid himself of bias toward the technology that he had discovered.

His removal from the Institute of Physical Problems was especially painful. It seemed to be a payback from Beria, but the real reason may have been more mundane. Kurchatov and the atomic bomb program badly needed Landau and his theory group of the IFP, and it was doubtful that Kapitza would have let a substantial section of his small institute be removed from his jurisdiction. For Beria, and for Stalin, the simplest solution appeared to be taking the whole IFP away from Kapitza.

The period from the end of the war, and especially from 1948 to Stalin's death in 1953, was another time of intensified repression in the Soviet Union. That included increasingly overt anti-Semitism. Kapitza was sensitive to the occurrence of any kind of anti-Semitism. His ancestors were of Russian and Polish stock. He subscribed to the tenet that anti-Semitism was a disease, and he would not tolerate any intolerance. In 1933, while he was still in Cambridge, Jewish German scientists started coming to Great Britain as refugees, and in his letters to his mother, Kapitza wrote about their plight. He noted the large number of famous scientists among the refugees, but was concerned especially with the fate of the unknown young ones who had not yet had the opportunity to make their names in science. Kapitza assisted Paul Ehrenfest and Leo Szilard in trying to find employment for the refugees in Western Europe and the United States. Again, he was in complete accord with Rutherford, who in spite of being most apolitical, devoted himself to helping the refugees. Rutherford was the president of the newly formed Academic Assistance Council and remained in this position to the end of his life.

Back in the Soviet Union, Kapitza remained vigilant in opposing anti-Semitism. He was horrified by the genocide during World War II and was one of the speakers on August 24, 1941, at the anti-fascist meeting of representatives of the Jewish people. The speeches were broadcast all over the world. Kapitza stressed that state-sponsored anti-Semitism, as was taking place in Germany, flared up when a country was undergoing a period of deep crisis and reactionary time. Such state-sponsored anti-Semitism relied on the lowest instincts of the people. He also referred to the pogroms during czarist times. Kapitza's speech and the whole anti-fascist movement must have been recalled with horror during the last years of Stalin's life when Soviet anti-Semitism took on appalling intensity. First, in 1948, was the assassination of Solomon Mikhoels, disguised as a road accident, which was followed, in 1952, by the execution of the rest of the leaders of the Jewish Anti-Fascist Committee and many other actions in between and after.

When Kapitza was exiled from his institute, he moved to his summer home in Nikolina Gora, one of the most exclusive suburbs of Moscow. A lesser scientist might have been lost under these circumstances, but not Kapitza. He ingeniously found a way to continue research in physics, and even to continue his letter writing to the high authorities. In Nikolina Gora, he formed a unique laboratory, which he called the Hut of Physical Problems by way of analogy with the Institute of Physical Problems. Lev Landau and Evgenii Lifshits were regular visitors. Eventually, the president of the Academy of Sciences and the new director of the Institute of Physical Problems allowed a technician to help Kapitza with his experiments.

Another occupation Kapitza found for himself was involvement with the preparation of young physicists. In 1947, a new Faculty of Physical Technology at Moscow State University came to life, which in 1951 became the exceptionally strong Moscow Institute of Physical Technology. Kapitza recognized the importance of educating well-trained physicists for the Soviet nuclear industry and elsewhere, and became one of its founding members. He was appointed chairman of the Department of General Physics, and he and Landau alternated giving the course of general physics ensuring the exceptionally high level of this course. The students crowded in for these lectures; the lucky ones were those who could attend during either of the two academic years 1947/48 and 1948/49 while Kapitza remained in this position.

Toward the end of 1949, huge festivities were organized to honor Stalin's seventieth birthday, December 21. Kapitza did not attend them, and the university responded by stopping him from teaching. The pro-rector wrote to him in a letter that "it is not possible to entrust the education of our scientific youth to someone who sets himself against the whole of our people in such a demonstrative manner."[43] He was officially dismissed from his job a few weeks later by the rector of the university, A. N. Nesmeyanov. The justification for dismissal was that Kapitza was no longer involved with teaching.

By this time Kapitza had stopped writing letters to Stalin, and on one occasion the dictator instructed Georgii Malenkov to call Kapitza and ask him why.[44] Stopping his letter writing and staying away from the festivities were the only actions Kapitza could take to express his resistance. Of course, Stalin might have ignored the slight, but he was so unused to even the minutest sign of resistance that he could not; thereby he helped, however involuntarily, Kapitza's action to take on real significance. Alas, the controversy between Stalin and Kapitza was not between equals as far as their potential to hurt the other was concerned. We have seen that Stalin's reaction to the snubbing by Kapitza was to have him dismissed from his teaching post. Curiously though, there was still someone, somewhere who tried to provide some solace: On June 1, 1950, the outstanding crystallographer Aleksey Shubnikov invited Kapitza to be a consultant at the Institute of Crystallography of the Soviet Academy of Sciences.[45] Incidentally, the Soviet Union was hermetically isolated from the external world in these years. Some outside organs watching the Soviet nuclear program for a while assumed that Kapitza's disappearance from view must mean that he was participating in the classified project. Misconceptions die hard; even in 1968, an article in the *New York Times* characterized Kapitza as "Stalin's chief scientific adviser."[46]

One of Kapitza's research topics in exile was the development of strong microwave beams. They were supposed to be so powerful that Kapitza considered their possible military use. The function of these high-energy beams would be to annihilate incoming attackers (this was reminiscent of President Reagan's SDI). Kapitza wrote in his letter of June 25, 1950, to Malenkov:[47]

> During the war I was already thinking a lot about methods of defence against bombing raids behind the lines [that would be] more effective than anti-aircraft fire or just crawling into bolt holes. Now that atomic bombs, jet aircraft and missiles have got into the arsenals, the question has assumed vastly greater importance. During the last four years I have devoted all my basic skills to the solution of this problem and I think I have now solved that part of the problem to which a scientist can contribute. The idea for the best possible method of protection is not new. It consists in creating a well-directed high-energy beam of such intensity that it would destroy practically instantaneously any object it struck. After two years work I have found a novel solution to this problem and, moreover, I have found that there are no fundamental obstacles in the way of realizing beams of the required intensity.

Kapitza soon renewed his letter writing to Stalin, and in a letter dated November 22, 1950, he mentioned this idea again and complained about the absence of assistance, without which he could not further develop this strategically important project. Kapitza could not be sure whether his letters reached Stalin, but in a phone call from Malenkov he learned that Stalin did indeed read his letters. By the way, Kapitza had to go to a location that had a telephone to receive Malenkov's call because his summer house did not have one.

On March 5, 1953, Stalin died, and on June 26, Beria was arrested, soon to be executed. On August 28, the Science Academy decided to create a new physical laboratory on the basis of Kapitza's makeshift unit at his summer home. Finally, on January 28, 1954, Kapitza was reinstated as director of the Institute of Physical Problems, remaining in this position for the rest of his life. His research interest for the next decades remained the same as what had formed in his summer home: high-performance electronics and plasma physics.

When Nikita Khrushchev came to power Kapitza started a correspondence with him. Their relationship was a more two-way interaction than Kapitza's had been with Stalin. A minor episode might illustrate Khrushchev's style and subjective judgment and that Kapitza did not lose his ability to take a principled stand in a controversy. At a Kremlin reception Kapitza wore a Ukrainian shirt rather than the prescribed formal attire and Khrushchev made a critical remark pointing this out. Kapitza's response was to inform Khrushchev that "surely diplomatic protocol recognizes national dress as appropriate for formal occasions."[48]

Kapitza must have felt satisfaction when Beria fell, and he wanted Khrushchev to learn about his own problems with Beria. In 1955, he sent copies of letters to the Soviet leader in which he had complained to Stalin about the way Beria treated scientists.[49] In another letter Kapitza scolded the Soviet scientific establishment.[50] However, conditions looked different from the outside. In the period between the mid-1950s and the mid-1960s the American scientists looked with envy at some of the achievements of Soviet science. There were problems to be sure, and the successes did not penetrate deeply. But the state of science education in the Soviet Union was an example for many American scientists when they urged the allocation of greater resources to science and science education in the United States.

One of the most important changes in Kapitza's life during its last two decades was that he was allowed to travel to the West, but this came only after Khrushchev's fall. In his memoirs Khrushchev discussed the reasons that even after Stalin's death Kapitza was not allowed to travel. It is incredible how erroneous and unsophisticated Khrushchev's evaluation was: "On the one hand, in the world [Kapitza] is recognized as one of the most significant scientist-physicists; on the other hand this scientist did not give us the possibility of obtaining the atomic bomb before the United States, or even if not before, he did not assist us in creating the Soviet atomic bomb."[51] Khrushchev consulted the leading mathematician academician M. A. Lavrentiev about Kapitza's trustworthiness. The Soviet leader was concerned that Kapitza might defect during visits to other countries or give away state secrets, even if unintentionally. Lavrentiev vouched for Kapitza's trustworthiness. Nonetheless, Kapitza was not allowed to travel, which the deposed leader later regretted.[52]

Petr Kapitza at the right listening to the presentation by Igor Landau, Lev Landau's son, at the Institute of Physical Problems in Moscow.

Source: Courtesy of Boris Gorobets, Moscow.

Petr Kapitza and Yulii Khariton.

Source: Courtesy of Alexey Semenov, Moscow.

Under subsequent leaders, Kapitza continued his letter writing. For example, he protested the practice of the Soviet censors of tearing out pages from publications. Sometimes whole articles in scientific journals and magazines were removed before they reached the libraries. Western scientific literature was scarce in the Soviet Union, but even those who had access to it could not be entrusted with some of the information it contained. When Kapitza received copies of the Italian translation of his book *Experiment, Theory, Practice*, the thirty-page preface had been torn out from each copy.[53] The book was published by an Italian communist publisher and the preface was written by an Italian communist philosopher. Kapitza, in his letter of April 22, 1980, to Yurii V. Andropov, then chairman of the secret police, protested: "I decided to write to you about the article torn out of the translated edition of my book because this was done by officials who are responsible to you."[54] By that time the secret police was called the Committee of State Security (Komitet Gosudarstvennoi Bezopasnosti [KGB]). Andropov responded that the KGB did not engage in such activity, and sent Kapitza a complete copy of the Italian edition, including the thirty-page preface.

Kapitza wrote to Andropov in defense of Soviet dissidents, too, including Sakharov; we have no information that the KGB chief would have been as accommodating in these cases as in the case of the Italian book. Incidentally, Kapitza tried repeatedly to intervene in Sakharov's case. Sakharov did not know of his actions and lamented that Kapitza did not defend him during his exile in Gorky. Later, Sakharov learned of Kapitza's actions.[55] Kapitza was an activist, but sometimes the absence of action spoke louder than his deeds. Thus, when in 1973, a letter was published in *Pravda* signed by forty fellow academicians condemning Sakharov's "anti-Soviet" activities, Kapitza's name was conspicuously missing.

Kapitza was a controversial individual, but he was an outstanding physicist and a brave fighter on behalf of his persecuted colleagues. Universal human values and Soviet-Russian patriotism together directed his actions while his demeanor was colored by autocratic attitude. He was a giant in a bygone era. C. P. Snow said of him: "If he hadn't existed, the world would have been worse: that is an epitaph that most of us would like and don't deserve."[56]

Evgenii Lifshits and Lev Landau in 1957.

Source: Courtesy of Boris Gorobets, Moscow.

5

Lev Landau: GENIUS and **Evgenii Lifshits:** MORE THAN LANDAU'S PEN

Lev D. Landau (1908–1968) initiated one of the great Soviet schools of theoretical physics. He began his studies at the University of Leningrad and completed them in Western Europe, where Niels Bohr was his mentor. In 1932, he established an outstanding theoretical physics department in Kharkov. In 1937, he moved to Moscow to Petr Kapitza's Institute of Physical Problems and remained there for the rest of his life. There were two interruptions: in 1938–1939, when he was incarcerated, and in 1941–1943, when the Institute of Physical Problems was in evacuation in Kazan.

Landau created in broad areas of physics; his pupils extended his abilities in covering virtually the whole range of theoretical physics. His first outstanding achievement was the interpretation of superfluidity. He was a superb teacher of talented theoretical physicists, and guided Landau and Lifshits's opus magnum, the series *Course of Theoretical Physics.*

Evgenii M. Lifshits (1915–1985) wrote much of the Landau-Lifshits *Course.* He liked to express himself in writing and was the perfect partner for the grapho-phobic Landau. He was an important theoretical physicist in his own right. His life was devoted to Landau, and to science. He was precise and demanding, did not have pupils, but he was mentor to the whole of the community of physicists. His legacy included the high professional level of the leading Soviet physics periodical, the *Journal of Experimental and Theoretical Physics.*

Lev Davidovich Landau was born on January 22, 1908, in Baku, Azerbaijan, then part of Russia. His father, David L. Landau, was an oil engineer. His mother, Lyubov V. Landau (née Harkavi), was an obstetrician-gynecologist. Landau studied first at home; then he went to school and completed his secondary education at the age of fourteen. For higher education, first he went to Baku University where he started two majors, physics-mathematics and chemistry. After the first year he dropped chemistry, and in 1924, he transferred to Leningrad University. There were famous professors in Leningrad, such as

A. F. Ioffe and D. S. Rozhdestvenskii in experimental physics and A. Friedman and Paul Ehrenfest in theoretical physics. Friedman was primarily interested in cosmology, and Ehrenfest had moved to Leiden, Holland, before Landau's time. Landau became friendly with the students Artem Alikhanyan and Dmitrii Ivanenko. The latter coined the name "Dau" for him. Soon other future stars joined their circle, including George Gamow.

Landau was very much taken by the new physics of Werner Heisenberg and Erwin Schrödinger, but he arrived on the scene a little too late to become one of the pioneers in the field. This also happened to Gamow, Edward Teller, Eugene Wigner, Robert Oppenheimer, and others who were to be superb physicists but had to contend largely with the pieces left over after the great discoveries of the first decades of the twentieth century.

Landau graduated from university at the age of nineteen. He already started publishing papers before graduation, and embarked on his doctoral studies at the age of eighteen in Abram Ioffe's section of the Leningrad Institute of Physical Technology (Leningradskii Fiziko-Tekhnicheskii Institut, [LFTI]). His friendships with Ivanenko and Gamow, as well as with Evgeniya Kanegiesser (the future Lady Rudolf Peierls) and Matvei Bronshtein, flourished. Kanegiesser characterized their circle of friends in a light verse in which the following referred directly to Landau:[1]

To tuneful songs, Landáu the clever
Who'll gladly argue anywhere,
At any time, with whom whatsoever,
Holds a discussion with a chair.

They played tennis and went swimming and to the movies to watch Hollywood films, but were mostly doing theoretical physics. Landau's unruly character had already manifested itself in his student days. Gamow, Landau, and their comrades made ruthless fun of Boris Gessen (see more about him in chapter 6), the author of a long entry about the (nonexistent) ether in the 1925 edition of the *Encyclopedia Sovietica*.[2] As a punishment, Landau was dismissed from his teaching job at LFTI, but he was allowed to continue his research in Ioffe's laboratory.[3] In 1928, Gamow, Ivanenko, and Landau published a joint paper about the universal constants—Planck's constant, h; the speed of light in vacuum, c; and the constant of universal gravity, G.[4]

Somewhat later, Landau and Ivanenko stopped being friends, and soon afterward became dedicated enemies. Ivanenko took up a leading role in the fight of philosophers and "purist" physicists against modern physics, "Western imperialistic and bourgeois" science and its penetration into Soviet science, the subservience of Soviet scientists before the West, and "cosmopolitanism." Simultaneously, Ivanenko fought for recognition of his discoveries that, according to him, others—foreign scientists—had expropriated. Ivanenko was a professor of physics during his entire career at Moscow State University and carried out excellent research.[5]

In 1929, the Soviet Ministry of Education sent Landau for a long study trip to Western Europe that included Germany, England, Switzerland, and Denmark. He was impressed by Niels Bohr's Copenhagen School, and considered Bohr his mentor, his only mentor, ever. Landau attended some of the Berlin colloquia at which Einstein and other giants were regulars. Landau appeared to his European friends to be a dedicated communist, and he was eager to disseminate his views. Even in his dress he displayed his politics by donning a red jacket. This attitude was not especially conspicuous, because in those years, many in Western Europe viewed the experiment of Soviet Russia with sympathy. They felt that Communism might be the only effective force for fighting poverty and stopping the Nazis. Even physicists who never so much as flirted with Communism, among them Edward Teller, became Landau's friends.

Teller's experience with Landau is of interest on two counts. One is that discussions with Landau greatly stimulated Teller in developing the idea that eventually became known as the Jahn-Teller effect. Teller repeatedly declared that it would have been fairer to call it the Landau-Jahn-Teller effect.[6] The other is that Teller recognized the extraordinary talent in Landau as a physicist. After World War II, when many scientists and politicians in the United States did *not* consider the Soviet Union capable of soon catching up with America in nuclear matters, Teller disagreed. In his opinion, the Soviet physicists were second to none. He warned that the free world should be prepared to witness great advances in the development of nuclear weaponry in the Soviet Union, which, as a totalitarian state, was capable of focusing its resources on solving selected problems even with a weak economy.[7]

Landau's political views started changing after his trips to Western Europe. In 1935, he still demonstrated his support for the Soviet regime by publishing an article in one of the central Moscow newspapers, *Izvestiya*, with the telling title, "Bourgeoisie and modern physics."[8] He noted that becoming a physicist in bourgeois societies was a question of who could afford it, whereas in the Soviet Union physicists hailed from all strata of society, including the working classes. There was indeed a drive during the first period of Soviet power to recruit personnel for the many newly established research institutes from those layers of society that used to be excluded from higher education. Landau called for "building up the best in the world physics institutions of higher education for training the best in the world physicist-researchers and for creating the richest and healthiest popular literature."[9]

Somewhat later, his disillusionment was manifested in his participation in the compilation of a leaflet that called for a mass movement for socialism, on the one hand, but condemned "Stalinist Fascism," on the other.[10] Landau's arrest and his horrible experience in the NKVD prison (discussed more later) only added to his alienation from Stalin's regime. For a long time, he ascribed all the ills of the Soviet system to Stalin and his associates but continued to adhere to Lenin and the ideals of the October 1917 Revolution. By the end of the 1950s, he had completely changed. He no longer had any illusions about the Soviet system and no longer restricted the application of the label "Fascist" to Stalin; now he called Lenin the very first Fascist.[11]

In 1957, the secret police (Committee of State Security—Komitet Gosudarstvennoi Bezopasnosti [KGB]) prepared a summary of Landau's anti-Soviet statements over the years, which had been detected by listening devices or by undercover agents.[12] Here, we present a sampling of Landau's statements: From 1947: "The patriotic line brings harm to our science. This will only further delineate us from the scientists of the West and add to the gap dividing us from progressive scientists and technologists."[13] From 1952: "No effort must be spared to stay away from entering the thicket of the atomic business."[14] From January 1953: "If it were not for the fifth point I would not involve myself with special assignment [nuclear weaponry], would only be doing science in which I am being left behind. However, my special assignment gives me some leverage."[15] Here, Landau was referring to the Soviet questionnaires in which the fifth point was about nationality, and being Jewish was considered a nationality in the Soviet Union.

From an agent's report dated April 9, 1955: "At the end of March 1953 Landau was summoned together with Ginzburg to a high official of nuclear matters, Avraami Zavenyagin, in connection with their special assignments. According to the agent's report, Landau bitterly accused Zeldovich 'who initiated all this sordid action.'"[16] From January 12, 1956, " [O]ur regime, as I have known it since 1937, is unambiguously a Fascist regime and it cannot simply change."[17] From November 30, 1956 (following the Soviet suppression of the Hungarian Revolution): "Ours [our soldiers] are soaked in blood to their waist. The Hungarians achieved something greatest. For the first time they destroyed, dealt a superb blow onto the Jesuit idea of our time. A superb blow!"[18] Here, Landau refers to the Hungarian Revolution of October 23, 1956. First, the Soviets withdrew from Budapest and let the Hungarians start building a multiparty democracy. But on November 4 and in the ensuing days, they ruthlessly suppressed the revolution. Landau equated the Jesuits and the Soviets.

After Landau had completed his visits to Western Europe, he continued his life in Leningrad, but was feeling increasingly frustrated. The father of Soviet physics, Abram Ioffe, and Landau somehow did not find the right synergy between them. It is true that Landau was not an easy personality. Later, after Landau was arrested, Kapitza characterized him in a letter to Stalin as follows: "He is provocative and cantankerous; likes to find errors in others, and when he does, especially in important senior people, such as our academicians, he irritates them with irreverence. This way he has acquired a lot of enemies."[19] Kapitza's words were meant to excuse Landau in case his arrest had been a consequence of some foul language or behavior. Kapitza's remarks sounded convincing, because they probably characterized Landau well.

In the mid-1930s, Ioffe facilitated the creation of a network of physics research institutes at a variety of places in the country, including Kharkov—until 1934 the capital of the Ukraine. Landau moved to Kharkov and was appointed to be in charge of the theoretical division at the Ukrainian Institute of Physical

Technology (Ukrainskii Fiziko-Tekhnicheskii Institut [UFTI]). This institute soon became the third in importance among the Soviet physics research institutes after the Physical Institute of the Soviet Academy of Sciences in Moscow (Fizicheskii Institut Akademii Nauk [FIAN]) and LFTI. Many foreign scientists came to Kharkov; some merely to attend international meetings; others came for years to work. Laszlo Tisza, for example, arrived at Edward Teller's recommendation. Tisza had become unemployable in Hungary after he was arrested and spent time in prison for having undertaken assignments for the Communist Party. He had already earned his doctorate, yet when he arrived in Kharkov he wanted to do doctoral studies under Landau. Tisza was one of the first to take Landau's famous exams called *teorminimum*, or "theoretical minimum"—a series of high-level tests in theoretical physics and mathematics. Altogether, forty-three physicists took the tests over the years. Tisza completed his studies under Landau and defended his Candidate of Science (PhD equivalent) dissertation in Kharkov. He then left the Soviet Union, after having a premonition about the difficult times to come. Tisza later had a long, distinguished career as a professor of physics at the Massachusetts Institute of Technology.[20] Tisza's dissertational work in Kharkov was the completion of a research project previously started by Leonid Pyatigorskii, whom Landau had excluded from continuing his project.

Lev Landau in the mid-1930s in the courtyard of the Institute of Physical Problems in Moscow.

Source: Photograph by and courtesy of the late David Shoenberg.

Pyatigorskii had suffered a tragic fate. As a ten-year-old child in 1919, he had witnessed the murder of his father and mother during an anti-Jewish pogrom in their village, not far from Kiev in the Ukraine. He himself was shot and his right arm had to be amputated. Pyatigorskii was gifted, became Landau's doctoral student in Kharkov, and passed the *teorminimum*. Pyatigorskii and Landau had been on opposite sides of the fence from 1935 when there was a big debate at UFTI about whether the institute should enhance its activities in support of defense efforts or continue to do primarily pure physics research. Pyatigorskii was a dedicated communist and pushed for defense-related research, whereas Landau was for maintaining the emphasis on pure physics. The institute was divided along this line, and eventually those who thought like Pyatigorskii gained the upper hand. When a few years later Landau was arrested, he supposed that his misfortune was due to Pyatigorskii. After he had been freed, he blacklisted Pyatigorskii, which led to Pyatigorskii's near-complete isolation. Pyatigorskii had to change his research topic, and until 1955 he could not defend his dissertation. It turned out that Landau's accusations were unfounded.

Incidentally, when in 1939 Landau resumed his work after prison, for a short while he did not completely break with Pyatigorskii; they were producing a book jointly. It was the first volume, *Mechanics*, of what later became the world-renowned *Course of Theoretical Physics*. Given Landau's grapho-phobia, the book had to be written by the one-armed Pyatigorskii. Later, Lifshits reworked the first volume, though it remained to a large extent unchanged. After that first book, Landau never had anything more to do with Pyatigorskii.

From the start of Stalin's Terror in 1937, UFTI fared very badly. Brilliant scientists were arrested, tried, sentenced, and executed in short order. Landau felt that he might be the next at any moment, and he turned to Petr Kapitza for help. Kapitza needed a bright theoretician and invited Landau to his Institute of Physical Problems (Institut Fizicheskikh Problem, IFP) in Moscow. Landau became an associate of IFP in February 1937. A little later, Landau helped two outstanding former colleagues at Kharkov, I. Ya. Pomeranchuk and E. M. Lifshits, to move to Moscow.[21]

The period 1937–1938 is known as the Great Purge, or Stalin's Terror. Landau was by then sufficiently disillusioned so as to allow himself to vent his discontent in his conversations with others. The Cambridge physicist, David Shoenberg, went to work in 1937 for a year in Moscow at Kapitza's invitation. Shoenberg had been Kapitza's student in Cambridge before Kapitza's confinement in the Soviet Union. Shoenberg became friendly with Landau; both were new arrivals at Kapitza's institute. Their friendship was becoming special because Shoenberg's other colleagues were turning increasingly taciturn under the prevailing atmosphere of fear and terror. Shoenberg observed a great shift in Landau's views. When they had met in the West, Landau sounded very "left-wing," in Shoenberg's words, and now "he became much more realistic." I spoke with Shoenberg about their interactions in February 2000 in Cambridge. Landau told his friend about "*konzlagers*," the German abbreviation for concentration camps. Shoenberg thought Landau was talking about Germany, not the Soviet Union, but Landau made the reference unambiguous when he said, "We invented them."[22] Shoenberg had comfortable living quarters at the IFP, where his meals were prepared for him, and he often invited Landau for lunch. Landau enjoyed gossiping in English, and Shoenberg found him rather indiscreet. One day Landau did not turn up for lunch, and Shoenberg learned that he had been arrested.

Shoenberg regretted his friend's fate, and not only for humanistic considerations. Shoenberg was doing experimental work with Kapitza. He was involved in determining the changes of the bismuth crystal under the impact of strong magnetic fields. He found a lot of curious results, and he was looking for their interpretation. He had shown his data to Landau, who became interested in them, did some calculations, and produced a formula that showed the relationships among the experimental results. Shoenberg was very pleased by Landau's assistance because all his measurements fitted Landau's formula beautifully. Unfortunately, Landau had not had time to show Shoenberg the details of his calculations or the way he obtained the formula before he disappeared.

The NKVD arrested Landau on April 27, 1938. The story of his getting freed as a consequence of Kapitza's guarantee is told in chapter 4 about Kapitza. Here, we limit our narrative to what happened to Landau. According to the information in NKVD archives, Landau was interrogated in June 1938; he signed the minutes on September 3, 1938, and the investigation was completed on November 21, 1938.[23] His case was transmitted to the public prosecutor on January 24, 1939, and it was transmitted from the prosecutor's office to the Moscow court on March 25, 1939. The case was handled in the secret political section. Some fragmented description of the investigation also appeared in secret documents. Thus, on occasion, Landau was made to stand for seven hours and was threatened but not beaten; they applied other means of pressure to extract a confession from him. Landau, however, had stamina and stayed quiet in the interrogations for long hours.

His fellow prisoners gave him advice about how to behave with the interrogators in order to not make them angry. The prison diet was based on the grain millet, which Landau could not eat. By the time he was freed, he was no longer strong enough to walk; he lay in bed all the time, and the only thing he could do was to think about scientific problems. Subsequently, he told his wife that he could be oblivious to the prison conditions—whether he was in solitary confinement or put into an overcrowded cell—because he was thinking about physics. He was not given paper and pencil, but given his graphophobia, this may have not been important. He worked out his problems in his head, performed complicated calculations, and stored his "manuscripts" in his head.[24] When he left prison, he had four research papers ready for publication. This is probably what saved his sanity.

During Landau's disappearance, he became a nonperson in Soviet society, in accordance with to the usual practice. Shoenberg experienced this first hand. When Shoenberg finished his work, he wrote up his paper, which he intended for the British journal *Proceedings of the Royal Society*. At that time, it was possible to publish results obtained in the Soviet Union abroad if the scientist published the same paper simultaneously in Russian. Shoenberg's manuscript was translated into Russian, and the next step was to secure official permission to publish the work. In his manuscript, he acknowledged Landau's assistance, and Landau's formula figured in the paper. In seeking permission to publish, Shoenberg was confronted by the assistant director of Kapitza's Institute. She was astonished that Shoenberg had expressed thanks to an "enemy of the people." She told Shoenberg that the acknowledgment had to go; Shoenberg refused and went to talk to Kapitza.

The director of the institute appeared quite sympathetic when Shoenberg told him the story. Shoenberg then narrated the rest of their meeting as follows: "[A]t that moment the Assistant Director marched into Kapitza's office, and Kapitza immediately turned to me as if he were in mid-sentence and said something like, 'So you understand, David, that you have to take that acknowledgment out.' He obviously had to show support for his Assistant Director. Then he gave me a hint that, as I was going back to England shortly, there was no reason not to put back the acknowledgment into the English version of my paper."[25]

This is how it happened that the Russian version of Shoenberg's paper did not mention Landau at all, and the mysterious formula appeared without Landau's name. When Shoenberg submitted the English version to the Royal Society, the referee thought that Shoenberg might be plagiarizing since there was no mention of where the basic formula came from. In the meantime, Landau's British friend, Rudolf Peierls, to whom Shoenberg showed the formula, understood that Landau must have worked it out. Peierls reconstructed the method and derived Landau's formula. Eventually, the formula with its derivation appeared as a one-page appendix by Peierls to Shoenberg's paper. The article became well known. Yet if Russian authors needed to refer to Landau's theory, they had to quote Peierls's appendix to Shoenberg's paper.[26]

As for Landau's existence in prison, eventually the secret police broke his resistance and extracted confessions from him. When Kapitza's efforts finally succeeded, Landau was released. The NKVD described their investigation and decision as follows:[27]

> LANDAU Lev Davidovich, born 1908 in Baku, until his arrest professor of physics, non-party, citizen of the USSR, was sufficiently proved to have participated in anti-Soviet group, harmful activities, and in an attempt to produce and disseminate an anti-Soviet leaflet.

However, taking into account

1) whereas LANDAU L. D. is an outstanding specialist in theoretical physics and may in the future be useful for Soviet science;
2) whereas Academician Kapitza P. L. expressed willingness to undertake guarantee for LANDAU L. D.
3) that the Minister of Internal Affairs of the Soviet Union, Commissar of State Security of the first rank, Comrade L. P. Beria ordered freeing LANDAU to be undertaken by academician Kapitza's guarantee,

IT IS RESOLVED:

Arrested LANDAU L. D. is to be freed, action in his respect to terminate, and his case to be deposited in the archives.

Landau was freed from prison exactly one year after his arrest. During the rest of his life, he barely talked about his one year in the hands of the NKVD. Landau's tragic experience in 1938–1939 was expunged from public awareness in the Soviet Union until its collapse. For example, in 1965, one of Landau's former pupils and longtime associate Alexei Abrikosov published a booklet about the life and oeuvre of his mentor; there is no mention of this "episode."[28] Another former Landau pupil, academician Isaak Khalatnikov edited a substantial volume of reminiscences about Landau; the book appeared in 1988.[29] It starts with a review by the late Evgenii Lifshits, again, without even a hint of the tragic turn of events in Landau's life in 1938–1939.[30] The N. E. Alekseevskii chapter—Alekseevskii worked at Kapitza's institute, just as Landau did—is specifically about Landau in the 1930s. Alekseevskii mentions that their meetings with Landau stopped at some point in 1938 but hints that Landau's changing family situation is the reason for this.[31]

After his liberation from NKVD captivity, Landau became more careful about what to say and when to stay quiet. He always remained deferential toward Kapitza.[32] According to one of the leading Soviet physicists, E. L. Feinberg, there were two Landaus. In his youth he was shy, and this made him unhappy. Later, he learned to be tough, and this made him happy.[33] According to Feinberg, the original shy Landau created a mask for himself that was harsh, and after long use the mask became his natural self. But even in this harsh behavior there appeared windows when his original self came to the surface. This was the more private Landau, in contrast with the less sympathetic public Landau. Being in the presence of one or two close friends or listening to some poetry that moved him would make Landau's softer self come out. According to Kapitza, "Those who closely knew Landau were aware that behind this briskness in statements there hid, in essence, a very kind and responsive person always ready to come to help to somebody who was unjustly offended."[34]

Landau's incarceration and his cruel treatment in the NKVD prison left indelible scars in his mind. He lived the rest of his life under the shadow of this experience., He was never legally rehabilitated during his lifetime, and

Kapitza's role as guarantor never officially ended, not after Landau's death in 1968, and not even after Kapitza's death in 1984. It was only in 1990, after the collapse of the Soviet Union, that his case was reviewed. On June 23, 1990, the office of the State Prosecutor General decided to discard the NKVD resolution of April 28, 1939 (when Landau was freed, conditionally), and to stop any further procedure in the absence of crime.[35]

As for Landau's work after his return from captivity, Kapitza was not disappointed. Kapitza discovered superfluidity of helium at the end of 1937, and he knew that it would take a theoretical physicist of Landau's caliber to provide an interpretation of the phenomenon. Without interpretation, the discovery could not be considered complete. In 1962, Landau was awarded the Nobel Prize in Physics for this theoretical contribution.

Stalin and Beria were not disappointed either in Landau's contribution to the work they considered crucial, namely, the development of the Soviet nuclear weapons. During the war years, 1941–1945, Landau investigated traditional explosives. He studied the trajectories of the explosion products and the nature of shock waves. This work later proved to be instrumental for the development of the nuclear bombs. Landau's engagement in nuclear weaponry was first with the atomic bomb and then the hydrogen bomb. Today, not only it is well known that the Soviet Union built its first atomic bombs as a result of information it gained through espionage, but the Russian authorities also admit it. However, the Soviet scientists and technologists labored on the atomic bomb as if they had not been provided the blueprints for their production. The Soviet leadership was extremely careful to restrict the circle of people who knew about the sources. and except for the two top physicists in the leadership of the project—Igor Kurchatov and Yulii Khariton—the others basically worked without such knowledge. At most, some may have wondered how it could happen that of the often great variety of possible solutions, Kurchatov always hit on the right ones.[36]

It appears that Kurchatov and the Soviet leadership did not become complacent due to the inflow of first-rate intelligence. The Soviet scientists had to work hard on solving every one of the problems in developing the atomic bomb despite the availability of intelligence. This served to camouflage the inflow of American information, and was also an excellent preparation for the continuation of the nuclear weapons program beyond the first atomic bomb. The next atomic bombs already utilized original Soviet solutions.

From the start of the Soviet atomic bomb program, Kurchatov understood Landau's potential value to the project. He kept urging the leading governmental organs to get Landau involved. His first attempts failed, but he did not give up. He and Khariton realized that "Landau combines the art of deep theoretical analysis of physical phenomena and the ability of finding the efficient approaches for quantitative calculations of extraordinarily complex problems.

As a result, relatively simple relationships are obtained whose application solves practical problems."[37]

In the period 1943–1945, Kapitza's IFP was not yet included in the nuclear program; hence Landau, working there, could not be involved. But because Landau was one of the leading scientists in nuclear physics, and because of his expertise in related areas, for example, shock waves, it was inevitable that sooner or later he would become part of it. First, he and his group, consisting of Lifshits and a few others, examined the theoretical aspects of detonations of ordinary explosives. By summer 1946, Landau had become a member of a commission of theoreticians charged with checking the previous calculations by Yakov Zeldovich's group at the Institute of Chemical Physics. Furthermore, they were branching out in new directions to describe the details of the events immediately following the nuclear explosions. In 1946, Kapitza left the nuclear program and was soon after removed from the directorship of the IFP. There was no longer any obstacle to involving the IFP and Landau in the atomic bomb program.

Landau's and Lifshits's contributions to weapons development have not been widely known, and they did not want them to become known either. Even when they spoke about these activities, they underlined that they were not building bombs, only carrying out calculations on the consequences of exploding them.[38] This was not quite so, because they investigated the processes within the bombs as well. They also dealt with the quantitative theory of nuclear chain reactions. The next step was the investigation of reactions that might play a role in thermonuclear detonations.

Landau and his group stayed at IFP in Moscow, although at one point his possible move to Arzamas-16 was considered. Ultimately, both he and his superiors thought that he would be better able to serve the project by staying at IFP. Landau had his own style of working, which was obviously most efficient for a theoretician, but was unacceptable to some bureaucrats. The NKVD general charged with supervision of the unit to which Landau belonged, observed Landau's absences for hours and days from the lab and contrasted this with Landau's high salary. At one point, he wanted to have Landau fired from the project. When the scientific director of this unit, academician A. I. Alikhanov, learned about the general's intention, he contacted the supreme leader of the atomic project, Lavrentii Beria. Alikhanov explained to Beria that Landau was working according to the needs of the project rather than following a strictly regulated but inefficient regime. Landau was left in peace.[39]

In the year 1946, another crucial event was the elections to the Academy of Sciences. The previous elections were held in wartime, in 1943, and their most conspicuous story was Kurchatov's election to full membership, skipping the corresponding member stage, as a result of Stalin's direct interference. Kurchatov considered Landau so important for the atomic bomb project that in 1946 he wanted to have Landau elected to full membership, and this is what happened. In the next elections, in 1953, Andrei Sakharov would also be elected, skipping the corresponding member stage. Based on these three examples, the jump in the elections to full membership might seem a rather frequent occurrence. This would, however, be a false impression because these three events were almost without precedent and obviously occurred due to the extreme importance of the nuclear weapons program. In Landau's case, his election was based more on expectations of his future contributions than rewarding services already rendered.

Landau was more than sanguine about skipping the corresponding membership and becoming full member directly; he even declared that should he be elected corresponding member, he would decline it. Stalin was the arbiter of such questions, and apparently the Soviet dictator supported Landau's candidacy for full membership. This is the more surprising in view of Landau's past incarceration, which was as much a consequence of the anti-Stalin leaflet as it was of unfounded accusations. Kurchatov's need for Landau's active involvement in the nuclear program must have played a role, and Stalin could act mysteriously.

Landau proved most valuable not only for the atomic bomb, but also for the hydrogen bomb project. He played a decisive role in reaching the conclusion that the so-called tube approach—the Americans called it the "classical Super"—would not lead to success. This gave green light to Sakharov's and Ginzburg's approaches to the successful Soviet thermonuclear bomb.

In 1953, as soon as Stalin disappeared from the scene Landau quit the nuclear weapons program. He told his friends: "This is all; I'm not afraid of him [Stalin] anymore, and I no longer work on this project."[40] As it was alluded to earlier, later, in spring 1955, some special problems necessitated that he be recalled. It was in connection with the hydrogen bomb of unprecedented strength that Khrushchev was enamored with. However, Landau refused to rejoin the project even for a temporary assignment.[41]

Landau could be very direct toward the authorities as well as toward his peers, and had a tendency for autocratic behavior. He could be cruel and unforgiving, even publicly, when he did not like a reported finding, regardless of the authority of the presenter. In some way, it was an expression of his democratic attitude that he did not distinguish between the targets of his criticism and that he considered those on the receiving end of his critiques as his equals. But even this democracy did not justify the occasional coarseness of his reactions. Clever people learned that it was better to approach Landau with finished, completed work that one could defend with assurance than with work in progress. It was also possible to convince Landau that his interlocutor was right, but he never uttered words admitting directly that he was wrong.

He was generous with his advice and did not expect acknowledgement for it, but he was not very careful to acknowledge what he may have learned from other people's unpublished communications either. He stated: "Some think that a teacher steals from his students, some, that students steal from their teachers. I think that both are right and participation in this mutual stealing is remarkable."[42] However, he never let his name be added to any authors' list in a publication unless his ideas had played a decisive role in the study and/or taken a substantial part in the calculations. The negative aspects in his interactions with junior colleagues were manifested when they turned to him with their ideas and he went ahead and solved the problem, and published the results on his own.[43]

There have been claims (see chapter 7) that Landau's criticism prevented the publication of important results that later were proven to be correct by foreign scientists. This meant loss of priority and recognition. This is a complex question. Landau did not forbid anybody to publish anything even if he disagreed—but it would have taken a very strong character to publish something that had earned opposition from Landau. He had such immense authority that he did not have to "forbid" publication; even his criticism sufficed to prevent it. The atmosphere in Landau's environment was authoritarian. Besides, Soviet scientists could not easily send their manuscripts abroad for publication, let alone for discussion with foreign colleagues. Exchanges were rare; most Soviet scientists hardly ever traveled; great authorities seldom came for visits, and when they did, there were carefully choreographed programs for them. These circumstances tremendously enhanced the importance of the opinion of such an authority as Landau. Igor Tamm taught his disciples how to handle Landau's criticism. He taught them to ignore general comments, such as "all this is nonsense," but to pay close attention to his specific remarks.[44]

Landau was a great physicist, and his acumen manifested itself not only in weapons design, but even more so in pure science. He was a theoretical physicist. One might tend to think that is already a restrictive delineation within physics, but in fact it is a vast area. Few physicists in modern times have been capable of contributing creatively to both experimental and theoretical physics—Enrico Fermi comes to mind as one. Landau was one of the last theoretical physicists who could navigate any area of theory with ease. It was unfortunate for him that most of his career was spent in isolation from much of world physics. This sometimes favored responding to what had been discovered elsewhere to continuously producing pioneering work in the forefront. This was yet more characteristic of many of his peers in Soviet physics. The high level of sophistication, including its higher mathematical character, makes it very difficult to describe Landau's achievements. He certainly earned the respect and admiration of his colleagues at home and internationally. On his fiftieth birthday, in 1958, he was presented with a symbolic two-stone tablet listing Landau's "Ten Commandments," his ten most important achievements in theoretical physics.[45] Then, the question arose, why not add yet another tablet with more "commandments?"[46]

Landau was not given to philosophizing. He liked to deal with concrete issues and sought order in everything, whether it was physics or everyday life. His mind was utterly analytical, and he liked to introduce classifications ubiquitously. In this, his rational approach to everything and his characteristic sarcasm blended inseparably.[47] He classified theoretical physicists, placing them on a steeply diminishing scale. His principal consideration was discoveries rather than knowledge, impact, or books. At the top of his list of twentieth century theoretical physicists was Einstein. Taking previous centuries into account, he put Newton in the same category with Einstein. The next category represented a fivefold reduction in the level of quality. Here, he put thirteen physicists, including Niels Bohr, Enrico Fermi, Werner Heisenberg, Wolfgang Pauli, Erwin Schrödinger, Paul Dirac, Max Planck, and Louis de Broglie; there is no unambiguous evidence of who the other five might have been. Landau placed his own name into the next category, but later he changed his mind and elevated his position closer to the category of Bohr, Fermi, and the others. There are lists of Soviet physicists as well, but only fragments of them have been made public; here Landau may have been reacting to subjective considerations.[48]

Lev Landau in 1958.

Source: Courtesy of Boris Gorobets, Moscow.

His classification of women was anything but politically correct, but Landau was far from political correctness even for his time. For women, he created six classes. In Class I he placed the German movie star Annie Ondra, a grey-eyed, Marilyn Monroe–type blond. "When looking at her; it is impossible to turn away. Class II is for pretty, slightly pug-nosed blonds. Class III, they are nothing particular, not quite terrible, but maybe not for looking at. Class IV, better not to look at; they are not dangerous for people, but scary for horses. Class V, interesting; they are not to look at; their parents need be reprimanded; Class VI, if repeated, they have to be shot."[49]

Landau and his future wife Konkordia ("Kora") Drobantseva met in Kharkov when he was twenty-seven years old. She was an engineer in a chocolate factory at the time. He had been shy with women and had never had any relationship before. When they married, they gave each other full freedom. Landau utilized his; he had several relationships with beautiful women; sometimes a relationship lasted for years, and it appears that it did not cause problems in his marriage. He and his wife had one son, born in 1946, Igor Landau, who studied at Moscow State University and became an experimental physicist in Kapitza's institute; later, he worked in Switzerland. He died in 2011 and was buried in Landau's gravesite at the Novodevichy Cemetery.

Landau classified sciences as "natural," "nonnatural," and "antinatural." He assigned some of the humanities to pseudoscience and thought that they cheated working people by yielding products of no value. Scientific communism, that is, the teachings of the philosophy and political economy of Marxism and Leninism, and similar disciplines gave Landau a fit. He considered them especially harmful and looked with contempt at the people who chose such disciplines for their profession. He classified intellectual workers in four categories. They could be imaginative and diligent, imaginative and lazy, lazy and dimwitted, or diligent and dimwitted.[50] His classification of research work showed both his elitism and his highly developed critical approach to his profession. The highest category was "gold fund." The next was "quiet pathology," to which he assigned 90 percent of publications. The authors of such works do not expropriate results of other authors, but they do not have their own results either. They are not engaged in pseudoscience, only quietly and unnecessarily rummage their own subject. "Philology" and "Exhibitionism" refer to pseudoscientific works, with aggressive claims for scientific results, whereas it is all merely self-advertising. The last class covered what could be described as "noise."[51]

Landau's grapho-phobia has been an interesting puzzle. Some felt it was congenital; others thought it was an acquired trait. Most probably, it was a mixture of both. According to one of his biographers, Boris Gorobets, Landau's grapho-phobia could be explained by his rationalism and egocentrism.[52] It was rational in that he was not very good at writing, whereas Lifshits excelled at it, so each should do what he did best. It was also egotism in that he did not want to waste his time and energy on writing down his thoughts; it sufficed that he had thought of them. Landau was a good lecturer; he did not have to think much in lecturing; his thoughts flowed seamlessly. But it has also been noted that they were often quite close to the text of the corresponding volume of the Landau-Lifshits *Course*. The first volumes of the *Course* had started with his pupils transcribing his lectures. Probably his grapho-phobia was much less pronounced in his youth and strengthened with age, not least because by then he could engage others to write down his thoughts. This was his practice not only for coauthored papers but also for papers produced by him as a solo author, which were written up by someone else, most often by Lifshits.

Landau's creative life was telescoped into three relatively short periods. His first paper appeared in 1926 when he was eighteen years old. From 1926 until the time of his incarceration in 1938 was a period of *twelve years*. Then there were *two years* before World War II began for the Soviet Union, lasting until the summer of 1945; but this period cannot be considered normal. From 1946, Landau was heavily involved with the nuclear weapons program. He quit in 1953, after Stalin's death. Another creative period of *nine years* followed, which lasted until his tragic automobile accident on January 7, 1962. Thus, his creative life lasted twelve + two + nine, a total of twenty-three years, between the ages of eighteen and fifty-four.

The tragic accident happened on a slippery winter road; the car in which Landau was traveling collided with a truck coming from the opposite direction. Neither the driver of the car nor the other passenger was hurt. Landau suffered such heavy injuries that the doctors who treated him did not expect him to survive. Most of all, he suffered head injuries. News of the accident filled the airwaves and all forms of the media. The country was traumatized. I was a student at Moscow State University at the time and experienced this myself. We first learned about what happened in our physics lecture; the effect was almost indescribable. People did not understand Landau's physics, but they understood his tragedy, and they could identify with him. He was not a politician; his portrait was not carried in May 1 demonstrations, but in contrast with the faceless politicians—in spite of their portraits everywhere—this was a real person. In addition, there was the great respect of the Russians for science.

Landau's colleagues, whether they knew him personally or not, sat vigil; everybody was trying to do something for him. Even the authorities overstepped their own limitations and lifted their rigorous rules to let medications and the chemicals for preparing medications be flown in from foreign lands without the usual red tape. Foreign doctors arrived for consultation. These extraordinary medical efforts kept Landau alive, but his next six years were ones of continuous suffering. He suffered physically from pain and he suffered spiritually from understanding that his mental capacities had shrunk. For a man of his sarcastic and critical nature, this must have been unbearable. It is a question whether he was lucky that due to his fame he was saved or whether he would have been better off if he had been allowed to die. When he asked his doctor whether there was any hope of ending his excruciating and constant pain, he added that he had never been afraid of death and that the condition he was in was no life for him. Yet this is how he lived for six long years until one day in 1968 he died.

The Royal Swedish Academy of Sciences found it prudent to announce Landau's Nobel Prize in Physics in October 1962, the year of the accident. The distinction was awarded for his theories of liquid helium. In 1954, Landau's possible Nobel Prize was reviewed for the first time in the Nobel Committee for Physics. During the 1950s, he had been nominated regularly, sometimes together with Kapitza, sometimes alone. In 1958, Wolfgang Pauli nominated Landau alone for the Prize, and Artsimovich and Leontovich nominated him and Kapitza together. In 1959 and again in 1960, Heisenberg nominated Landau, but there was no nomination for him in 1961. When they were nominated together, Kapitza was nominated for the discovery of the superfluidity of helium and Landau for the theoretical interpretation of the phenomenon. Bohr had been submitting such nominations for years. However, the Nobel Committee for Physics hesitated, because superfluidity had also been discovered by two Cambridge physicists, essentially at the same time as by Kapitza.

In January 1962, as soon as Bohr learned about Landau's accident, he changed his mind about the usual joint nomination and decided to submit one solely for Landau. There was no ambiguity of priority for Landau's work; hence, he had a better chance of receiving the award alone than coupled with Kapitza.[53] Bohr was highly experienced in Nobel Prize matters, and his nomination carried weight. Besides, the members of the Nobel Committee could not have been insensitive to Landau's tragedy. He was awarded an unshared Nobel Prize.

There was no question about Landau's traveling to Stockholm to receive the award. The Swedish ambassador in Moscow handed it to Landau in the morning of the same day when the festive award ceremony was taking place in Stockholm for the rest of the awards and in Oslo for the peace prize. Protocol did not call for his saying anything during the ceremony when he was given the medal, the diploma, and the check in the conference room of the Central Hospital of the Soviet Academy of Sciences. Yet Landau said a few sentences in English in response to the Swedish ambassador's speech. Only a limited number of people were allowed to be present during this ceremony. Landau's doctors did not want a crowd. Kapitza was there; the Soviet Nobel laureates had also been designated to be present though two of them were deleted from the list to make room for some officials of the Science Academy.[54] There was a reception in the evening of the same day at the Swedish Embassy in Moscow at which Landau was represented by his wife and Kapitza; this time all the Soviet Nobel laureates attended.

At the same time, the traditional Nobel banquet was taking place in the famous Blue Room of the Stockholm City Hall. Landau would have been in stellar company. The chemistry prize was awarded to Max Perutz and John Kendrew for the structure of globular proteins. Yet more exciting was that the physiology or medicine prize went to Francis Crick, James Watson, and Maurice Wilkins for the discovery of the double-helix structure of DNA. John Steinbeck received the Nobel Prize in Literature.

Each category of the Nobel Prizes is designated one two-minute speech at the Nobel banquet. In lieu of Landau, the Soviet ambassador in Stockholm delivered the two-minute address that had been carefully prepared in Moscow by apparatchiks rather than by scientists, and it was exactly what could have been expected from officials sitting in Moscow offices. It read as if Landau had asked the ambassador to ascribe his success to the labors of all Soviet scientists and to the achievements of Soviet science. Further, Landau—ostensibly—thought that the award would facilitate the cooperation between Soviet and Swedish scientists and scientists of the whole planet.[55] Luckily, the Soviet authorities did not charge any of their office people with giving Landau's scientific lecture in his stead; he never gave his Nobel lecture.

The tombstone over Landau's grave is an exquisite bust of the physicist resting on a tall metallic stand. The bust is the creation of Ernst Neizvestny, who was Kapitza's choice. Kapitza was the director of the institute where Landau worked, and a considerable share of the expenses of the tombstone was paid by the Academy of Sciences.

(a)

(b)

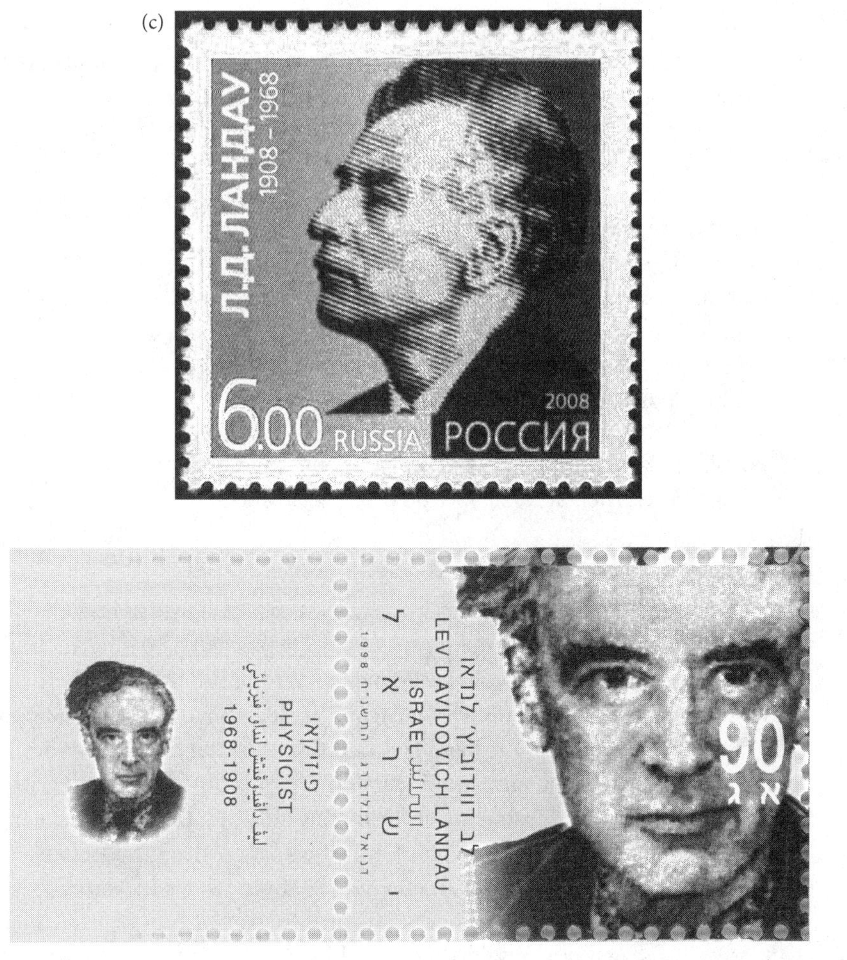

Commemorating Lev Landau on postage stamps of (a) Azerbaijan (2008), (b) the Ukraine (2010), (c) Russia (2008), and (d) Israel (1998).

Upon hearing about Landau's death, Nobel laureate Hans Bethe referred to him as "one of the greatest theoretical physicist of our time."[56] Two-time Nobel laureate John Bardeen sent a telegram to the Soviet Academy of Sciences stressing Landau's "profound contribution to elementary particle physics, to low temperature phenomena, to the theory of solid state, and to other branches of physics."[57] At Landau's burial ceremonies, there were quite a few speeches. Evgenii Lifshits spoke on behalf of Landau's pupils. He ended his speech with the following sentence: "As we are bidding goodbye to him, we are bidding goodbye to the best part of our lives."[58] Lifshits was Landau's right hand—literally, Landau's writing hand—and during the subsequent years, Lifshits often recited the poet David Samoilov's lines, obviously meaning Landau without explicitly referencing him:[59]

> It's all over. The eyes of genius have closed.
> And when the skies have darkened,
> As if in a now deserted building
> Our voices have become audible.
> Let us drawl, drawl the hackneyed word,
> Let us speak languidly and vaguely.
> How they feast us and treat us graciously!
> They do not exist. Anything is permitted.

Lifshits was delicate enough not to offend anyone personally who might have considered themselves Landau's direct followers.

***

Evgenii Mikhailovich Lifshits was born in 1915 in Kharkov. He had a younger brother, Ilya; both boys became physicists, both excelled as theoreticians, and both became full members of the Soviet Academy of Sciences. Their father was a renowned medical doctor, a gastroenterologist. He had a broad education and outlook on the world and traveled often, taking his family with him. It was most unusual, but in the Lifshits family, they spoke English with the children and had an English teacher for them until 1937. They also had a music teacher, and both boys at some point considered becoming musicians. Their mother, Berta Lifshits (née Mazel', meaning "happiness," as in Mazel' tov) was very proud of her sons' achievements. According to an anecdote, Berta's friend called her a liar when she said that both her two sons were members of the Science Academy because it was so unbelievable.[60] She survived her husband, who died in 1934; she died in 1976.

Evgenii was first taught at home and entered school starting with the sixth grade. He was not very sociable as a child, but he did not make enemies. From 1929, he attended a chemical college for two years. He started his university studies in 1932 at the Faculty of Physics and Mechanics and graduated a year later in 1933. He immediately began his graduate studies as Landau's student at the UFTI and completed his project and examinations in 1934. His project was related to the existence and properties of electrons and positrons. Landau, who had not believed in the existence of positrons, may have hastened the completion of Lifshits's graduate studies. Some physicists remembered Landau repeating the saying, "Dirac-durak," meaning in Russian "Dirac-nitwit."[61]

Lifshits then charged ahead after becoming a Candidate of Science (PhD equivalent), and in five years he acquired the higher degree; he became a

Doctor of Science. There was one short period in Lifshits's life, which looks like derailment, in spring 1938 when for a few months he worked in the Institute of Leather in Moscow, followed by about a one-year stint at the Kharkov Institute of Chemical Technology. By September 1939, however, he was back on the right track at Kapitza's Institute of Physical Problems (IFP). The timing of the detour coincided with Landau's imprisonment by the NKDV. It was not just that Lifshits was without a mentor and had to seek alternative employment. His main worry was that he might get entangled in the case. Under normal circumstances he need have not feared such a development, but those were abnormal times.

From the second half of the 1930s, Landau and Lifshits were engaged in creating their famous *Course of Theoretical Physics*. Apart from the first version of *Mechanics*, on which Landau cooperated with Pyatigorskii, it was always Landau and Lifshits. Landau highly valued Lifshits's writing ability. On occasion, he called him "a great writer, the Leo Tolstoy of physics," in which one might even spot a bit of patronizing. There have been comments that Landau "may have exploited" Lifshits.[62] However, they found the ideal partners in each other. Lifshits was Landau's pupil, and remained his professional pupil to the end. Lifshits has been called the "Apostle of Landauism with no compromise."[63]

There has been a popular joke that "there is not a line written by Landau and not a thought by Lifshits" in the *Course*, but this is a very unfair characterization of what the series represents.[64] The two discussed every chapter and every paragraph, and afterward Lifshits wrote it up and then returned it to Landau, who read everything. Another discussion followed, of the minutest details, and Lifshits prepared what was usually the final version. There have been translations into twenty languages of the set or its individual volumes. The complete set is 5300 pages long. The *Course* is Lifshits's principal legacy, but not the only one.

Although Lifshits did not have his own pupils, he was still a great pedagogue; his courses were coveted, and he had plenty of opportunity to teach. He taught courses at several institutions of higher education. He loved lecturing. In the immediate postwar period, new institutions mushroomed, and there was a need for good lecturers. Lifshits had a clear mind and a precise style of lecturing. He lectured as only scientists who have a full understanding of the subject matter can.

His biographers described a characteristic episode. When Dirac gave a lecture to a huge audience in Moscow, most of those present could not have followed Dirac's lecture in English (maybe some native English speakers could not have either). Lifshits volunteered to translate, but Dirac did not want the interruption of preparing a near-simultaneous translation. So he gave his lecture, and it was followed by Lifshits's "translation." He gave a brilliant presentation of what Dirac had said, he used Dirac's formulas preserved on the board, and he faithfully followed the sequence and contents of Dirac's lecture. It was much more difficult than giving one's own lecture, but Lifshits's presentation proved impeccable.[65] In contrast, he was not very good in live scientific discussions. It was a standard outcome that when people turned to Lifshits with questions or for clarification, instead of entering an exchange, he readily gave them the exact place in the *Course*, the volume, section, and often even page number, where the answer could be found. Incidentally, Dirac was known to lecture as if reading from his own book—he was not reading, he remembered everything verbatim. When his students then asked for clarification of a problem, his response was to repeat exactly what he had already said.

Evgenii Lifshits lecturing in the early 1980s.

Source: Courtesy of Boris Gorobets, Moscow.

Whatever Lifshits did, he did with perfection. For his college courses, the preparation he needed to do was obvious. It was less so when he was invited to give guest presentations because he did not always know the level and background of his audience. He therefore prepared three versions of his talks, at three different levels, and at a moment's notice he could change from one to another. He never missed an opportunity to deliver a talk about Landau. His attachment to Landau went beyond science; it was a great friendship, and it endured the trying times of the period between 1962 and 1968, from the moment of Landau's accident to Landau's death. And it did not end with Landau's passing, either. It is curious to think that Lifshits lived a fully active twenty-three years after Landau's incapacitating accident; yet most people think of them as a unit to the end. This image has formed due to Lifshits's faithfulness to Landau and his legacy.

One of the few areas of Lifshits's independent activities was his work as editor of the *Zhurnal eksperimentalnoi i teoreticheskoi fiziki* (*Journal of*

Experimental and Theoretical Physics [*JETP*]). In 1931 this periodical was a spin-off of the *Zhurnal Russkogo fiziko-khimicheskogo obshchestva* (*Journal of the Russian Physical-Chemical Society*), which had been appearing since 1878. In 1955, Petr Kapitza was appointed editor-in-chief of *JETP*, and he invited Lifshits to be his first deputy, delegating to him the running of the journal's day-to-day operations. It was an excellent choice and an excellent partnership. Kapitza was not interested in the details, but he was a big name. Lifshits immersed himself in all the operational details; he had a vast knowledge of topics and authors as well as potential reviewers; and he loved spending his time and efforts on the journal. Even physically, it was a most convenient arrangement. Lifshits's home, his work place at the IFP, and the editorial office of *JETP* were situated around the same courtyard. He could go from one place to another without a winter coat even in cold weather.

He was in charge of the journal for forty-six years, and it was estimated that during this period some one hundred thousand manuscripts went through his hands. He was direct. candid, and dedicated to bringing up the level of the journal, and he succeeded. It became a sign of prestige to appear in the pages of this journal, and it became the leading physics publication in the Soviet Union.[66] He often made corrections and suggested changes to authors. It even happened that he rewrote parts of the manuscript and some considered his contributions so substantial that he could have been a co-author of their papers. Yet he respected the intentions and the desires of the authors. He liked to say, "The editor suggests, but the author decides."[67] Another of his favorite aphorisms about editorial work was, "The principal editorial instrument is deletion."[68] *JETP* was the first Soviet physics journal to appear in the West in a full English translation. This provided an added incentive to its authors. With the appearance of the English translation they were entitled to a small honorarium in Western currency. It was given to the authors of *JETP* in form of so-called certificates to be used in special shops, *Berezka* ("birch tree" in diminutive),[69] where products that were mostly unavailable elsewhere could be purchased.

Lifshits received recognition for his work, but it came rather late. In 1962 he was awarded jointly with Landau the Lenin Prize for the *Course*, which was high on the list of Soviet awards. Lifshits had participated in the work on thermonuclear bombs under Landau's direction and in 1954 was awarded the Stalin Prize for his contribution (Landau received Hero of Socialist Labor). Lifshits's election to the Soviet Academy of Sciences did not come too early either. Although he had been nominated in 1953 and 1958, he was not elected corresponding member until 1966, and full member in 1979. Thus, he was an academician only during the last six years of his life. Looking at the dates, he joined the first tier years after Landau's accident; and the second tier, many years after Landau's death. It seemed that while Landau was alive, he did not consider it a high priority to get Lifshits into the Academy. He did not hinder his election, but he did not go out of his way to facilitate it either, and his silence was interpreted as lack of support. It seems that for Landau, it was a convenient arrangement to have Lifshits in the position he was in, a respected professor, but not a member of the Academy.

For his election as corresponding member in 1966, Lifshits received strong support from Vitaly Ginzburg. Ginzburg was not yet a full member—he was elected in 1966—but he had been a corresponding member since 1953. For Lifshits, the situation was becoming critical because at fifty-one years of age, he was getting close to the age limit: at the time the directive was not to elect new corresponding members aged fifty-five-years or older. Ginzburg looked around and noticed that among the candidates for corresponding member was L. V. Keldysh, thirty-five years old and having a good chance of being elected. Ginzburg asked Keldysh to withdraw his candidacy in favor of Lifshits, which he agreed to do. Lifshits was elected in 1966 and Keldysh was elected the next time, in 1968.[70]

For scientific output, Lifshits altogether had close to fifty research papers, including about twenty as sole author.[71] He published very solid articles, on average much more substantial than is customary nowadays. He worked out the theories of various phenomena, such as ferromagnetism, the molecular interactions between bodies in condensed phase, symmetry changes in crystals, and various problems of low-temperature physics, among others. During the last period of his life, cosmology became his favorite subject both for research and lecturing.

Politically, Lifshits was very reserved. He read a lot, including literature that was frowned upon in the Soviet Union, but his interest in clandestine literature did not translate into action. He never initiated or joined any kind of protest. He never signed any letter or statement against those who actively involved themselves in dissent, and he never signed anything critical of Andrei Sakharov.

When he got involved in an action that the reigning regime disliked, it was not because he wanted to stage a political protest; rather, because he was absolutely honest in all scientific matters. In 1955, there was a commemorative session at the Academy of Sciences honoring the fiftieth anniversary of Einstein's theory of relativity of 1905. Lifshits gave one of the talks and spoke about the age of the universe and its expanding character. This was heresy in the eyes of the watchdogs of ideological purity in the Communist Party. In hindsight, it is tragicomic that the communist ideologists elected to fight the latest achievements of science whereas these achievements had nothing to do with politics. Of course, there was reason the Communist Party meddled in the affairs of the Science Academy. It could not tolerate any sphere of Soviet life that stayed outside of their jurisdiction, and they were afraid of any Western influence. In the long run, they were fighting a losing war, but they were winning the battles in the short run, causing all kinds of difficulties for many people, including the best scientists. The drastic measures of the early 1950s taken against scientific views that were contrary to the perceived Communist ideology were gone. But unruly scientists could still lose such "privileges" as teaching in the universities of Moscow and foreign travel, to mention just two among many. Lifshits never gave a second thought to such possible consequences of his taking a stand in science, and luckily for him, his peers, including Ginzburg, Zeldovich, Tamm, Landau, and others, stood by him.

Lifshits died in 1985, and the era of Landau and Lifshits ended. Isaak Khalatnikov, the director of the Landau Institute of Theoretical Physics, said: "Physicists were afraid of Dau [Landau] and they were shy before Evgenii Mikhailovich [Lifshits]; while they were around, all made an effort to behave with propriety. Now they are no more."[72] E. M. Lifshits is buried at Kuntsevskoe Cemetery, a branch of the Novedevichy Cemetery.

Vitaly Ginzburg.

Source: Courtesy of Valentina Berezovskaya, Moscow.

Vitaly Ginzburg's grave and tombstone in the Novodevichy Cemetery.

Source: Photograph by and courtesy of Larissa Zasourskaya, Moscow.

6

Vitaly Ginzburg

AMATEUR ASTRONOMER

Vitaly L. Ginzburg (1916–2009) was a member of the Mandelshtam-Tamm School, and also considered Lev Landau as his teacher. Ginzburg contributed to the development of the Soviet hydrogen bomb with an important suggestion, but never had full clearance to conduct classified work. He achieved outstanding results in many areas of theoretical physics, was recognized by an array of scientific awards and toward the end of his life, by the Nobel Prize for work he had done more than half a century earlier.

His life was characterized by a measure of constancy though he faced unending "temporary difficulties." He was often obliged to keep quiet about events in Soviet society that he would have preferred protesting against, but he never made statements or signed documents contrary to his beliefs. He was a prolific researcher and author who wrote hundreds of papers and several books in which he described the science of his time. His writings made predictions for the future development of physics and astrophysics. He also wrote about great individuals of whom many were his personal friends.

I had a memorable meeting with Vitaly Ginzburg on a beautiful late-September day in 2004 in Moscow. We had corresponded about my visit, and I recorded a conversation with him. He was one of the previous year's Nobel laureates. I had already met with Alexei Abrikosov, one of his co-laureates, in January of the same year, in Lemont, Illinois. On the day of my meeting with Ginzburg, September 21, we both attended the session of the Presidium of the Soviet Academy of Sciences in its magnificent old building. He was a member of the Presidium, and I was there as recipient of an honorary doctorate from the Russian Academy of Sciences. I was invited to sit at the head table with the president, vice presidents, and the secretary general of the Academy, facing the rest of the hall. At the end of the ceremony I was asked to say a few words. During my presentation, I scanned the audience trying to guess who Ginzburg might be. The ceremony was followed by the business part of the meeting, and the president indicated that I was free to leave; on similar occasions their foreign guests invariably elected to leave. I stayed because I was interested in seeing what was going on. Besides, I had nowhere to go and since following the meeting, Ginzburg and I had arranged to go to his office together at the Physical Institute of the Academy of Sciences (Fizicheskii Institut Akademii Nauk [FIAN]) named after the Russian physicist P. N. Lebedev.

The meeting of the Presidium was lively; the discussions were frank and sometimes charged with emotion. One of those who spoke was Ginzburg, so I now could identify him. His topic was the obligatory religious instruction in state schools. He was very much against what he called the clericalization of Russian society and found it especially upsetting to see religious instructions being introduced into the school curriculum. He was against it even if it was an "elective" subject.

When the meeting was over, an Academy car took us to FIAN, where Ginzburg had worked since 1940. We settled in his office, which was so cluttered with books and papers that it seemed smaller than it really was. From the first moment he made me feel comfortable, and the feeling must have been mutual because at the end of the recording when I asked him whether he would like to add anything, he said, "I enjoyed our conversation because I found that we speak a common language."[1]

Vitaly Lazarevich Ginzburg was born on October 4, 1916, in Moscow. He had a strong feeling about his family, his Jewish roots, his science; and he had strong feelings about issues that concerned him. He was born late, when his father was fifty-three years old. His mother died of typhus when Vitaly was four years old. When I asked him to send me some pictures of himself, one of the few he sent me was his four-year-old self. He did not have brothers or sisters and did not have friends during his childhood either. He did not go to school until he was eleven years old—there was no obligatory school attendance at the time, and the elementary schools were not much trusted in the early Soviet period. It was not at all uncommon for families of the intelligentsia to keep their children at home and to organize their instruction privately. He did not have an easy childhood due to the upheaval in the world around him, in addition to having lost his mother. He revealed what he suffered from most when he said: "What I had in excess was loneliness."[2]

Ginzburg's paternal ancestors originated from the German town Günzburg on the Danube River in Bavaria, about fifteen miles to the east of Ulm. A Jewish community existed there as early as 1566.[3] Ginzburg's father, Lazar Ginzburg, was born in a small town in Belarus; he must have been very gifted and very determined to receive a higher education because at the time this was exceptional for Jews. He studied first in St. Petersburg, then in Riga, and graduated as an engineer. He specialized in water purification and held several patents. His office and laboratory were in the family home. Ginzburg's mother, Augusta Vildauer-Ginzburg came from Latvia; she had studied medicine and was a practicing physician. When she died young, at age thirty-three, her unmarried sister Rosa Vildauer came to live with Lazar and Vitaly. She had prepared to become a dentist, but interrupted her studies and eventually, on account of her knowledge of foreign languages, worked in a company importing foreign-language literature to the Soviet Union. The rest of Ginzburg's maternal family was murdered during the German occupation in World War II.[4]

When Ginzburg was fifteen years old, he left school. According to the practice then followed, he was expected to gain some work experience before applying to university. He was fortunate to find a job as a laboratory assistant in an X-ray laboratory engaged in structure analysis. This introduced him to research and to physicists who would later resurface in his career. It was at this time that Ginzburg chose physics for his studies, and a book by Orest Khvolson, *Fizika nashikh dnei* (*Physics of Our Days*), further sparked his interest in the subject. When he reached the age of seventeen, he applied to university after taking a private crash course to make up for the two years of school education he had missed. He succeeded in getting accepted, but always regretted not spending those two years in a school atmosphere rather than cramming the material into three months of intensive studies.

Ginzburg's entrance examination at the Faculty of Physics of Moscow State University qualified him for a correspondence course rather than the regular one. He was not among the best applicants; but there were additional disadvantages. He did not have a proletarian background; neither was he at that time a Komsomol member (the organization of young communists; Ginzburg would join in 1937). Ginzburg stressed that anti-Semitism did not play a role in his difficulty in getting admitted to Moscow State University in 1933 (the situation would be much different after World War II).[5] The next year, he transferred to the second course of the regular instruction, but here again, he missed subjects such as astronomy and chemistry that were part of the regular first-year curriculum but not of the correspondence version. Ginzburg especially regretted that he did not learn English properly. Later, he had no difficulty reading the scientific literature in English, but he had not acquired good English-language communications skills. According to others, however, his English was good; what Ginzburg missed was having rhetorical skills in English, something he possessed in Russian.[6] Our communications were in Russian, including the conversation we recorded in September 2004.

His having skipped astronomy at the elementary level seems to have haunted Ginzburg in later years, because he did achieve important results in that field. However, his results were always in "sophisticated" astronomy concerning quasars, pulsars, and about the sky of radio waves, X-rays, and gamma rays. What he remained unfamiliar with was the ordinary stellar sky; he never learned the stars and their constellations. It did not bother him too much, and he never considered himself a professional astronomer; rather, he stressed his amateur status even though he was far above that. He concluded that "a lack of elementary knowledge in one or another field," and he meant astronomy and physics, "is not yet an obstacle for obtaining interesting and important results in these fields."[7] What was more important, Ginzburg stressed, was the ability to distinguish between—as a popular Russian children's poem said—"what is good and what is bad." This quote is from his Nobel lecture in 2003 in Stockholm.[8]

Ginzburg immersed himself in his studies of physics but could not remain oblivious to the fights between two groups of physicists that increasingly reflected the political-ideological struggle then taking place in the Soviet Union. This struggle happened in 1935–1936, just prior to Stalin's Terror, with the most tragic consequences. There was one group of professors who followed with active interest the development of modern physics—the theory of relativity and quantum mechanics. Their approach to physics invited the wrath of another group that considered these developments heresy to Marxism-Leninism. This is where the scientific debate received an ugly political coloring and could no longer stay within the framework of arguments and counterarguments. The main culprit for the "modernists," or those who "bowed to Western influence"—depending on which of the two sides was looking at him—was none other than the dean of the Faculty of Physics, B. M. Gessen. He was both a physicist and a philosopher and had broad interests and knowledge. He was also an old Bolshevik, meaning that he had been a follower of Lenin back in the old days. This did not protect him; it might even have worked against him in Stalin's struggle to consolidate power. Gessen was arrested in August 1936, tried, and sentenced to death in December of the same year; he was shot on the day of his sentencing.

Once Ginzburg became a regular student, his studies went well and he soon faced the question of specialization. His heart was attracted to theoretical research, but he had doubts about his abilities, and chose experimental optics instead. Gessen's fate may have also served as a warning for young physicists to stay away from theory, which could be interpreted as ideology. The chairman of the department of optics was Grigory Landsberg, associate of the better-known Leonid Mandelshtam. Mandelshtam and Landsberg discovered what has become known as Raman spectroscopy, after the independent discoverer of the phenomenon. Mandelshtam and Landsberg observed that when light is shined onto molecules, the frequency of the scattered radiation is the combination of the frequencies of the light photons and the molecular vibrations. Hence, the Russian name of the Raman scattering: "combination scattering" and, accordingly, "spectroscopy of combination scattering." An associate professor in this department, Saul Levi, was Ginzburg's tutor; Ginzburg found him very helpful. Levi was a Jew from Lithuania, a refugee from Germany.

During the 1937 Terror, he was dismissed from his job for being a refugee from Germany. Luckily, he and his wife managed to leave the Soviet Union and ended up in the United States.[9]

When in 1938, Ginzburg graduated with his master's degree equivalent, Landsberg invited him to stay on for graduate work for his Candidate of Science degree (PhD equivalent). He was lucky because for a short while graduate students were exempt from conscription. Had the exemptions not been available then or during the war, Ginzburg would have not tried to avoid serving; but staying out of the army allowed him to go ahead with his studies. Thus, he continued his experimental work in optics. A photo showing the graduate student Ginzburg amid his complex-looking optical experimental setup appeared in a newspaper with a caption saying that graduate student Ginzburg pledged to defend his dissertation by a certain deadline signified by a jubilee of the University.

Ginzburg was, however, growing restive in the loneliness of the darkroom conditions necessary for his optical experiment. For some experiments he had to create conditions of "vacuum," that is, very low pressure, in his apparatus. This takes a lot of time even if there is no leak in the apparatus. I know from my experience that creating these conditions often involves long hours of waiting. At the same time, Ginzburg had ideas for interpreting his observations, but he lacked the background in theory to develop them. At this point, he decided to consult with Igor Tamm, and this proved to be decisive for his career. Tamm was at this time the chairman of theoretical physics at Moscow State University. He welcomed Ginzburg and helped his young colleague to put together his first papers in theoretical physics. Ginzburg wanted to publish them in the prestigious *Doklady Akademii Nauk* (Proceedings of the Science Academy), for which he needed a member of the Academy to sponsor his manuscripts. Tamm was not yet an academician, but Ginzburg found Vladimir Fock, who was, and he provided the necessary assistance. This was how Ginzburg changed from an experimentalist to a theoretician.[10]

Ginzburg made an early theoretical contribution to the description of Vavilov-Cherenkov radiation (known in the West as "Cherenkov radiation"). This effect concerns the radiation emitted by a moving charge in various media. The experimental discovery was made by Pavel Cherenkov under the direction of his mentor, Sergei Vavilov. It was then interpreted theoretically by Igor Tamm and Ilya Frank (see chapter 1). The three, Cherenkov, Frank, and Tamm shared the Nobel Prize in 1958 (by then Vavilov had died, so there was no dilemma due to the three-person restriction of the Nobel Prize). Back in the late 1930s, following Tamm and Frank's work, Ginzburg introduced some refinement to the theoretical interpretation of Cherenkov radiation, and provided its quantum mechanical description. His results did not add much to Tamm and Frank's interpretation, but his approach was useful for the treatment of important similar phenomena. Lev Landau, who was the chief theoretician of another physics institute, the Institute of Physical Problems, did not place much value on the quantum description of Cherenkov radiation. He found it superfluous to spend effort merely improving something rather than producing something entirely novel. His approach was quite elitist, whereas Tamm's and Ginzburg's were more down-to-earth. The differences between the two leading theoreticians, Landau and Tamm, became apparent when the two groups occasionally held joint seminars.

Vitaly Ginzburg conducting an experiment in optics.
Source: Courtesy of Victoria Dorman, Princeton, New Jersey.

Ginzburg completed his dissertation very soon, and defended it in 1940. He then moved from the University to FIAN, where Tamm had in the meantime organized a theoretical department. The same people at the university who had caused the Gessen tragedy worked against Tamm as well, and eventually he had to give up his university involvement. At FIAN, Tamm became Ginzburg's mentor during the next step in his scientific career, the Doctor of Science degree. It was quite unusual that it did not take Ginzberg more than two years to complete this degree, especially considering that just in the middle of this period the war started for the Soviet Union. Germany attacked the Soviet Union on June 22, 1941, and advanced quickly into Soviet territory, and soon FIAN, along with many other research institutes, was evacuated to Kazan. Ginzburg defended his D.Sc. dissertation in 1942, in Kazan.

Some considered Ginzburg an "impudent" student.[11] This was a mistaken notion, although he understood why such impression of him could be formed. He was easily excited by interesting scientific problems and participated in debates with zest. At that time he had problems with public speaking; speaking in front of others was torture for him. Tamm encouraged him to develop his ability to communicate his thoughts to others, and it was helpful that he welcomed Ginzburg's ideas. Eventually, Ginzburg developed into a good speaker.

Subsequently, Ginzburg often wondered how his career might have developed had not Tamm but Landau been his mentor. Landau as a rule did not show enthusiasm; on the contrary, he often cooled other people's excitement. Ginzburg regretted, however, that he never took Landau's famous multipart exam, the *teorminimum*. Ginzburg felt himself an outsider in theoretical physics—when he administered examinations, he sometimes had to ask questions that he did not know the answer to, for example, about the derivation of formulas. Yet he was successful, which he explained as follows: "Firstly, there is some hunch, an understanding of physics, tenacity, a grip on combinations and associations. Secondly, there was a great wish 'to think up an effect,' to do something. Why? I think that it came from an inferiority complex."[12] He felt good about having ambition; to him there was "a distinction between a 'good' ambition, on the one hand, and ambition, in general, and vanity, on the other hand."[13] He defined "good" ambition as "an aspiration to do my work, to do it well, and to have it acknowledged and acclaimed."[14] But he considered it important that he never wanted to be successful at someone else's expense.

During his student years, Ginzburg was a distant observer of political developments. Then came the war, and he enthusiastically supported the defense of the Soviet fatherland, like everybody else. When the Russians refer to World War II as the Great Patriotic War, they express the sentiments of that time exactly. During the war, Stalin shrewdly invoked feelings of patriotism even among those people who opposed communism; he masterly played on nationalistic and religious dedication of the masses. After the war he did everything he could to crush those sentiments and added a vicious anti-Semitism to his pathological policies. However, during the war, patriotism and communism got blended in many people's minds.

Under the wartime conditions, Ginzburg decided to join the Communist Party. This happened during the worst crisis, when the Germans reached the Volga, and taking this step in 1942 could not be considered a career move. He became a candidate for membership in the party in 1942 and a bona fide member in 1944. Indeed, his party membership did not alleviate the difficulties that came about at more than one level during the immediate postwar years. Ginzburg remained in the party until 1991, when he resigned from it. By then he had become rather active in forming and voicing political views that were not in line with the Communist Party. Between 1989 and 1991, he was an elected member of the Congress of People's Deputies, just like Andrei Sakharov. They both were delegated by the membership of the Academy of Sciences.

Ginzburg married fellow physics student in 1937 Olga Zamsha; they had one daughter born in 1939, Irina. They divorced in 1946. Ginzburg maintained a good relationship with daughter Irina, who married and had two daughters. At the time of Ginzburg's death, he had one great-grandson and one great-granddaughter—the children of one of Irina's daughters—living in Princeton, New Jersey.

Ginzburg returned from the evacuation to Moscow with the rest of FIAN and continued at FIAN for the rest of his life. He had an additional affiliation at the University of Gorky. This town has since regained its original name Nizhnii Novgorod; it is east of Moscow, between Moscow and Kazan. The University of Gorky organized a new department of radiophysics, and because during the war Ginzburg conducted research on the propagation of radio waves in the ionosphere, he was invited to be professor and chair.

He started commuting between Moscow and Gorky toward the end of 1945 and soon he met in Gorky his future second wife, Nina Ermakova, who was originally from Moscow but now lived in a nearby village. She lived in exile, ostensibly for her participation in a conspiracy to kill Stalin. The supposed plan was to shoot the Soviet leader from her apartment window. Later, it was shown that her apartment did not even have a window facing the road where Stalin used to ride; but this did not matter. Ginzburg and Ermakova married in the summer of 1946, and she became Nina Ginzburg. When Ginzburg died in 2009, they had been happily married for sixty-three years. Between 1946 and 1953 they applied annually for permission to let Nina return to Moscow, but to no avail. Only after Stalin's death they could finally live together in Moscow. From 1954, in March every year, Ginzburg and his wife made it a point to remember the death of the dictator with gratitude. His wife's exile prevented Ginzburg from getting the highest security clearance, and this is why he was not invited to join the secret location for the development of nuclear weapons. In hindsight, Ginzburg considered this his lucky break. Yet, even his partial involvement with the weapons program may have saved his life.

During Stalin's last years, in the period between 1946 and 1953, there were two developments in the Soviet Union that proved life threatening for Ginzburg. One was the emergence of state-sponsored anti-Semitism, and the other was the ideological struggle against modern science. The two developments were not entirely independent of each other. Often, when referring to Jews, the euphemism "cosmopolitans" was used, meaning people who worshipped the West. It was easy to accuse even non-Jewish scientists of cosmopolitanism because of their practice of making references to other people's work in their publications. The two accusations found confluence in Ginzburg in the widely distributed *Literaturnaya Gazeta* (Literary Newspaper), which published an article titled "Against Servility." In it, Ginzburg and a number of respected biologists were attacked for their ostensible opposition to the teachings of the powerful charlatan Trofim Lysenko. Lysenko was responsible for overseeing the Soviet agricultural scientific programs intended to increase crop yields. However, his unyielding belief that acquired characteristics were inherited resulted in overwhelming crop failures and mass famines.

In the same article, Ginzburg was also criticized for insufficiently referring to the works by the Soviet theoretical physicist, Dmitrii Ivanenko. The article was signed by an academician of the agricultural academy, V. S. Nemchinov, who could have hardly been versed about the importance of Ivanenko's research for Ginzburg's papers. Later, Ginzburg found out that the article was written by a person of much lesser stature, but to give it more weight they made the better-known scientist sign it. It was quite natural to suppose that the part of the attack concerned with physics had been inspired by Ivanenko, who denied these charges. Yet Ivanenko, a professor at Moscow State University, was an active participant in the ideological struggles in which not only Ginzburg, but others, notably Igor Tamm, were also accused without foundation. Upon the attack on Ginzburg in *Literaturnaya Gazeta*, eleven academicians wrote a letter to the magazine in Ginzburg's defense, but the letter was not published. Ivanenko's influence was substantial. It was also at this time that Ginzburg was denied the title of professor, supposedly again at Ivanenko's initiative.[15] This article might have doomed Ginzburg, except for his involvement in the nuclear weapons project, which I discuss more later. It protected him from severe consequences that might have included arrest and a harsh sentence. He would have certainly figured heavily in the attack that was already being prepared against physics and physicists accusing them of bourgeois idealism. It had been carefully orchestrated to happen in March 1949. It was finally halted given that there was an unambiguous conflict between ideology and the atomic bomb. The Soviet leadership, and in particular Lavrentii Beria, chose the bombs over ideological purity.

During his Gorky professorship and his wife's continuing exile, Ginzburg had to spend a lot of time traveling. To overcome boredom, he developed a special pastime, which he called "brainstorming." Its purpose was to come up with some idea and to do so within a limited amount of time, say, half an hour. For example, once in 1964, he was on a long train ride from Kislovodsk in the northern Caucasian Mountains to Moscow. He was alone, no partners and no books, and he decided to brainstorm. In choosing the topic he considered his longtime involvement in low-temperature physics and in astrophysics. He decided that this brainstorming would be about the question whether these two areas might be connected. In other words, he started brainstorming about the possibilities for superconductivity and superfluidity in space, and came to interesting conclusions. Within the set time period he concluded that superfluidity was possible in neutron stars and that superconductivity was possible in the atmosphere of white dwarfs. Upon returning to Moscow, he engaged the cooperation of one associate each to research these two conclusions, and they published interesting findings shortly after.[16]

Vitaly Ginzburg with his second wife.

Source: Courtesy of Victoria Dorman, Princeton, New Jersey.

This brainstorming was a remarkable exercise; one has the impression of being present in the process of the mind of a genius at work. Genius is loosely defined as the ability to relate things that others find unrelated. But once Ginzburg suggested looking into the possibility of the relationship between superconductivity and superfluidity under the extreme conditions of space, even for an outsider this consideration seemed logical. Superconductivity and superfluidity had been discovered under extreme conditions that had to be created in the physical experiments. In space, these extreme conditions are already present, so the supposition that these extreme properties may also be there is no longer surprising. Ginzburg noted: "To formulate a question is frequently equivalent to doing half the work."[17] This resembles the statement attributed to Leo Szilard about the most important part of a project being the *recognition of the problem*.

For a critical period, starting in 1947–1948, Ginzburg was involved in the Soviet hydrogen bomb project. The work had already been going on for some time when in 1947 Igor Tamm was invited to join. Tamm had not been considered entirely reliable due to his Menshevik background (see chapter 1) although by this time he had long since stopped any involvement in politics. Tamm formed a special group at FIAN which included Ginzburg, Sakharov, and a few other young theoretical physicists. As noted earlier, Ginzburg had some problems obtaining security clearance because of his wife's exile, but for the time being, the security organs tolerated his participation. There were limits, though, to this tolerance; Ginzburg's closest friend, the also-outstanding theoretical physicist E. L. Feinberg, was not given clearance because his wife had once lived in the United States.

The idea of developing the hydrogen bomb had been around, but it was not yet clear whether it would be possible to achieve that goal. This was also the dilemma in the American program. Once the decision by President Truman to develop the hydrogen bomb was made, it was a sobering recognition that nobody knew yet whether it was possible. Looking back, it was this uncertainty that necessitated the involvement of additional people in the Soviet project and led the security organs to relax the secrecy requirements, at least for the time being.

The inclusion of Tamm's group turned out to be most fortunate for the Soviet hydrogen bomb program. There were to be three fundamental ideas for the solution, and Tamm's associates proved to be instrumental in bringing about all three. The euphemistic description the "first, second, and third ideas" originated with Sakharov at a time when their physical essence could not yet be communicated, but these labels remained in usage even afterward. The first idea, by Sakharov, was the layer structure, *sloika*, and the third idea was the radiation implosion. The third idea was also by Sakharov, although Yakov Zeldovich and others participated substantially in the development of this approach. The second idea came from Ginzburg, and it concerned the thermonuclear fuel. Ginzburg suggested to use lithium(6) deuteride, ^{6}LiD, for obtaining tritium, ^{3}H, in the reaction,

$$^{6}\mathrm{Li} + \mathrm{n} \rightarrow {}^{3}\mathrm{H} + {}^{4}\mathrm{He} + 4.6\mathrm{MeV}$$

where ^{6}Li is a light lithium isotope, n is neutron, and ^{3}H is the heaviest hydrogen isotope, tritium, often labeled as T.

The third idea came later, but the first two ideas proved sufficient to build the first deliverable hydrogen bomb, and the work intensified. In 1950, Tamm, Sakharov, and some others were directed to move to the secret location of the Soviet nuclear weapons laboratory, Arzamas-16. For this, a yet higher security clearance was needed that Ginzburg did not have, so he stayed in Moscow, where his involvement continued for some time.

In 1950, Ginzburg was made head of a small support group at FIAN participating in the work on the hydrogen bomb to augment the work at Arzamas-16. Even in Moscow the project was classified enough to have a guard sitting before the door to the room where Ginzburg and the others labored. Then, in 1951, as the project was moving ahead with the promise of success, Ginzburg was dismissed even from his limited assignment. It came about so abruptly that he was even barred from access to his own papers. On the other hand, somehow, by inertia, he was left alone and not bothered during the anti-Semitic state actions culminating just before Stalin's death. If Stalin had lived longer, mass repercussions against Jews would have taken place. The best Ginzburg might have hoped for would be to end up in a *sharashka*, a special kind of labor camp where skilled prisoners conducted scientific and technological research.

On August 12, 1953, there was a big success with the test of the Soviet hydrogen bomb in which Ginzburg's second idea figured heavily. As was described in previous chapters, it was not yet a bona fide hydrogen bomb, but the test was a triumph at a critical time for the project—immediately following Stalin's death and Beria's disappearance from the Soviet leadership. Ginzburg was no longer involved directly in the project and did not receive the highest award, Hero of Socialist Labor, and the perks that went with it. But he was still distinguished with sufficient recognition, the Order of Lenin and the Stalin Prize, which made his life easier. Even more significant, he was soon elected corresponding member of the Soviet Academy of Sciences. His contribution to the nuclear weapons program was a decisive factor in his promotion.

In my conversation with Ginzburg, he posed the question about the general morality of scientists working on weapons. He was especially concerned with his own participation in the work on the Soviet hydrogen bomb. According to Ginzburg, scientists generally speaking do have responsibility for their participation in creating weapons of mass destruction. However, a lot depends on the specific situation, and this is why he liked to insert the qualifier "generally speaking" when describing this responsibility.[18] For instance, he fully approved of Albert Einstein's involvement in the initiation of the work on the atomic bomb in the United States. Einstein's action was justified by the danger of the possibility that Nazi Germany might also have developed the atomic bomb, and might have developed it first. Ginzburg was disturbed by Werner Heisenberg's claims of moral superiority over Einstein.[19] Ginzburg thought that Heisenberg would have served his interests better by staying quiet about this topic, because he was a leader of the German Uranium Project. The Germans did not succeed in their project. They committed several rudimentary errors, and to Ginzburg it seemed highly doubtful when they explained them by active or passive resistance to creating an atomic bomb.[20]

According to Ginzburg, those Soviet physicists, whom he knew, including Sakharov and Tamm, justified their participation in the nuclear project by the necessity of counterweighing the American monopoly. He accepted the notion that having more than one power in possession of the hydrogen bomb had a stabilizing effect and served as a deterrent. This was the policy of mutually assured destruction (MAD). Ginzburg never tried to mask his own role and responsibility in creating the Soviet hydrogen bomb, which was, however, more limited than Tamm's and much more limited than Sakharov's. It is also true though that at the time of his participation, and for some time afterward, it never occurred to him that the Soviet Union might use such a weapon as a means of aggression. He and his colleagues did not have any doubts at the time that it was their duty to work on the project. This was very different from the sentiments of the American physicists participating in the nuclear program.

Suffice it to mention the fierce debates in 1949–1950 and that the majority of leading physicists opposed the development of the hydrogen bomb in the United States.

During our conversation in September 2004, Ginzburg admitted that he and his colleagues did not understand Stalin's real aspirations; hardly anybody did. There was, however, one physicist working on the bomb who understood Stalin's aims, and this physicist was participating in the project out of fear. He kept silent about his fears at that time, and Ginzburg did not want to pass judgment on him, but he would not identify him either. We now know that Lev Landau was this physicist; he considered himself to be a "learned slave" and quit working on the nuclear project right after Stalin's death. Looking back, Ginzburg had no doubt that Stalin was a ruthless bandit. He would have employed even the most terrible weapons without hesitation if he thought he needed them to accomplish his goals and could get away with it. Ginzburg added, "It is the luck of humankind that Stalin and Hitler did not possess atomic bombs first."[21]

Of course, it was just as well that under Stalin the physicists did not initiate a discussion of whether or not it was morally justifiable to develop the hydrogen bomb. The restrictions on the freedom of scientists went much beyond their inability to question the propriety of Stalin's decision about weapons development. The scientists were not allowed to maintain contact with Western colleagues, even on purely scientific issues. Their ability to publish in international journals was severely curtailed. The foreign-language scientific publications of the Soviet Academy of Sciences ceased to exist after the war. Much of the excellent production of Soviet scientists remained in oblivion for quite some time before Western publishing companies started publishing complete translations of Soviet scientific journals, including Ginzburg's works. Ginzburg's best known paper, co-authored with Landau, was published in 1950, and it served as the basis of his 2003 Nobel Prize. Fortunately, Petr Kapitza's former doctoral student, David Shoenberg in Cambridge (see chapter 4), subscribed to some of the Russian-language journals and even translated some papers. The Ginzburg-Landau paper was among them. Thus Shoenberg was instrumental in disseminating Ginzburg's results as well as those of other Soviet physicists among their Western colleagues.

It was obvious that Ginzburg was troubled about the responsibility of the scientist participating in the creation of weapons. In our conversation, he returned to this question repeatedly and stressed that this responsibility depended on the goals for which such weapons were being developed. He was ready and willing to sanction the creation of weapons serving the protection of one's country from aggressors and terrorists. He was unambiguous that he specifically meant Israel in this context.[22]

Ginzburg did not aspire for awards and position, and in this respect the totalitarian regime did not have much leverage over him. There was one aspect, though, they could hurt him, and they did. It was always very difficult for him to get permission for foreign travel, especially together with his wife.

There were some periods when travel was possible, and on the whole he traveled more than most Soviet scientists. For example, in 1947, he was a part of an expedition to Brazil organized by the Soviet Academy of Sciences. They conducted radio observations on board the Soviet ship *Griboyedov* on the occasion of the total solar eclipse on May 20. During the second half of the 1960s, he was allowed to travel to the West and was even permitted to visit the United States on three occasions, in 1965, 1967, and 1969. He was elected to full membership of the Science Academy in 1966 and thereafter even his wife could sometimes accompany him. The high point of this period was in 1967, when he spent a few months in Cambridge, England.

But in most cases, his applications for permission to travel were declined, citing his possession of state secrets from his work on classified projects. This was only an excuse; some people were allowed to travel more easily although they possessed more secrets than Ginzburg. The lowest point was when the authorities did not let him attend an international physics meeting in Kiev(!), that is, within the borders of the Soviet Union. Landau was also forbidden to go, but Landau declared that he would go anyway, and the authorities, wanting to save face, relented. Ginzburg did not make such a fuss but later regretted his inaction, and felt humiliated for as long as he lived.

Ginzburg had been Tamm's deputy as the theoretical department head and following Tamm's death in 1971, he was appointed to be head of the department, which was named after Tamm. Ginzburg's appointment was inevitable because at the time the two ranking members of the department—the two full members of the Academy—were Ginzburg and Sakharov. Sakharov had already been involved in his "dissident" activities, so he could have not been appointed. Ginzburg was the department head for seventeen years. By the end of his tenure the department had grown considerably. It included sixty associates and seven members of the Academy—three full members and four corresponding members.[23] It carried great relative weight in FIAN because the theoretical department had as many Academy members as the rest of the Institute.

Ginzburg greatly respected Sakharov both as a physicist and as a human being, but Sakharov being a member of his department also caused problems. The two worked out a tacit agreement for maintaining their lives in a mutually acceptable manner.[24] On his part, Sakharov kept his social and political activities separated from the department, and never asked department members to sign his petitions and other documents. The department considered Sakharov one of its members during his entire predicament, including his years of exile to Gorky. Once or twice annually, department members traveled to Gorky to keep Sakharov informed about developments in physics, and Ginzburg visited him twice. It gave him a strange feeling to be traveling to Gorky to visit an exiled person, reminding him of the time that his wife was exiled there three decades before.

When the political changes came at the very end of the 1980s, Ginzburg was still sufficiently fit to enjoy his greater freedom, and he attended meetings and wrote articles as part of his enhanced activities. By the time his Nobel Prize was announced in 2003, his possibilities were more limited because of age and general well-being. I met with him within a year of his new life as a Nobel laureate, and detected no elation in his demeanor, as sometimes characterizes people in similar situations. Also, he was not one of those laureates who would claim that they had never thought about the possibility of winning the award or that it was a complete surprise. He had thought about it, but after a while decided that he had been passed over. He had a few reasons for thinking so and he spoke about them freely.[25]

October 7, 2003, was a memorable day for Ginzburg. He was sitting at his desk in his office at FIAN. He knew that it was the day when the Nobel Prize in Physics would be announced, but he did not give it much thought. He was writing a letter to his daughter. Suddenly, the telephone rang and a voice in English told him, "This is from Stockholm; you have received the Nobel Prize." He might have thought that somebody was making a joke. However, the caller further told him that he shared the prize with Alexei Abrikosov and Anthony Leggett. At that moment Ginzburg understood that it was for real. Abrikosov's name brought home this reality. He did not know anything about Leggett. He knew that there had been nominations for the three of them, Abrikosov, Lev Gorkov, and himself as early as the early 1970s. Landau might have been in their group, but he had already received his Nobel Prize in 1962.

In his Nobel lecture, Ginzburg alluded to the fact that it was a long wait for him to receive his Nobel recognition, but he did this with taste and humor.[26] He said that he was eighty-seven years old; the prize almost would have had to be given posthumously if they had waited any longer. He then added a twist by saying that there were no posthumous Nobel Prizes and such recognition would have meant very little to him anyway, because he did not believe in life after death. He was especially glad to have received the Nobel Prize because the two persons whom he always considered his teachers had also been given this distinction, Igor Tamm in 1958 and Lev Landau in 1962.

The title of Ginzburg's Nobel lecture was "On Superconductivity and Superfluidity." His prize was given for the development of the "Ginzburg-Landau theory of superconductivity," which in the Russian literature is known by the name of the "Ψ-theory of superconductivity" (Ψ being the capital *psi* of the Greek alphabet). In fact, in Western literature, this theory had often been referred to as the Landau-Ginzburg theory of superconductivity. Ginzburg explained to me that this was incorrect, but that previously he had not thought it prudent to correct this usage lest people think that he was placing himself in front of Landau as a physicist.[27] This was not the case, and Ginzburg considered Landau to be a greater physicist. However, the Ψ-theory was primarily Ginzburg's work; he just consulted with Landau on a few occasions. Landau's contribution was that Ginzburg followed Landau's theory of phase transitions published in 1937. The introduction of the Ψ wave function in the description of superconductivity meant the expression of order; in other words, Ψ plays a role of an order parameter. The importance of the theory was enhanced by the discovery of high-temperature superconductivity; Ginzburg estimated that during the decade following the discovery of Georg Bednorz and Alexander Müller, there were about fifty thousand papers published on the topic, that is, about fifteen papers daily![28]

The science Nobel laureates in December 2003 in Stockholm, from left to right, Peter Mansfield, Vitaly Ginzburg, Peter Agre, Anthony Leggett, Roderick MacKinnon, and Paul Lauterbur (Alexei Abrikosov is missing from this picture).

Source: Courtesy of Anthony Leggett, Urbana, Illinois.

Ginzburg had long been interested in the institution of the Nobel Prize. It was not just because of his desire to receive it. On a broader scale, he was especially interested in why several outstanding Soviet and Russian physicists had never been honored by this award, in particular, L. I. Mandelstam and G. S. Landsberg, who had discovered in 1928 what has become known as the Raman Effect. Raman made his discovery also in 1928 and was awarded the Nobel Prize for it in 1930. For a long time, Ginzburg and many others in Russia believed that Mandelstam and Landsberg had not been awarded the Nobel Prize because of the anti-Soviet sentiments of the award givers. This was also the party line. In reality, however, Mandelstam and Landsberg hardly received any nominations, the only exception being a nomination from the Soviet physicist Orest Khvolson (the author of the young Ginzburg's favorite physics book). There was insufficient publicity for the discovery by Mandelstam and Landsberg; whereas Raman, as soon as he made his discovery, sent out a bulletin about it to numerous Nobel laureates, asking for their support.

Ginzburg also cited the case of V. N. Ipatev, an important chemist who after the 1917 revolution dedicated himself to developing the chemical industry in the Soviet Union. However, in the 1930s he, like many others, became the target of government attacks at the conclusion of a foreign visit. Next time he was abroad, he did not return to the Soviet Union. In 1931, when Ipatev was still in the Soviet Union, two German chemists, Carl Bosch and Friedrich Bergius, received the Nobel Prize in Chemistry for their contributions to high-pressure methods in chemistry. Ipatev could have been included in this prize. There was an article in the *Herald of the Russian Academy of Science* in 1997 arguing that Ipatev had been overlooked because he had collaborated with the Soviet government.[29] By then, however, the archives of the Nobel Prize documents of the years around 1931 could be investigated, and there was not a single nomination for Ipatev. Not even a Soviet chemist had sent any nomination for Ipatev, whereas many Soviet chemists received invitations to submit nominations.

Abrikosov had been very active in disseminating the results of Soviet researchers in the areas mentioned in the motivation formulated by the Nobel Prize: "for contributions to the understanding of superconductivity and superfluidity." Ginzburg's paper with Landau appeared in 1950, but Ginzburg had been working in this field since 1943. Ginzburg was aware of nominations that included Abrikosov, Ginzburg, and Gorkov. Their names were linked in 1966 when they jointly received the Lenin Prize. It was soon after that, according to Ginzburg, that Abrikosov and Gorkov decided to get nominated for the Nobel Prize. They told Ginzburg that they wanted to include him as well. So, all three wrote up their own contributions. The question was who should submit the nomination, because self-nomination is not allowed. Abrikosov suggested asking Ilya Frank, who by then was a Nobel laureate. However, Ginzburg did not want Frank involved because he, Ginzburg, had helped get Frank nominated, and their request would have looked like asking for payback. Abrikosov did ask Frank, however, and Frank submitted a nomination. Subsequently, the three were nominated several times. In 2003, Gorkov was omitted from the award. Ginzburg was surprised; he called Gorkov "a wonderful physicist," and had no explanation for the omission. However, he found this omission less conspicuous than what happened in 1997 when three physicists, Steven Chu, Claude Cohen-Tannoudji, and William Phillips, received the prize for the laser cooling of atoms. At that time there was a lot of discussion in Russia that the Russian physicist V. S. Letokhov was omitted.[30]

Ginzburg had an extraordinary overview of physics. He devoted a tremendous amount of time to reading, and his interest defined his weekly seminars. By 2001, he had held altogether 1700 seminars. Ginzburg's seminar was famous; people from all over Moscow used to attend it at FIAN, and when out-of-town physicists were visiting Moscow, they also joined it. The seminars were friendly; Ginzburg encouraged participation. There was none of Landau's sarcasm or impatience there.

In the early 1970s, Ginzburg began compiling lists of challenges for physics. It was a roster of unsolved problems that he thought should be attacked and, hopefully, solved. The list kept expanding. It was not merely a list; Ginzburg augmented it with commentaries that showed his deep understanding of the problems. It provided assistance to physicists who were looking for ideas and projects. As Ginzburg was concluding his Nobel lecture on December 8, 2003, in Stockholm, he added a section about the especially important and interesting problems of physics and astrophysics in the beginning of the twenty-first century.[31] Now he listed thirty problems with controlled nuclear fusion as number one. His interest in collecting the most important problems of contemporary physics prompted me to ask Ginzburg about his personal choice of a project or projects if he could be twenty-five years old again.

Ginzburg initially declined to answer because he was against making such choices. He preferred a broad base, and his scientific career was an example of the benefits of being engaged in numerous areas.[32] He was well aware of the fact that the results of scientific research were not long lasting in the sense that they are soon overtaken by yet newer findings—this was the nature of scientific progress. Creativity in science differs from creativity in the arts. There are products of the arts that are forever associated with the names of their creators, but in science, most discoveries are built into the edifice of science. Only the most conspicuous discoveries find their way into textbooks, and this is what only the greatest of the greats may hope for; all the others disappear in oblivion.

Ginzburg was highly prolific. He published a large number of scientific papers, and by the end of the 1980s he estimated the number of his publications as approaching one thousand—something very rare, especially among physicists.[33] For the fifteen-year period 1961–1975, he was among the five top-cited Soviet scientists, with close to seven thousand citations. But he saw clearly that much of his writing in physics would become obsolete as science progressed. The thought of longevity of scientific production clearly made him wonder. He told me about another scientist's approach to this problem.

When Yakov Zeldovich was nearing his seventieth birthday, he devoted two years to collecting and commenting on his scientific production. He was assisted by some of his associates.[34] Zeldovich thought that his two-volume compilation would be useful for posterity. Not long after Zeldovich had completed the two volumes, he died. Because he had been a foreign member of the Royal Society (London), Ginzburg, being another foreign member, was asked to write a biographical memoir about Zeldovich. These bibliographical memoirs are detailed treatises, and Ginzburg made much use of Zeldovich's two volumes in producing his composition for the Royal Society series.[35]

Vitaly Ginzburg and Istvan Hargittai in September 2004 in Ginzburg's office at FIAN (by unknown photographer).

When Zeldovich gave Ginzburg his two volumes as a gift, he said, "You will soon be 70 years old," and suggested to Ginzburg that he follow his example and compile his own two volumes.[36] Ginzburg liked the idea but was reluctant to take up the huge task of organizing such a compilation. Besides, Ginzburg's production, at least considering its volume and the number of his publications, was much larger than his friend's. He decided instead to compile his scientific autobiography, emphasizing the works he considered most important.[37] This autobiography shows that he worked in many areas in theoretical physics, and he could easily change his focus of research from one area to another.

When I talked with Ginzburg, our first topic was about a general characterization of his life, which appeared to me to carry considerable constancy in that he had lived all his life in Moscow and FIAN remained his only workplace. In response, he quoted what people used to say in Soviet times. "Question: What is constant under the Soviets? Answer: Temporary difficulties."[38] Of course, behind the apparent constancy, there were genuinely difficult times in Ginzburg's life. He wondered whether he could have performed better in science under less trying conditions. He was not sure, but this question had occurred to him, and a decade and a half earlier he had written: "[I]f I had lived under better conditions, I would have probably been happier and have rested and seen more. But the integral of my scientific activity, if I may say so, most probably would not be larger than it is."[39]

At some point in our conversation, Ginzburg overcame his initial aversion to my question. Were he twenty-five years old again, but in possession of his accumulated experience and knowledge, he said he was sure he would again become a theoretical physicist. Then, he added: "I have a prayer.... As you know, Jewish men have such a prayer in which they thank God that he did not make them into women. In my prayer, I am thanking God for having made me into a theoretical physicist. This does not mean that I have anything against experimental physicists. In my eyes they have the most difficult job possible. They have to sit at some apparatus all their lives. What the great luck of the theoretical physicist is that he can easily change his topics all the time."

After further contemplation—it was obvious that he had warmed to the challenge of the question—he gave me this response about his hypothetical choice of research project for a twenty-five-year old Ginzburg:

> I always dealt with many problems. This is why I cannot give you a specific response to your question about what would be my choice today if I were 25 years old again. Problems in theoretical physics would be sufficient to keep busy a thousand Ginzburgs. If then, taking a closer look at theoretical physics, I do have some fixed ideas. From 1964, I have been interested in high-temperature superconductivity. Today, the question is about room-temperature superconductivity. Could we make a superconductor that it would be possible to utilize at room temperature for which, for example, water-cooling would suffice? This is what I find to be a most interesting problem. It may not be the greatest challenge in physics today, but I would probably select this one to pursue it further if I could suddenly become young again.

I found Ginzburg, even at the age of eighty-eight, to be young in spirit, and this is how I remember him. For many, he has remained an inspiration.

Alexei Abrikosov and other members of the Theoretical Division of the Institute of Physical Problems, 1956. From left to right in the front row: Abrikosov, I. M. Khalatnikov, L. D. Landau, and E. M. Lifshits; and in the back row: S. S. Gershtein, L. P. Pitaevskii, L. A. Vainshtein, R. G. Arkhipov, and I. E. Dzyaloshinskii.

Source: Courtesy of Boris Gorobets, Moscow.

7

Alexei Abrikosov

"UNMANAGEABLE"

Alexei A. Abrikosov (1928–) was born into an elite Russian family and received the best possible training in Lev Landau's theoretical physics school. He had the ability to communicate with experimentalists and early on made important discoveries about superconductors. He was elected in his thirties to be corresponding member of the Soviet Academy of Sciences, but then had to wait over two decades to become a full member. His marriage to a French woman—his second wife—made him suspect in the eyes of the authorities. His third marriage brought stability to his life.

Abrikosov was active in the communist youth movement but never became a party member. His father, a famous pathologist, talked him out of joining the party based on his own bitter experience. Abrikosov was half-Jewish, which almost prevented him from getting the job he desired. He was also not allowed into classified nuclear weapons work because he had an uncle who lived in the West.

Abrikosov emigrated to the United States when the Soviet Union was dissolved. When he received the Nobel Prize, he declared that he was first and foremost an American.

Alexei Alexeevich Abrikosov was born on June 25, 1928, in Moscow. His parents were medical doctors, pathologists. His father was Russian and his mother Jewish; her name was Fanny Davidovna Wul'ff. This did not make Abrikosov Jewish, as he would be considered according to Judaism, because in Russia the father was the determining factor for one's nationality (religious denomination was not registered whereas being Jewish was counted among nationalities). So Abrikosov was Russian, and only when his family background was investigated did his mother become a consideration—as happened when he finished his doctoral studies. Abrikosov had been Lev Landau's pupil at the Institute of Physical Problems (Institut Fizicheskikh Problem [IFP]). Landau recognized Abrikosov's talent, and wanted to keep him in his group after graduation. This was in the very early 1950s. The director of IFP, A. P.

Aleksandrov (Petr Kapitza was in exile) did not object to Abrikosov's appointment. However, the local representative of Lavrentii Beria's Interior Ministry responsible for security matters decided that Abrikosov was not to be trusted. As a pretext, the anti-nepotism rule was cited, referring to Landau's patronymic, Davidovich, and to the patronymic of Abrikosov's mother, Davidovna, as if this were evidence of their being closely related.

Abrikosov went to work at the Institute of the Physics of Earth and completed a good piece of theoretical research about the internal structure of Jupiter. However, an independent event unexpectedly changed his fortune. I heard this story from Abrikosov himself.[1] Later I was told it repeatedly by Russian colleagues whenever Abrikosov's name came up in conversations. It has become part of the folklore. In 1952, the Stalinist leader of Mongolia, Khorloogiin Choibalsan, was visiting the Soviet Union. He arrived gravely ill and died in Moscow. One of the signatories of the medical report of his death was Abrikosov's mother. The communication was printed in *Pravda*, the principal newspaper of the Communist Party, and received wide publicity. The fact that she had been allowed to perform the autopsy of this most important leader greatly impressed the representative of the Interior Ministry. He changed his mind about Abrikosov being employed by the IFP, and Abrikosov was duly transferred there. One of Abrikosov's colleagues at the IFP, the future academician Aleksandr Shalnikov, suggested to Abrikosov that he express his gratitude by hanging a portrait of Choibalsan in coffin on the wall of Abrikosov's office.

By the time Abrikosov joined the IFP in 1952, many of its activities had been transformed into nuclear-weapons-related research. Landau at this time was a full-fledged member of this project, and he wanted to involve Abrikosov, who was ready to join. Six decades later, however, Abrikosov appeared happy remembering that he had never participated in any weapons-related project.[2] It was, again, due to interference by the security organs that he was not allowed to join the project. Abrikosov learned the reason only decades later.

Abrikosov's father had a younger brother about whom the family never spoke, and Abrikosov grew up without knowing that he had an uncle. This uncle, Dmitrii Abrikosov, was a diplomat under the czarist government, and in 1917, when the revolution broke out, he was serving at the Imperial Russian Embassy in Japan, first as a member of the embassy staff, later becoming the acting ambassador. Japan had not yet recognized the Soviet state. When Japan recognized the Soviet government, Dmitrii Abrikosov had to leave his post. He never returned to the Soviet Union. He became a private citizen and stayed in Japan until the end of World War II. He then moved to the United States, but did not live for a long time after. He left behind a manuscript, which was deposited in the Archival Collections of the Columbia University Libraries. The first part of his manuscript was published under the title *Revelations of a Russian Diplomat*,[3] after a historian researcher discovered it in the archives. When Abrikosov eventually had the opportunity to read the memoirs, he found them fascinating. The security organs had known that Abrikosov had an uncle abroad long before Abrikosov learned about his existence.

Although Kapitza was in internal exile, even in his absence his principal research interest—superfluidity and superconductivity—was kept alive at the IFP. These areas became Abrikosov's topics of theoretical investigation in Landau's group.[4] Abrikosov had a good career in Soviet science, but he thought his career could have been better had he not married a French woman in 1970. The story of this marriage—not terribly remarkable under ordinary circumstances, though it was under the Soviet regime—has also become part of the folklore.

It was Abrikosov's second marriage. He had divorced his first wife after twenty years of marriage. They had one son. He met his future second wife during a conference in India, where a French colleague arrived with his beautiful wife. She came from a mixed French-Vietnamese family; her father was a college professor with a dominating personality. His daughter, Anni, wanted to gain her independence by marrying early. Alas, she did not feel independent enough even in her marriage. She may have been overshadowed by her husband, or may have felt confined for other reasons.

After getting to know each other in India, Anni and Abrikosov next met in 1968 in the Soviet Union, during a French-Soviet meeting of theoretical physicists in the the Caucasian Mountains. One of the social programs was hiking. Abrikosov participated, but Anni's husband did not. Abrikosov and Anni reached the summit first, and Abrikosov wrote a complimentary remark about her in the journal placed there in which the conquerors of the summit could leave comments. She was much moved by his compliments. The next year they met in Paris, where Abrikosov had traveled on business, and on that occasion they decided to marry.

To do the necessary paperwork, Abrikosov had to extend his stay in Paris, and this was not taken lightly by the Soviet Embassy. The Soviet officials implied that he was a traitor to his Fatherland. Abrikosov issued an ultimatum: either his stay was legally extended or he would not return to the Soviet Union at all. When the embassy people told him that this was not the proper way to repay the tuition-free education he had received in the Soviet Union, he responded: "I also taught there for many years."[5] Soon, a compromise was reached and Abrikosov traveled back to the Soviet Union together with Anni. Upon his return, he was put into quarantine as far as foreign travel, especially to the West, was concerned. In 1975, he was allowed to visit Finland, but Finland was a special country. Although it was part of the West, Finland had signed an agreement with the Soviet Union stipulating that if Soviet citizens in Finland wanted to stay out of their country, Finland would force them to return to the Soviet Union.

Abrikosov's second marriage produced another son. The marriage lasted seven years. Unfortunately, Anni did not find independence under the conditions of Soviet life, although she learned Russian quickly. She became homesick for France and her former family, and when she went back for a visit, she never returned. Eventually, Abrikosov married his third wife. They have been married for thirty-seven years as of 2013. They had one daughter who went to college in the United States, graduated from medical school, and became a physician.

Returning to the subject of Abrikosov's career, he defended his higher doctorate in 1955. In 1964, he was elected corresponding member of the Soviet Academy of Sciences, and in 1987, he was elected full member. In 1966, Abrikosov, Vitaly L. Ginzburg, of the Lebedev Institute of Physics of the Academy of Sciences (Fizicheskii Institut Akademii Nauk [FIAN]), and Lev P. Gorkov, also of the IFP (as Abrikosov was), were jointly awarded the Lenin Prize "for the theory of superconductivity in strong magnetic fields." The Lenin Prize was a very high recognition in the former Soviet Union. It was different from the State Prize (previously called the Stalin Prize), of which one could receive several. The Lenin Prize could only be received once.

At this time, both Landau and Kapitza were alive, but Landau had already been injured in a car accident in 1962, and he was no longer a factor in scientific life. Kapitza had changed the direction of his research after he returned from internal exile in 1954. Abrikosov, Ginzburg, and Gorkov—all three independent researchers—continued what Kapitza and Landau had started. When in 1965 the theoreticians of IFP moved away to form a new institute, Abrikosov and Gorkov both followed. The new institute became the L. D. Landau Institute of Theoretical Physics of the Soviet (later Russian) Academy of Sciences. There, Abrikosov became head of one of the divisions. Abrikosov's Soviet career was capped in 1988 by his appointment to director of the High Pressure Physics Institute in Troitsk, Moscow Region. He stayed in this position for three years before he emigrated to the United States.

In 1991, the three Lenin Prize winners, Abrikosov, Ginzburg, and Gorkov, were jointly awarded the Bardeen Prize, making them the first recipients of the award. The Bardeen Prize was established to commemorate John Bardeen, the only two-time winner of the Nobel Prize in Physics. Its principal sponsor is the physics department of the University of Illinois, and it is a coveted international recognition.

In 1992, Gorkov, also a full member of the Russian Academy of Sciences, also immigrated to the United States. He became a leading scientist of the National High Magnetic Field Laboratory at Florida State University in Tallahassee. Although, like Abrikosov, Gorkov had become a US citizen and a member of the National Academy of Sciences of the United States, he has remained in active contact with the Russian Academy and the Landau Institute. It must have been a great disappointment to Gorkov in 2003 that he was left out of the Nobel Prize in Physics, of which his former co-awardees of the Lenin Prize and the Bardeen Prize were co-recipients with Anthony

Leggett of the University of Illinois. Both Abrikosov and Ginzburg were puzzled by Gorkov's absence in the Nobel Prize announcement.

When Abrikosov was a young man, he was a member of the Communist Youth Organization (Komsomol), as almost everybody was in his generation, and without which one could not get a higher education.[6] He was exceptionally active and successful in the movement. His assignment was to talk with people at election times to make sure that everybody went to vote. It was a meaningless exercise, because the 99.9 percent majority was assured in the Soviet-style uncontested elections with one-candidate for one-place, but still, people had to participate to maintain the appearance of democracy. Abrikosov was an "agitator," and he was good at it. He was soon promoted to so-called propagandist. This was, in fact, adult education, and Abrikosov had to teach the history of the Soviet Communist Party to grown-ups; this was to a large extent Stalin's biography. To facilitate learning, he found some books for his "pupils" to read, and at study sessions they would discuss these books. When the regional party people did inspections, they were pleased by the progress of Abrikosov's group and by the way he conducted the training. They decided to further elevate him, and he was appointed deputy party secretary for propaganda. This was a strange state of affairs, because Abrikosov was not even a party member. The solution to the problem seemed obvious; they decided that Abrikosov should join the party. Although Abrikosov did everything with enthusiasm, at this point he cooled his heels, and decided to seek his father's advice.[7]

Alexei Abrikosov Senior was a well-known scientist, full member of the Soviet Academy of Sciences, and the vice president of the Soviet Academy of Medicinal Sciences. He was best known for having performed the autopsies of Lenin and Stalin. He was charged with embalming Lenin's body, which is still on display in the Lenin Mausoleum. At the age of sixty-five, he became a party member, and his son wondered why he had waited so long, and what had made him finally join the Party.

Abrikosov described his situation to his father, and asked for advice. He told his father that it was his intention to join the party, because it was the ruling party (there was no other party) and, of course, he believed, as everybody else did, that this system would last forever. He told his father that he was afraid that if he did not join, the party would fall into the hands of some nasty people. To him at that moment, it seemed to be his duty to join.

The father said, "Do you understand that joining the party means that you would become a soldier of the party?" Abrikosov was prepared for that. Then the old man said, "You probably don't know what it really means. Do you know how I resigned from my position as director of the Institute of Morphology, which I'd created myself?" Abrikosov thought that his father had resigned because he was old and sick, but his father contradicted him: "It was nothing like that. One day I was called to the regional party committee and I was asked whether I knew that all nationalities in the Soviet Union were equal. Of course, I did. Then they told me that half of the co-workers in my Institute were Jewish, whereas in the whole population the Jews amounted to a few percent only. They told me that this situation was unjust to the other nationalities and that I should correct the situation." This story is consistent with what we have learned since about the brutal anti-Semitic campaign in the Soviet Union during the last years of Stalin's life.

By then, Abrikosov's father had been a party member for some time. When he joined, he did not think much of it, and he had an easy-going nature, so when he was invited, he complied. It was before Stalin's anti-Semitic policies were begun. Abrikosov's father never openly revolted, and he did not at this time either, but he did not take any action. When he was called the second time, he explained to the party people that he had examined the files of all his collaborators and found no excuse to let any of them go. They were very good, and they had to be, because that was his criteria for hiring them in the first place. The party people told him that if he could not take action, they would do it themselves. The old man went home and wrote his resignation. He told his son that he must be prepared for such things if he became a party member. Abrikosov immediately understood that he would not become one.

However, he had to find a good excuse to decline the invitation to join the party. He came up with a shrewd explanation: "I told them that when I am doing something, I am doing it with zest. If I would join the Party, I would make a very successful career in the Party even if I had no intention of doing so. That would mean that eventually I would have to give up science. I felt—I told them—that I was at a branching point and it was at this point that I had to stop."[8]

Abrikosov was never entirely happy living in the Soviet Union, and he was not satisfied with Gorbachev's policy of perestroika ("restructuring") either: "I understood that under the pretext of perestroika, they were destroying the socialist economy without replacing it with something else. I anticipated that the first victim of the situation would be basic science."[9] By then he was the director of the Institute of High Pressure Physics. To ease the difficult economic situation of his institute, Abrikosov had tried to establish industrial connections. He thought that the institute could produce and sell diamond instrumentation for industry. He wanted these activities to be legitimate but could not legalize them because private enterprise was still frowned upon, officially, that is. At the same time, Abrikosov saw some of his friends moving to America and becoming successful there, so he decided to follow them. He did not feel it to be a great sacrifice to give up his position of director because, as he put it, "My reputation was always based on my science and not on my position."[10]

He observed an all-around deterioration in the most diverse areas of life and activities in the Soviet Union. If the country had always been a paper tiger in the sense of the economy but a giant in selected areas of the sciences, among them nuclear weaponry and the space program, it was on its way to also becoming a paper tiger in those areas. The earlier Soviet advantage had disappeared, and the United States prevailed. According to Abrikosov, the arms race and programs like the Strategic Defense Initiative (SDI) contributed to the collapse of the Soviet Union. He no longer saw a future for himself in what had already become Russia. He was in his early sixties when he started thinking seriously about moving to the United States.

Alexei Abrikosov (on the right) and Lev Landau (on the left) with two foreign visitors between them, in Moscow.

Source: Courtesy of Boris Gorobets, Moscow.

One of his concerns was how to get a good job at such an age in America. Another was how his colleagues would view his departure. In this regard, he thought the sooner he left, the better because with the situation further deteriorating, they would try to prevent him from leaving by exercising moral pressure. This was no trivial matter because his appointment to the directorship had been preceded by a democratic election at the institute; he became director by popular demand. Thinking about the move toward the very end of the 1980s, he wanted to take his future into his own hands, and for this, he first wanted to visit the United States.

Following his second marriage, he was never let out of the country to travel to the West. He used to have a big map of the world in his office on which all the cities from which he had received invitations were marked with little flags, to which he had never been permitted to go. Now he went directly to the secret police, the KGB, and requested that they let him travel. It is not known what he told them, but he received permission; probably the KGB was also sensing the changes and might have been impressed by Abrikosov's determination as well as his position. On his visit to the United States, Abrikosov called on some of his colleagues and inquired about the possibility of his finding a job there. One of them told him that because of his high position in Russia, it would not be an easy task.

Abrikosov did not like uncertain situations. Heeding the fates of his friends who went to America and started looking for a position after they had arrived, he was determined that he would not move before he knew what he would be doing. So he sent out feelers, and after a while resigned himself to having to stay in Russia. Eventually, news came that there was an opening for him at Argonne National Laboratory near Chicago. On his next trip abroad, which happened to be to Venezuela, Abrikosov arranged for a week-long visit to Argonne, and there was a mutual interest in his moving there. The deal was that he would get a permanent position, and not just any position but one specifically created for him by the U.S. Department of Energy, as Distinguished Scientist. Abrikosov moved and has been at Argonne since 1991. He has not been back in Russia, and when I talked with him, in 2004, he had no intention of going back, even for a visit.

He no longer has a base in Russia, although he is still full member of the Russian Academy of Sciences and receives his enumeration there as a full member. His former associates at his institute consider him a captain who was the first to abandon his ship when it faced the danger of sinking. He and his first wife had been members of a closely knit circle of friends. When they divorced, those friends stayed friendly with his wife, and not with him, so he has no friends there anymore. His third wife is Russian, nineteen years his junior; their home in Lemont, Illinois, looks like a Russian island, down to the last porcelain figurine in their china cabinet. Their meals consist of traditional Russian dishes.

Abrikosov talks about Russia and the conditions there with contempt. When the news came about his Nobel Prize in October 2003, President Vladimir Putin sent him a congratulatory telegram, and the Russian Embassy in Washington invited him to attend a reception given by the Russian president in Moscow. Abrikosov declined; instead, he attended a reception by President George W. Bush in Washington, at which Abrikosov declared, "I am first and foremost an American citizen."[11] He had indeed become an American citizen in 1996 when he was sixty-eight-years old; he remained a Russian citizen, too.

Back in 1993, within a couple of years after his departure from Russia, Abrikosov was interviewed by a Russian journalist. In this interview Abrikosov made a powerful statement about science and scientists in Russia. Although it is impossible to know how truthfully the journalist conveyed his words, Abrikosov never publicly protested what was printed: "I am convinced that it is meaningless helping science there, in Russia. It is possible to raise the salaries of scientists, but it's not possible to bring over instrumentation and equipment.

As of today, there is only one recipe for the preservation of Russian science: all gifted scientists should be assisted to leave Russia and ignore those who stay there. My expression may seem unnecessarily brutal to you, and many disagree with me. But life has proved me right.... Any Russian scientist who has the opportunity departs from the country. This is also better for Russia."[12]

Among those who disagreed was the American Nobel laureate Roald Hoffmann of Cornell University, himself an immigrant. He protested Abrikosov's statement: "This is a bizarre and defeatist illogic." He called Abrikosov's statement "over-pessimistic and self-serving rationalization of emigration." He expressed his hope for the future strengthening of science in Russia. Hoffmann's statement first appeared on July 25, 1993, in the *New York Times* and then in *Pravda*.[13] In the same issue of the *New York Times* that carried Hoffmann's letter, there was a yet stronger and more personal condemnation of Abrikosov's statement. It came from Alexander Migdal, then a Princeton University physics professor, himself a recent immigrant from the Soviet Union. Migdal's father, the late Arkadii Migdal was one of Landau's pupils, and a full member of the Science Academy.

Alexander Migdal also became a theoretical physicist, and before his immigration he worked for twenty years with Abrikosov in the Landau Institute. Migdal corrected an erroneous translation of one of Abrikosov's statements in the *New York Times*. Abrikosov was quoted as saying that "he got hungry doing science," which Migdal felt could be interpreted to mean that people went hungry if they were engaged in doing science. Although Soviet life was famous for food shortages, Migdal pointed out that the members of the Academy had special privileges to shop in exclusive stores. He interpreted Abrikosov's remark as getting "hungry *to do* science." Migdal referred to the fact that many high-positioned scientists no longer involved themselves in bona fide research work. Regarding Abrikosov's suggestion that all talented scientists should be helped to leave Russia, and the rest ignored, Migdal wrote: "This smells like Stalinism! Stalin entertained himself by moving whole nations around the empire. But what about those who love Russia, those who would rather share the troubles of their country than leave it, those who feel responsible for the laboratories and the institutes they lead and those simply too young to have an international reputation?"[14]

There may be different opinions about such questions, but Abrikosov's statement had considerable impact, because he had been a star in Soviet science and he had become one at an early age. His paper on the seminal discovery of Type II superconductors appeared in 1952 when he was twenty-four-years old, and he was awarded the Nobel Prize for it in 2003, at the age of seventy five. To get a sense of his discovery, one has to go back in the history of science to the discovery of superconductivity itself.

The Dutch physicist Heike Kamerlingh-Onnes spoke in his 1913 Nobel lecture about his discovery of superconductivity. He observed that on cooling mercury to near absolute zero temperature, the resistance to electrical conductivity drastically diminished—this vanishing took place abruptly rather than gradually. Further, he observed that the state of superconductivity could be destroyed by the application of a strong magnetic field. In 1933, German scientists reported the inverse phenomenon that in the superconductors there was practically no magnetic field present. In an independent investigation in the mid-1930s in Kharkov, Lev Shubnikov discovered the same phenomenon. Shubnikov and his pioneering achievements almost disappeared into oblivion. In 1937, when he was thirty-six years old, he was arrested on false charges during Stalin's terror, tortured, and after being forced to admit the crime of conspiring against the Soviet state—which he never committed—he was executed. He was rehabilitated in 1956. In 2003, in his Nobel lecture, Abrikosov quoted Shubnikov's discoveries and said, "I would like to pay here a tribute to Shubnikov, whose data gave me real inspiration. I never met him but I heard about him from Landau, who was his close friend."[15]

For a long time, what later were called Type I superconductors, were the only superconductors that were known. But the observations about the relationship between superconducting and magnetic properties kept puzzling the physicists. Even the experimental discovery of Kammerlingh-Onnes found no theoretical interpretation for decades. The phenomenon was finally explained by John Bardeen, Leon Cooper, and J. Robert Schrieffer in 1957. Even before then, people understood many aspects of superconductors. In 1950, Vitaly Ginzburg and Lev Landau published a theory of superconductivity, which was based on Landau's theory of the so-called second-order phase transitions, which he constructed in 1937. It was a relatively simple but at the same time general theory, which could describe phase transitions in many different substances. The Ginzburg-Landau theory made many useful predictions, but they required experimental verification.

One of Abrikosov's great merits as a physicist was that he was always on the lookout for interesting experimental findings. One of his colleagues at the IFP, Nikolai Zavaritskii, was an experimental physicist. They had known each other since their university studies, and the two always discussed Zavaritskii's experiments. At some point, Zavaritskii started to do experiments checking the predictions of the Ginzburg-Landau theory. Zavaritskii's scientific advisor at the Institute, Aleksandr Shalnikov, had done similar experiments years before. Shalnikov had been one of Nikolai Semenov's (see chapter 8) star associates in Leningrad, and in 1935 he moved to Moscow to join Kapitza's new Institute of Physical Problems. At the time of Shalnikov's experiments, there was no theory with whose predictions he could have compared his measurements. Zavaritskii was Shalnikov's PhD student, and now everything came together to enable such comparisons. Zavaritskii found that the Ginzburg-Landau theory described his experiments to perfection.

These experiments were done on thin films, which Zavaritskii prepared by evaporating a metal onto a glass plate. Everybody was satisfied seeing the perfect match between Zavaritskii's experiments and the Ginzburg-Landau prediction, everybody, that is, except Shalnikov. He was a perfectionist and was not satisfied because he did not find the preparation of the thin film in Zavaritskii's experiment well defined. When Zavaritskii evaporated the metal onto the glass plate, the metal atoms reached the glass surface. It was at room temperature, which was sufficiently warm to allow the metal atoms to move around, and they formed microcrystallites. This meant that the metal cover of the glass plate was far from uniform, and the film was poorly characterized. Consequently, the conditions on the glass plate were not reproducible, and this was an important caveat, because for a scientific finding, it is a rigorous requirement that it be reproducible!

Shalnikov suggested keeping the glass plate very cold, at liquid helium temperature. The atoms reaching the glass surface would then stick to it; they would not move around and would not aggregate into microcrystallites. Rather, a smooth and uniform film would form—a reproducible structure. Shalnikov also warned that the film should not be heated until the measurements had been completed. Although this experiment was not easy to accomplish, Zavaritskii carried it out, and he made all the measurements for such low-temperature films. The results astonished everybody, because the measurements did not fit the Ginzburg-Landau theory at all. It seemed as if Shalnikov's caution had destroyed a beautiful result. In reality, Shalnikov's caution saved his colleagues from future embarrassment, which would have meant different results under different experimental conditions. In fact, these careful experiments paved the way to their yet more important discovery.

Abrikosov and Zavaritskii started to discuss what actually happened. Of course, one could always say that the theory was wrong, but the theory was beautiful by itself, and it explained many other properties correctly. These experiments and discussions prompted Abrikosov to work out a new theory. He called the superconductors for which he had worked it out "superconductors of the second group." Eventually, they became known as Type II superconductors, and it was then that the other superconductors, already being known, received the name Type I superconductors. Eventually, it turned out that the Type II superconductors were the widespread kind, while the Type I ones seldom occur.

After the Bardeen-Cooper-Schrieffer (BCS) theory of superconductivity was published, Gorkov showed that the Ginzburg-Landau theory was a limiting case of the BCS theory meaning that its validity was limited to a certain set of specific conditions. In other words, the Ginzburg-Landau theory was consistent with the BCS theory, but the BCS theory provided a more general approach. The equivalence of the BCS theory and the Ginzburg-Landau theory—under certain conditions—was understood later.

Abrikosov and Zavaritskii published their papers separately, but they appeared in the same issue of *Doklady Akademii Nauk* (Proceedings of the Academy of Sciences)—Zavaritskii, his experimental data, and Abrikosov, his theory.[16] In my conversation with Abrikosov, he stressed that often a discovery comes in an unimportant form. But then, the significance of the discovery becomes more and more recognized. At the beginning, the Type II superconductors were considered to be something exotic. They referred to films that were prepared in a very special way. At that time, it would have been impossible to predict that all superconductors discovered since 1916 were Type II superconductors. In contrast, today the question arises, why should we distinguish between Type II and Type I superconductors since nearly all superconductors are Type II. The fact remains that their understanding started with Zavaritskii's very artificial thin films and Shalnikov's criticism.

Following this initial success, Abrikosov went on to make further important theoretical discoveries, including some structures of what is called today the "Abrikosov vortex lattice." I am not going to describe these structures as they are beyond the scope of this discussion, but one aspect of this work should throw some light on the working atmosphere in Landau's group. Abrikosov was Landau's pupil and Landau encouraged his pupils to become independent. The watershed in becoming independent was the passing of a set of exams, the *teorminimum*. Abrikosov was one of those who passed it.

Landau did not like to have people dependent on him for their choice of research projects; once he saw a project, he liked to solve it himself. This was another reason he encouraged his pupils to become independent. What Abrikosov learned from Landau on his way to independence was that he should "read the journals, attend the seminars and what is most important, discuss with experimentalists. [He has] done that all [his] life. [He has] read the papers, listened to people and listened to them very attentively; [he has] developed long ears."[17]

Landau's group at the IFP was small; at one point it consisted of himself, Abrikosov, Gorkov, Evgenii Lifshits, and Isaak Khalatnikov (the future director of the future Landau Institute). But Landau's influence was much broader than the size of his research group suggests. His former pupils worked in the leading positions of many other research institutions. Most of them regularly attended Landau's weekly seminars at the IFP. The attendees participated in these seminars not merely as passive listeners; everybody had to give a talk according to a timetable. Abrikosov acted for a long time as the secretary of these seminars; he kept the record of speakers. Some reported on their own work, others were given assignments to discuss published papers. *Physical Review* was Landau's favorite journal and he marked the articles he wanted to hear about. The potential reporters were shown the selected papers, and they could choose one of them to discuss at one of the following seminars. This is how Abrikosov remembered those sessions:[18]

> The nucleus [of the seminar] was about fifteen people, but others could come and sit in; Landau did not object to their presence. Preparing the

reports was a heavy duty. Landau was very critical and if a person did not prepare well his talk, meaning mainly that he did not fully understand the paper which he was supposed to talk about, then Landau was furious. Although I was the secretary of the seminar, I also had to prepare reports. Once I was reporting on a paper that was especially difficult for me because I was not familiar with the field, and it meant a tremendous effort for me, but I remembered that paper for the whole of my life. I have benefited from the ideas of that paper and accomplished much more in that topic than the original author did.

Landau enjoyed tremendous authority before his pupils. Again, in Abrikosov's words, "We, Landau's pupils, trusted him more than we trusted our own judgment, and we never even attempted publishing anything about which we did not have his agreement."[19] This may be the key to understanding what happened in 1953 when Abrikosov first told Landau about what later became known as the Abrikosov vortex lattice. Initially, Landau did not comprehend the way Abrikosov was discussing it, and this is why Abrikosov stopped developing the idea. Later, after Richard Feynman had published something similar, which Landau liked, Abrikosov went ahead, developed his theory, and published it. But by then, his priority in creating an entirely new theory had been damaged. He has never forgiven his mentor for letting this happen.

It could indeed be interpreted that in this case Landau prevented Abrikosov from gaining full priority of an important theoretical discovery. But, of course, Landau did not do anything of the kind. It was Abrikosov's deference to Landau's authority that really made him stop continuing the project at that early stage, and the work, accordingly, suffered a sad delay. The story probably took on a life of its own when it became part of a controversy, as Abrikosov liked to complain about this episode—it obviously kept bothering him. But Landau cannot be blamed for Abrikosov's failure to convince him about the validity of his idea when they first discussed it. The question arises, since Landau was a universally recognized genius, how could it happen that he did not see the point, even if Abrikosov was not sufficiently convincing? To this question Abrikosov responded: "Landau had a very special mind. Because he understood things much deeper than others, some things were not so easy for him to understand. He often noticed contradictions, which the original author did not even think about."[20]

When Abrikosov told me this story, he appeared rather reserved and did not accuse Landau of, for example, forbidding him to work on and eventually publish his new idea. I understood that Landau's authority was so great that the lack of his enthusiasm sufficed to make Abrikosov refrain from developing his idea further at that point. On other occasions, Abrikosov may have given a less subtle description of what had happened. This is perhaps why rumors started circulating that Landau had prevented Abrikosov from becoming the discoverer of an important effect. This was going on when Landau was already dead, and the story infuriated Evgenii Lifshits, Landau's closest associate and coauthor. In 1978, Lifshits wrote a letter to John Bardeen who was not only a famous American physicist but also a giant in the research interests of Landau's and Abrikosov's. Lifshits wrote that Landau could have not hindered Abrikosov's studying this topic and publishing his results, and asked Bardeen to inform the international community about this. Bardeen complied.

For the record, this is what Abrikosov said about this matter in 2003 in his Nobel lecture: "I made my derivation of the vortex lattice in 1953 but the publication was postponed since Landau first disagreed with the whole idea. Only after R. Feynman published his paper on vortices in superfluid helium, and Landau accepted the idea of vortices, he agreed with my derivation, and I published my paper in 1957."[21]

Abrikosov was never shy about letting others know about his achievements. Vitaly Ginzburg recognized that Abrikosov's activities in promoting recognition might have helped them receive the Nobel Prize. He said, "I might even say that in some sense our prize was due to Abrikosov. I think that he was very active in disseminating his results, sending out reprints, and so on."[22] Aspects of their nomination were narrated in the chapter 6, about Ginzburg.

Alexei Abrikosov and his third wife in 2004 in their home in Lemont, Illinois.

Source: Photograph by and courtesy of the author.

Abrikosov had always had heroes, but they changed over time. In this, as in many other aspects of life, he had a utilitarian approach. In his old age, Niels Bohr became his hero because Bohr continued doing physics almost until his death. According to Abrikosov, "There is illness in old age, on the one hand, the abilities of a person decrease, on the other hand, his self-esteem increases; he may think that what he can do now is only great work. Of course, this is not the case and such people usually do not produce anything. Bohr had a different attitude. He worked for fun, not fame, and with that attitude, he could work any time. I am doing exactly the same."[23]

For all his flamboyance, Abrikosov was careful not to get involved in political dissidence. He respected political activists, such as Sakharov, but considered them to be fighting for a lost cause. He never wanted anything to do with lost causes; he wanted to do physics, and any political involvement with "lost causes" would have interfered with it. He was suspicious in the eyes of the authorities, not for any political action, but because of his marriage to a foreigner. They found him "unmanageable."[24] This is how Abrikosov summarized his personal philosophy: "I was independent. My mother was clever and able, but she was also very hard, even dictatorial, and for this reason, from my very childhood, I wanted independence. I learned to fight for my independence. At the same time, I was smart and I did not do foolish things. In science, I could have a choice and Landau encouraged it, so I chose my own way. Think thoroughly about various things and choose your own way. For a career, for publications, it is harder, but it is also more enjoyable."[25]

At the age of eighty-five, with failing eyesight and in fragile health, I wonder whether Abrikosov's quite skeptical outlook on life has deepened. He appeared taciturn in our latest exchange of e-mail messages. It is remarkable that having come to the United States in his mid-sixties, he could still make a career in his new environment. However, once he has passed his creative period, he and his wife might find themselves quite lonely. I wonder if he thinks about what it would have been like to spend his twilight years in Russia. For some reason, Lyndon Johnson's words (and I forget their source) when he decided to return to Texas after his presidency come to mind. He wanted to go there to exchange his loneliness for a place where "they know when you're sick and care when you die." In 2004, Abrikosov appeared very hardened to me in his determination not to go back to Russia, ever. I wonder if he has ever wavered in his determination.

PART III

Chemists and Chemical Physicists

Nikolai Semenov in the laboratory.

Source: Courtesy of Alexey Semenov, Moscow.

8

Nikolai Semenov

MR. CHAIN REACTION

Nikolai N. Semenov (1896–1986) was both an extraordinary researcher and a successful organizer–science administrator. He applied modern physics to chemistry, discovered the branched chain reactions in chemistry (for which he received the Nobel Prize), formed a large school, created a network of research institutes, and was peripatetic and popular. His Institute of Chemical Physics played a crucial role in the Soviet nuclear weapons program. As a scientist, Semenov was intermediate between experimental and theoretical, and could be called the theoretician of the experiment. He had his own political difficulties caused by overzealous dogmatists. He was compliant and learned to coexist with an oppressive regime rather than oppose it, and he took pride in representing it.

Nikolai Semenov was the first person with whom I ever recorded a conversation. This happened in September 1965, at the request of the science editor of Radio Budapest. He had called me when he was looking for someone with a background in chemistry or physics and knowledge of Russian. I had just returned from Moscow with a fresh master's degree equivalent and had started working. I did not think much of the task. I had all the arrogance of an ignorant person who goes to interview a foreign Nobel laureate scientist without knowing much about the scientist's achievements or anything about the art of interviewing. The Radio sent a technician with me and a tape recorder, which looked like a medium-sized suitcase. I was very lucky, but I did not know then, that Semenov was a seasoned interviewee and a natural story teller. However ignorant I was, he saved the interview. Thirty-five years later, when I was compiling the *Candid Science* volumes of my collected interviews, I remembered the Semenov interview. I purchased a copy of the original tape from the Archives of Radio Budapest and included the English translation of the conversation in the first volume of the series.[1]

Nikolai Nikolaevich Semenov was born on April 15, 1896. His father had been a professional military man but resigned his commission after his marriage. He served as manager of a large government-owned estate in the Volga basin, not far from the town of Saratov, some 525 miles southeast of Moscow, where Semenov was born. His mother was well-educated, taught mathematics, and had a good affinity for the arts. The Semenovs lived in small places, where there was no secondary school, but by the time he reached the age for secondary education, they had moved to Samara, some 675 miles east/southeast of Moscow (in Soviet times, this city was called Kuibyshev). Semenov went to a "real" (as opposed to "humanistic") gymnasium, with emphasis on science and mathematics.

As a schoolboy he came down with typhus and had to stay at home for quite a while. He was already ambitious and did not want to be left behind in his studies, so he read a lot of books. His favorites were about chemistry, and he could hardly wait to become healthy again to go to the nearest drug store and purchase the ingredients for chemical experiments. He read that sodium and chlorine are the components of table salt, which surprised him because sodium was a flammable and malleable metal, and chlorine was an extremely reactive gas. In his first experiment, he burned a piece of sodium in chlorine gas; the reaction produced a precipitate, which he recrystallized. The product was a white powder, which Semenov poured over a slice of bread, and he found that it was the best table salt he had ever tasted. He was puzzled by the striking difference between the properties of table salt, on the one hand, and those of the chlorine gas and metallic sodium, on the other. His puzzlement touched upon the most fundamental question of chemistry: the nature of chemical bonding and how such bonding leads to new substances and properties.

Already in the secondary school, Semenov decided that he would become a *good* chemist. He read that the future of chemistry was in physics, so in 1914, he began his studies as a physics major at the University of St. Petersburg. He found a mentor who impressed him enormously, Abram Ioffe, and started working for him from his sophomore year.

Ioffe graduated from the St. Petersburg Institute of Technology, and between 1902 and 1905 he did his doctoral studies in Munich with W. C. Röntgen, the discoverer of Röntgen rays, called X-rays in the English language. Ioffe earned degrees from St. Petersburg University in 1913 and in 1915 (by then, it was called Petrograd University). He eventually founded the Leningrad Institute of Physical Technology, which then became the starting point for other research institutes in Leningrad, Moscow, and elsewhere in the Soviet Union. He was mentor to many of the great generation of Soviet scientists. He stayed active throughout his life and at the age of seventy-five got his last appointment: director of the Institute of Semiconductors.[2]

Abram F. Ioffe's grave in the museum section, called "Literary Bridges," of the Volkovskoe Cemetery in St. Petersburg.

Source: Photograph by and courtesy of the author.

St. Petersburg, then Petrograd, finally, Leningrad, was an excellent venue for a young and ambitious scientist to start his career. In addition to Ioffe, there were other outstanding professors of physics; one of them was the pioneer cosmologist Aleksandr Friedman. Among Semenov's fellow students, there were many future internationally renowned scientists. One of them, the future academician I. V. Obreimov, noted that even as an undergraduate, Semenov's scientific acumen and organizational skills were already apparent. He could, though, become tired of things quickly and lose his enthusiasm, but he soon rid himself of these traits in striving to better himself.[3]

Semenov did not limit his studies to the requirements of the curriculum and was keen on educating himself. He was among the youngest participants in two informal gatherings of budding scientists who wanted to be informed about the development of their science. Friedman held informal evening discussions on a variety of problems in theoretical physics that were not properly treated in the formal courses. Then there was the Sunday circle, in which selected papers from the scientific literature were discussed. Paul Ehrenfest founded this circle in 1908 and was in charge of it until he left St. Petersburg in 1912. Between 1915 and 1921, Obreimov and Semenov ran the circle, although with some interruptions.

Semenov conducted experiments and published papers during his student years. In 1917, he graduated from the university, and continued doing research until the spring of 1918. Then he moved to his parents' home in Samara. The civil war was raging, and the young scientist joined the White Army; this remained a taint in his biography. Had he been captured by the Red Army during his brief service, he would have been summarily executed. He soon deserted the White Army and was happy to return to academia. He got a job doing research and teaching in Tomsk, in Siberia. He was successful, did independent work, and published papers. By 1919, he was back in Samara with his parents.

Semenov met an extraordinary woman in Samara, Maria Boreisha, seventeen years his senior who had just been appointed professor of philology at Samara University. She was married to a renowned medical doctor, Alexei Liverovskii. They had four children; the eldest was only a few years younger than Semenov. Nikolai and Maria met at a literary event; they shared a keen interest in literature and, especially, in music. He loved to sing and, by coincidence, his favorite songs were her favorite ones, too. They fell in love with each other; she divorced her husband and moved with her youngest child to Petrograd, where she was from, and Semenov followed. They got married and were living a happy life, with Maria always at the center of any group that they found themselves in. But Maria soon became ill with cancer, to which she succumbed in 1923. According to family legend, when Maria Boreisha was on her deathbed, she willed to Semenov to marry her niece, Natalia Burtseva.[4] This is what happened, and Semenov and Burtseva stayed married from 1924 until 1975; they had two children, a boy and a girl.

In Petrograd, Semenov worked in Ioffe's new institute. Semenov was initially in charge of a research laboratory studying electronic phenomena. Soon a group formed around him, and he was pursuing several research directions. His interests blended fundamental questions and applications, and included the ionization of atoms and molecules; the phenomena of condensation and adsorption; and electricity and electric techniques. There were future leading researchers among his first pupils: V. N. Kondratev, Yu. B. Khariton, A. I. Shalnikov, A. F. Valter, and A. I. Leipunskii. In 1922, Semenov was promoted to deputy director of the entire institute.

The 1920s were an exciting time for science in the Soviet Union. What was happening in science might be compared to the Soviet avant-garde art of the same period. It was characterized by enthusiasm, innovation, and the power to mobilize youths who would have not had the opportunity to get engaged in these activities under other circumstances. Semenov talked about this in our 1965 conversation and conveyed the atmosphere of that time:[5]

> There were tremendous changes in our country. There was a revolution, civil war was on its way and so was foreign intervention. Many thought at that time that under the circumstances scientific research would be suspended. Due to Lenin's foresight, however, things took a different turn for science. He convened the scientists scattered all over the country, created conditions for independent work, and secured the necessary funds for international monographs and journals.

This was a heroic era indeed. It was moving and uplifting to see the thirst for science of our impoverished, tortured, and liberated people. We received letters from the most remote corners of the country. If somebody read or created something of interest, he let us know at once. The scientific institutes themselves turned often to the public for recruiting new coworkers. Although we could hardly pay them, people were joining us *en masse*. Let me just mention one example, [Nikolai] Chirkov, who is now a professor, had abandoned animal husbandry for science, and he was not alone. Everybody was willing and ready, even if living merely on bread and water, to participate in creating the new science. It is also true, though, that the young researchers were given greater independence than is customary today. But we were all very young. Even Ioffe, the oldest among us, was not yet forty.

I was put in charge of an independent laboratory in 1920, and was already directing sixty coworkers by the early thirties. Progress was breathtaking and new institutes were mushrooming. Among them Pavlov's physiological institute, Ioffe's physical institute, institutes for radio-physics, radioactivity, and others. The technical physics institute was very strong and I was in charge of its Department of Physical Chemistry. The reactions abroad were interesting. For quite a while they could not appreciate what was going on in our country. Only when the spectacular results appeared, the atomic bomb, the atomic power station, the first Sputnik, only then did we finally receive their recognition.

However, it was a long, very long way before we reached that destination. In the 1920s, England and Germany were the great scientific powers. Our Soviet science was, to some extent, provincial, in spite of the many excellent scientists. American science was in a similar situation at that time. Today Soviet and American sciences are on the top. Without belittling the American achievements, their development was, no doubt, greatly accelerated by the tremendous import of outstanding European scientists. We, on the other hand, had to do everything ourselves but, I think, we have solved our task with great success.

Semenov was bold in charging young and inexperienced, but obviously gifted, people with exciting projects. In 1921, he asked the seventeen-year-old student Yulii Khariton to join his research group. After Khariton received his Diploma (master's degree equivalent), in 1925, Semenov suggested to him that he and one of his associates, Zinaida Valta, investigate the oxidation of phosphorus. The topic became immensely successful, and even decades later Khariton marveled at Semenov's power of intuition.[6]

The experimental setup and the results in Khariton and Valta's initial work are described in chapter 9, about Khariton. The experiments measured the pressure and temperature conditions at which phosphorus would show luminescence (emission of light) in the presence of oxygen. The essence of the observations was that at a given temperature, at very low pressure, nothing happened. Then, keeping the temperature constant but gradually increasing the pressure of oxygen, at certain minimal pressure, phosphorus lit up, and the luminescence continued up to a certain value of elevated pressure. Above that pressure, again, there was no luminescence.

The observations were unexpected and called for communicating them in a research article in 1926. For the time being, it was a valuable empirical study. Khariton and Valta did not think much of carrying it further or of looking for a theoretical interpretation. Khariton was preparing for his work at the Cavendish Laboratory in Cambridge. Valta was also moving away from the institute to join another work place. So it was only after Khariton spotted the internationally renowned German scientist Max Bodenstein's devastating criticism of their observations that Semenov returned to the topic. The physical chemist Bodenstein did not believe the report and was convinced that the Russian authors had committed an error, and he said so without mincing his words. The criticism hurt because the report had come from a virtually unknown institution and had been written by unknown authors. Many in the West thought that the conditions in the Soviet Union were not yet ripe to produce trustworthy science.

It often happens that there is a new experimental observation, which is then given different theoretical interpretations. Usually, the controversy is not about the experimental facts, but about their meaning. Here, however, this was not the case because the experimental observation was judged suspect. Khariton informed Semenov about Bodenstein's opinion in a letter of March 13, 1927, and it sounded a justified alarm. Bodenstein's criticism put the reputation of Semenov's laboratory under question.[7] Khariton did not hesitate to declare that Bodenstein was not right, and pleaded with Semenov to think about the problem.[8]

Semenov understood that he had to repeat the experiment and play around with the experimental conditions before he could state unambiguously that the Khariton-Valta observations were correct. This was done, and Semenov fully *reproduced* Khariton and Valta's findings. The next task was to *interpret* the experimental findings. Bodenstein's disbelief only raised the stakes—it underlined that the observations were unexpected and, accordingly, carried the promise of a discovery.

Semenov spent the next few months working on a theory for the interpretation of the observations. He soon understood that the oxidation of phosphorus was a chain reaction, but it differed from previously described chain reactions. This in itself was an important notion. It made it also easier to understand why Bodenstein was puzzled by Khariton and Valta's observations. Bodenstein was a pioneer in the chemistry of chain reactions and, in particular, of chain reactions in photochemical processes.

In a photochemical process, chemical change occurs under the impact of light; in other reactions the reaction is initiated by heat. To understand the interpretation of the reactions discovered by Khariton, Valta, and Semenov, we first have to understand the chain reactions. The reaction between hydrogen and chlorine, $H_2 + Cl_2$, can serve as example. According to the interpretation by German physical chemist Walther Nernst, for this reaction to begin, there must be an active center in the reaction volume. In the mixture of chlorine gas and hydrogen gas, one of the molecules may dissociate under the impact of light. Let us thus suppose that one of the chlorine molecules dissociates, resulting in two reactive chlorine atoms. When one of these chlorine atoms collides with a hydrogen molecule, a hydrogen chloride molecule forms, plus a free hydrogen atom,

$$Cl + H_2 \rightarrow HCl + H.$$

At this point, the hydrogen atom may carry on the reaction,

$$H + Cl_2 \rightarrow HCl + Cl.$$

This continues as a chain until two hydrogen atoms, or two chlorine atoms, meet and form an H_2 or a Cl_2 molecule, causing the termination of that particular chain. Thus, there is a distinct beginning and a distinct end in this chain reaction.

The oxidation of phosphorus proved to be a more complex process in which active centers can form not only under the impact of external light (or heat) but also by the large amount of energy liberated by a collision of two particles. This energy can break an oxygen molecule, producing two reactive oxygen atoms,

$$O_2 + \text{energy} \rightarrow 2O.$$

and every oxygen atom may serve as the beginning of a new chain. This process is accompanied by the emission of light, and it often leads to an explosion.

If the pressure is very low in the reaction mixture of phosphorus and oxygen, there will be very few branches before the active particle reaches the wall, and the reaction stops. On the other hand, with increasing pressure, the probability of having enough branches increases, and the reaction continues. This mechanism provided an explanation for the observation that had been a puzzle. For the observation that at a certain higher pressure the oxidation of phosphorus stops again, there was a similarly simple explanation. At higher pressures, the large number of oxygen atoms increases the probability that they will bump into each other, causing two oxygen atoms to combine into an oxygen molecule, which would terminate the chain. Hence, the upper limit of pressure.

The peculiarities of the phosphorus oxidation reaction led Semenov to realize that this process was not simply a chain reaction; it was a *branched* chain reaction. In the chain reaction, like the one between hydrogen and chlorine, there is always only one active center for the continuation of the reaction, so it continues along the same chain. In the oxidation of phosphorus, there are sometimes two, and in rare cases even three, active centers in the events of the reaction; Thus the reaction continues not merely in one chain; it might continue along two, sometimes even three, chains. The chains soon multiply, hence the branched chain reaction, which might lead to *explosion*.

Ingenuity in setting up the experiment, the unexpected observations, and the frustration caused by criticism and disbelief on the part of an authoritative foreign scientist stimulated Semenov to arrive at a new theory. Not only did his discovery solve the puzzle of phosphorus oxidation, it also provided interpretation for many other chemical reactions. It would also be applicable to nuclear chain reactions, but they had not yet been discovered. It will be a few more years before the concept of nuclear chain reaction occurs to Leo Szilard, and he also arrives at the concept of critical mass in nuclear chain reactions reaching explosion.[9] It will be yet a few more years before Otto Hahn and his colleagues discover nuclear fission of uranium, leading to the first experimental identification of a substance suitable for nuclear chain reactions.

Semenov early on recognized the importance of researching explosives. He assigned a strong work force to investigate them. His possibilities kept broadening; he was moving up, first within Ioffe's Institute, when he became responsible for the whole sector engaged in chemical physics. Then, in 1932, a separate institute was formed under Semenov's leadership, the Institute of Chemical Physics. In 1929, Semenov was elected corresponding member of the Soviet Academy of Sciences. He was young, but his election did not come conspicuously early. He spent a mere three years as corresponding member, and in 1932, he was elected to full membership of the Soviet Academy of Sciences—the pinnacle of recognition for a scientist.

Nikolai Semenov and his associates in the early 1930s at the Institute of Chemical Physics in Leningrad. On Semenov's right, Viktor Kondratev; on Semenov's left, Yulii Khariton; and on Khariton's left, Aleksandr Shalnikov.

Source: Courtesy of Alexey Semenov, Moscow.

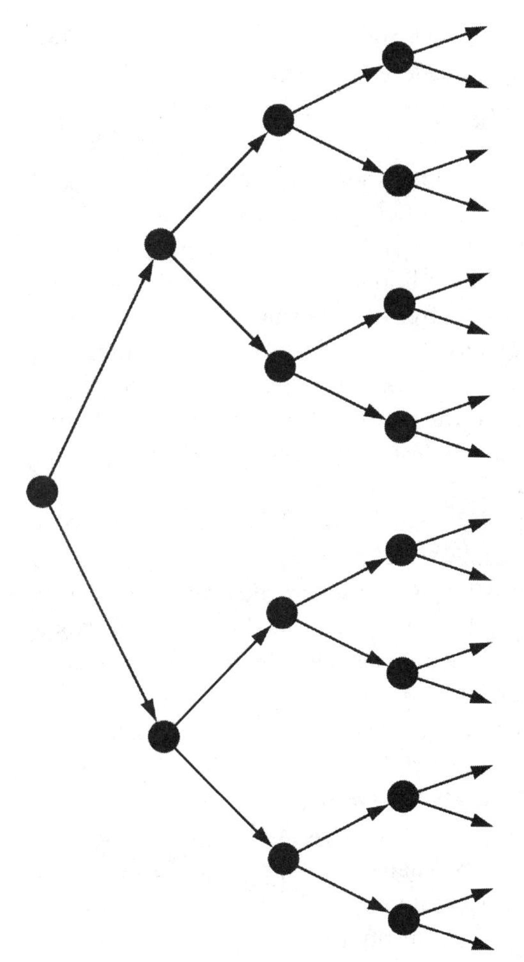

Scheme of a branched chain reaction.

The discovery of branched chain reactions, and their theoretical interpretation in 1928, was a milestone result. Semenov felt the need for generalization. In 1931–1934, he wrote a monograph *Tsepnie reaktsii* (*Chain Reactions*). It was published in Russian, and in the following year, it appeared in English translation in Oxford.[10] Semenov worked on his book with exceptional intensity, although he could devote only his nights to this project. When the book came out, he gave a copy to Khariton in which he wrote: "To dear Yulii Borisovich, you were the first who pushed my thoughts toward the chain reactions."[11] This was the highest praise, but Khariton was more moved that Semenov used the polite form of address, that is, Khariton's first name and his patronymic, not just his first name. This was an expression of respect and showed that Semenov now considered Khariton to be an equal in their adventures. Incidentally, Semenov and Khariton used the polite form of address with each other throughout their lives, even though they became close friends and were related by close family ties as

Semenov's son would marry Khariton's daughter.[12] Among the people of their generation, the familiar way of mutually addressing each other was very rare.

Semenov always found reasons to invoke the concept of chain reactions. He found plenty of examples in chemical and nuclear processes, burning, combustion, and oxidation. He could also discuss biological processes in terms of chain reactions and, in particular, the process of aging. In the framework of normal operations of the organism, the processes of oxidation and reduction provide equilibrium conditions for the organism. However, when there is even a minute shift away from equilibrium, the normal operation of the organism suffers, and transformation from minute modifications to changes on the scale of the organism happens. According to Semenov, these are chain processes (although they might also be called "cooperative effects").[13] Years after Semenov's death, he was still being referred to as "Mr. Chain Reaction."[14]

Semenov wanted to build up a big center to tackle his research problems and wanted the best possible personnel for it. He sent his senior associates to visit various locations in the Soviet Union and invite gifted young people to work in his institute. In a parallel effort, they established interactions with a number of military organizations to identify talented scientists and to become versed in the research needs of national defense.

Semenov proved to be an outstanding organizer, and his institute grew as if it were "fed by yeast," to use Khariton's description.[15] Semenov reminded Khariton of Ioffe, Semenov's mentor, who also proved to be an excellent organizer. Semenov did not tell people what to do, what kind of projects to start, but he supported their initiatives if they seemed promising. Also, he supported the formation of new groups, new laboratories, and even new institutes that would be independent of his own. Such institutes remained working closely with Semenov. They appeared in a number of cities throughout the Soviet Union, and they were organized by Ioffe's or Semenov's former pupils.

In 1945, Semenov was appointed professor at Moscow State University, where he founded the Department of Chemical Kinetics. He became its chairman and held this appointment till the end of his life. However, he never thought of himself as a bona fide pedagogue, and did not care much for lecturing to students. He was not a very good educator and was not even very interested in his children and grandchildren. But there were occasions when he rose to the challenge of properly handling them. His son, Yurii, was about ten years old when Semenov caught him smoking. Semenov had been a chain smoker all his life, but he did not want his son to become a smoker, especially not at such an early age. He did not say a word, only invited Yurii into his office and offered him one of his own very strong cigarettes. The boy became ill, and never again smoked in his entire life.[16]

By the mid-1930s, conditions in the Soviet Union worsened considerably as far democratic atmosphere and even the security of ordinary people were concerned. In 1937, Semenov might have been arrested as a member of a counterrevolutionary group, however absurd such an accusation might have been. M. P. Bronshtein of Ioffe's Institute defended his doctoral dissertation in 1935 under the title "Quantization of Gravitational Waves." He was arrested in 1937 on frivolous charges; Ioffe attempted to save him but could not. Bronshtein was tried, sentenced, and executed.[17] Twenty years later, in 1957, his sentence was annulled for lacking any foundation. Only in the 1990s did it become known that Bronshtein had been accused of being a member of a Fascist-terrorist organization that, ostensibly, included a large group of other brilliant scientists, such as Vladimir Fock, Lev Landau, and Semenov. Fock and Landau each spent some time in prison; however, Semenov was fortunate to have avoided arrest.[18]

Alas, other alarming developments were making Semenov's life difficult and hindering his charging ahead with his goals. They were the doing of Nikolai Akulov, Professor of Physics at Moscow State University.[19] In April 1941, he published an article in the *Zhurnal Fizicheskoi Khimii* (Journal of Physical Chemistry) in which he accused Semenov of plagiarizing the theory of two Danish scientists when he had advanced his theory of branched chain reactions.[20] When Akulov was told that it did not look good to popularize the priority of foreign scientists, he changed his accusations and ascribed the priority in creating the theory of branched chain reactions to a Russian physical chemist. Semenov was accused also of fawning before foreign scientists. Back in 1934, he had dedicated his monograph about the chain reactions to the Swedish Svante Arrhenius and the Dutch Jacobus van 't Hoff.[21] His accusers declared that Semenov was a cosmopolitan.[22]

Semenov got tangled in a long and unpleasant struggle to clear his name, however improbable the accusations appeared. He asked the Soviet Academy of Sciences to form a commission to investigate the accusations and to publish its findings. This effort had to be suspended during the war years, 1941–1945, and the commission was not created until the spring of 1945. It was chaired by A. N. Nesmeyanov, rector of Moscow State University. The commission was very authoritative; members included such internationally renowned scientists as the mathematician A. N. Kolmogorov, theoretical physicist V. A. Fock, electrochemist A. N. Frumkin, and physical chemist A. A. Balandin.

The commission had numerous meetings, and at one of them Semenov was given the opportunity to state his position. He called his discovery "one of the greatest discoveries of Soviet science." He found it contemptible that Akulov attempted to ascribe this discovery to foreign scientists or to a Russian scientist of prerevolution times.[23] Here Semenov shrewdly played on the political sentiments of the time and turning Akulov's own approach in his accusation against Akulov. Furthermore, Semenov stressed, "These falsifications of history don't help enhance the glory of *our science*; on the contrary"(italics added).[24] At the end, Semenov was exonerated. In 1954, Akulov was fired from

Moscow State University, and he moved to Minsk, Belarus, where he had been elected full member of the Belarus Academy of Sciences.

Still in 1951, Akulov published his own book about the theory of chain reactions.[25] Semenov, Khariton, and Yakov Zeldovich offered a critique of this book in a letter to the president of the Soviet Academy of Sciences, who by then was none other than A. N. Nesmeyanov. In 1952, there was a public discussion of Akulov's book at Moscow State University. Semenov gave a presentation in which he criticized many of the statements in the book. The struggle with Akulov lasted thirteen long years and took up a lot of Semenov's time and energy. Akulov managed to relate his criticism of Semenov to the fight against cosmopolitans. The attack could not be effective against the *physicist* Semenov, who, along with his many associates, was deeply involved in the creation of the Soviet nuclear might at the time. However, Akulov supposed that it might work against the *chemist* Semenov since chemists were more vulnerable to ideological accusations than the physicists during the last years of Stalin's life.

The Soviet Union stayed neutral with respect to the belligerence between the British-French Alliance and Nazi Germany for almost two years after World War II broke out on September 1, 1939. The Soviet-German nonaggression pact made the German aggression in the West easier. When Germany attacked the Soviet Union on June 22, 1941, it had the advantage of the element of surprise—this in itself was puzzling because the attack should have been expected. The Germans were advancing fast toward Moscow and Leningrad, and the research institutes were evacuated to Kazan. Semenov played an active role in initiating the evacuation. In Kazan, Semenov and his coworkers continued their work on explosives with increased intensity. At the request of the military, in early 1942, Semenov went to a meeting in Moscow, the purpose of which was to determine how to solve the puzzle of a German innovation—the combined charges being used by their artillery. Semenov, accompanied by Khariton, visited a classified research institute; Khariton stayed on in this institute.

There was an aspect of the work in Semenov's Institute that even before the Soviet Union became involved in World War II had special importance for the nuclear future. The understanding of branched chemical chain reactions facilitated the understanding of the similar chain reactions in nuclear phenomena resulting from the bombardment of uranium with neutrons. Nuclear fission was discovered at the end of 1938 and was understood at the beginning of 1939. The Soviet physicists were ready right away to move on with further studies of this phenomenon and its applications. Khariton and Zeldovich of Semenov's institute understood at once that the reactions induced by bombarding uranium-235 with slow neutrons were branched chain reactions that might lead to explosion. But because both were busy with their work at the institute, they could not simply move to a new project, however intriguing and promising it appeared to them. They decided to work on it after hours on their own time. Thus Khariton and Zeldovich kept working on their official projects during the day, but come evening, they switched to their calculations on the fission phenomenon and the branched nuclear chain reaction. They kept Semenov informed about their "side project."

At the time this work was going on, Semenov's Institute was under the jurisdiction of the ministry of oil industry. When Semenov learned about the "hobby" of his two brilliant associates, he at once became enthusiastic and proposed to his ministry to give this new project proper weight. He never received a response; thus, the work continued on a low scale. Next, they prepared a report for the Academy of Sciences about the feasibility of an atomic bomb. The Academy proved more responsive than the ministry and it created a uranium commission. Its members included a number of academicians, plus Igor Kurchatov and Khariton who were not yet members of the Science Academy. With the German attack though, the uncertain project of the atomic bomb was shelved. Khariton, Zeldovich, and all the others were directed to war-related research of immediate urgency.

The evacuation to Kazan lasted from September 1941 to December 1943. During this period, Semenov and his associates' research focused on burning, explosions, and specific weapons improvements. At the end of the evacuation, Semenov did not return to Leningrad from Kazan, but was directed to continue his activities in Moscow. His coworkers and his Institute followed him to Moscow before the end of the war. Khariton and Zeldovich along with some others eventually became leading members of the nuclear weapons program (see chapters 2 and 9). In addition, a substantial research program was developed in Semenov's Institute assisting the nuclear project.

Semenov's wartime experience contributed to his decision to join the Communist Party. In fall 1945, he was accepted as candidate to membership, and in 1947 he advanced to full membership. He held various positions in party organizations, including being a candidate for membership on the Central Committee of the Party between 1961 and 1966. It was a rare distinction for a scientist, but he never became a full member of the central committee. Semenov had a more brilliant career in the Science Academy. In 1957, in the year following his Nobel Prize in 1956, he was elected to be in charge of the chemistry section of the Academy and to be a member of the Presidium of the Academy. He remained a presidium member to the end of his life, but served as chair of chemistry only until 1963, when he was elected vice president of the Academy. He stayed in this position until 1971. He remained director of his institute through his life and also chaired the council of directors of the research institutes at the research center in Chernogolovka in the Moscow Region. His position there involved determining general directions more than running things on a day-to-day basis.

Semenov always had big ideas and plans for bringing them to fruition. In this regard, his functions in the Party and at the Academy were most helpful. He appeared to cling to his roles and acquired additional ones, even when he was an octogenarian. When he was eighty-two years old he became the chair of the Scientific Council on the Utilization of Solar Energy. At the age of eighty-four he was appointed to serve as chair of the Council on the Philosophical and Social Problems of Science and Technology. At the age of eighty-five he was elected chair of the Council on Chemical Physics. In the same year, he became the editor-in-chief of the Russian journal *Khimicheskaya Fizika* (Chemical Physics). Finally, at the age of eighty-seven he was appointed as chair of the Council on Fuels. None of his other positions ended; he was simply adding one after the other.

Even in his ninetieth year, he stayed in charge of all the institutions and organizations in whose charge he had been when his abilities were at full capacity. He just could not quit, and there was nobody in his environment to tell him that he should step down. He thus prevented a whole generation of his brilliant pupils from becoming leaders in their own right. He acted this way not out of spitefulness; on the contrary, he was truly kind and caring; he just could not leave the scene of his influence. His institute and later institutes were his life. He was proud of the institutions he created and was not bashful about it.

Semenov's greatest organizational achievement was the creation of the scientific center in Chernogolovka, thirty-five miles northeast of Moscow. Chernogolovka today is a town with over twenty thousand inhabitants, and a large number of research institutes. The origin of this science center dates back to the mid-1950s.[26] At the time, several research areas in Semenov's Institute were fast expanding. They investigated a great variety of explosives and the Institute needed a proving ground; they could not run tests of explosion in the middle of Moscow. The solution was to build an affiliated laboratory for experimental work outside Moscow. The authorities offered Semenov to choose the most appropriate location for his new base. Semenov and a few of his associates visited the village of Chernogolovka at the end of November 1955, and decided that it fit their needs perfectly. There was a proving ground of over two thousand hectares previously belonging to the Soviet Air Force. Semenov wrote a formal request on December 3, 1955, and the Soviet government issued an order on February 28, 1956, about the creation of this experimental base for the Institute of Chemical Physics, serving also as the nucleus of the future science city.

The Soviet regime did not tolerate private initiative, but it was possible to carry through a private initiative if one had the ability, power, and connections to transform it into an initiative as if coming from higher authorities. Some initiatives could even benefit from the contradictions of Soviet society: it was easier for a totalitarian regime than for a democratic one to focus its efforts on a few selected projects. The tremendous success of the development of Soviet nuclear weaponry and likewise the Soviet rocket industry were examples of such an approach. On a lesser scale, this also happened to polymer science and industry, and Semenov was one of its architects. Toward the end of the 1950s, the polymer industry still was not developed in the Soviet Union. But the founder of Soviet polymer science, academician V. A. Kargin, succeeded in convincing Semenov that there should be polymer chemistry in Soviet science as well as industry. Semenov became enthusiastic, and requested to be received by the supreme Soviet leader, Nikita Khrushchev, to whom he explained the importance of polymer science and technology. Khrushchev grasped the significance of what Semenov told him and asked the scientist to prepare a report within two weeks for the highest governing body of the country, the Presidium of the Central Committee of the Communist Party.[27] Semenov prepared a detailed factual report, gave his talk to the party leaders, and answered their questions. As a consequence, fast development of polymer science and industry in the Soviet Union was begun.[28]

Semenov was popular, and it came naturally to him. On the occasion when he was returning from Stockholm from the Nobel Prize ceremonies, his family members and his associates were meeting him at the airport. There was a luxurious car waiting for him and his family, and there was a bus in which his colleagues traveled. Semenov's grandson asked him: "Will you ride with us or with the people?" Semenov rode with the people.[29]

He always needed to be in communication with people; this was his way of operating: having discussions and talking and developing ideas. The ranks and titles of his partners in discussion hardly influenced him. But he needed the people around him. Sometimes his associates, after they had been sitting with him in a meeting for many hours, would beg him to let them go, but he would not budge. He used people as a bookshelf full of books. At any moment he might need a piece of information that one of those around him might possess.[30]

Nikolai Semenov (on the right) talking to Yakov Zeldovich with Yulii Khariton looking on.

Source: Courtesy of Olga Zeldovich, Moscow.

Semenov and Petr Kapitza were close friends from early youth; they both came from Ioffe's School. Reading through their correspondence, one is moved by their genuine affection and caring for each other. They were similarly gifted, original, and revered; they were also very different personalities. Kapitza was an aristocrat; Semenov was at ease in the most diverse circumstances. It would be difficult to imagine Kapitza with nineteen orders on his jacket—the number I counted on a photograph displaying Semenov's.[31] Yet Kapitza proudly displayed his two Hero of Socialist Labor stars on his formal dress when he received the Nobel Prize from the King of Sweden.

When Khariton compared Kapitza and Semenov, he admired Kapitza's courage, which Khariton must have considered reckless because he called it "almost ill-advised."[32] Kapitza took considerable, even life-threatening risks in his correspondence with Stalin. Kapitza never hesitated to raise his voice in the interest of the persecuted. He knew that such actions should be taken quietly and not publicly in order not to cause embarrassment to Stalin and his colleagues. It was a tacit mutual agreement, and all of Kapitza's correspondence with the leaders of the Soviet Union remained classified for a long time. Khariton wondered whether similar correspondence by Semenov might one day also surface, but I doubt that there is much to surface.

Kapitza felt good about being in charge of a relatively small institute, where he was familiar with every aspect of its operation. In contrast, Semenov was in charge of a research empire.

In 1979, he was on a visit in France on the occasion of receiving his foreign membership in the French Academy of Sciences. Two other chemists were receiving the same honor at the same time, Derek Barton from Great Britain and Robert Woodward from the United States. Woodward wanted to talk with Semenov, which was not without complications because Semenov did not speak English, though he could read without any problem. On such occasions, his second wife, Natalia Semenova, helped him in conversation while the marriage lasted. His third wife, Lidia Shcherbakova, did likewise after 1975. Woodward invited Semenov for a visit to America, and in addition to Harvard University he wanted Semenov to visit his ranch. He boasted to Semenov that he had hundreds of cows on his ranch. To this, Semenov reposted that he wanted Woodward to visit him and his institute in Moscow, where there were five thousand people working under his command.[33]

Continuing the Kapitza/Semenov comparison, there was a difference in their interactions with the powers that be. Kapitza did not yearn for official positions, with two exceptions; they were the directorship of his Institute of Physical Problems and being in charge of the oxygen project before, during, and after the war. He was a contemplative type; he read a lot, worked out his own philosophy, and worked more in the background than in any visible way. Kapitza's lack of desire to expand his institute, sphere of influence, or personal clout in any other way had a liberating effect on him. As we have seen, he bravely acted on his principles.

In addition to the positions mentioned earlier, Semenov was also a member of the Supreme Soviet—the Soviet parliament. His positions meant a lot to Semenov, and he used them and his many awards to facilitate his efficiency in negotiations with various officials. He knew the limitations of what he could do and what he could achieve, and he was ready to play the game with the state and party bureaucrats according to their rules. Considering his positions, it is, then, the more puzzling that he deemed it necessary to behave in a most accommodating manner—bordering on self-humiliation—toward the authorities. Besides, he frequently used rhetoric in expressing support for the party and the Academy leadership that others considered superfluous, especially coming from such an international authority in science. He publicly praised Stalin's "teachings" about Marxism and linguistics.[34] Even though he was vice president of the Soviet Academy of Sciences, Nobel laureate, and Hero of Socialist Labor, when he had to talk with someone in the permanent office of the Academy, he was observed to be uneasy, subservient, and eager to oblige.[35]

This did not necessarily mean that he enjoyed doing it. When Khrushchev was "retired" and Lysenko no longer had his protector, Semenov prepared the article "Science and Pseudo-Science" with the help of his philosopher son and three biologist colleagues of his biologist daughter-in-law. One of these biologists, Sergei Kovalev, later became a well-known supporter of Andrei Sakharov. Semenov maintained good relationship with Kovalev and tried to talk him out of his "anti-Soviet" activities.[36] When Semenov's anti-Lysenko article was ready, he wanted to publish it in the central party newspaper *Pravda*. This did not happen; the article appeared under the title "Science Does Not Tolerate Subjectivism" in the popular magazine *Nauka i Zhizn* (Science and Life).[37] At this time, at the end of 1965 and early 1966, there were indications that the Soviet Union might return to Stalin's approaches. A group of intellectuals compiled a letter to the Soviet leadership calling for them to abstain from such a move. This became known to the public. What remained classified was that three academicians, A. P. Aleksandrov, Semenov, and Khariton, forwarded a letter to the party leader Leonid Brezhnev expressing their views even in stronger terms about the catastrophic consequences of a return to Stalinist policies.[38]

Semenov's constant need to expand created the sensation in him that he was at the mercy of the authorities. He did not hide the fact that he considered his institute more important than his life, and he would not have made such a statement lightly.[39] He was grateful to the Soviet government for the support that made his ambitious plans possible. Semenov made his principal discovery in the 1920s when his country was in a very difficult situation yet support for science was strong, it would seem, out of proportion. He was a good politician; he knew how to deal with people and understood that a good politician would have to deal with people in different situations. Given this, it was inconsistent when he was puzzled by other peoples' ambition. He was free of anti-Semitism, and this is why one of his Jewish associates was surprised when, at the time of Academy elections, Semenov asked him: "Tell me why so many Jews want to become corresponding members of the Academy?"[40]

Semenov was awarded the Nobel Prize in Chemistry in 1956 jointly with the British chemist Cyril Norman Hinshelwood "for their researches into the mechanism of chemical reactions." Hinshelwood's contribution focused on the mechanism of the reaction above the upper pressure limit in the oxidation of phosphorus. Semenov's Nobel Prize was distinguished by the fact that it was the first ever awarded to a Soviet citizen. Russian scientists had not been spoiled by Nobel Prizes, though it is also true that before the Soviet Union came into existence there were two Russian Nobel laureates, both in the category of "Physiology or Medicine." Ivan Pavlov received one in 1904 "in recognition of his work on the physiology of digestion, through which knowledge on vital aspects of the subject has been transformed and enlarged." Ilya Mechnikov shared the prize in 1908 with Paul Ehrlich "in recognition of their work on immunity." When the Russian writer I. A. Bunin received the literature prize in 1933, the news was met in the Soviet Union with indignation as he was an emigrant living in Paris.

Nikolai Semenov at his 70th birthday dancing with his daughter Ludmilla.

Source: Courtesy of Zhanna Smorodinskaya, Moscow.

After the war, there were many nominations of Soviet scientists and writers for the Nobel Prizes. This was in part inspired by the widespread recognition of the sacrifice of the Soviet people in World War II. From 1946, there were nominations, among others, for Petr Kapitza in physics and Nikolai Semenov in chemistry. There was, however, hardly any interaction between the Nobel Prize organizations and individual Soviet scientists; even when individuals were invited to submit nominations, few responded. In 1948, these interactions ceased entirely. For example, in 1949, four invitations were sent out to Soviet scientists, including Kapitza and Semenov, soliciting nominations. They never reached the addressees, however, and ended up in the archives of the Soviet Ministry of Foreign Affairs. This ministry should not be blamed exclusively for breaking down the communications. When it queried the Soviet Academy of Sciences whether or not they would be willing to deal with the matter, the Academy responded that it had no intention of making nominations for Nobel Prizes.[41]

The Nobel Prize institution was quite determined to include Soviet scientists in their considerations. As part of their efforts, they were willing to compromise on their otherwise rigorous principle of accepting nominations only from individuals who had prepared their nominations without involving other people. The Swedish institution understood that if they wanted to succeed, they could not avoid the interference of embassies, ministries, the Communist Party, and even the secret police. There was an enthusiastic chemistry professor in Stockholm, Lars Gunnar Sillén, who efficiently facilitated the flow of information between the Nobel Prize institutions and the Soviet side.[42] Sillén had been trying to get documentation on Semenov's scientific activities since 1952. Sillén spoke some Russian, which he had learned in high school from his Russian gym teacher.

In 1954, the secretary of the biology division of the Soviet Academy of Sciences, the biochemist Aleksander Oparin, well known for his theory of the origin of life, attended a meeting in Stockholm. Upon his return to Moscow, he reported to the president of the Academy, Aleksandr Nesmeyanov, on the discussion he had had in Stockholm with two members of the Nobel Committee for Chemistry. Arne Tiselius and Arne Fredga told Oparin that "they would like to see Soviet scientists participate in the Nobel movement."[43] Nesmeyanov informed the Central Committee of the Communist Party about Oparin's report, and by the end of 1954 detailed documentation of Semenov's scientific activities was sent to Stockholm. The documentation arrived too late for the deliberations of the 1955 prize, but could be used for the following year.

That politics could not have been left out of this affair was also shown by the fact that the topic of Semenov's possible Nobel Prize came up in the conversation between the Swedish prime minister, Tage Erlander, and Nikita Khrushchev during their meeting in the summer of 1956 in Moscow.[44] It was a far cry from the Soviet attitude toward the Nobel Prize institution during Stalin's last years. There was a turnabout, and it was also in 1956 that the Presidium of the Central Committee of the Communist Party let the Soviet

Ministry of Higher Education nominate two Soviet citizens for the Nobel Peace Prize.[45]

There was a sad coincidence between Semenov's Nobel Prize, announced on November 1, and the Soviet suppression of the Hungarian Revolution, October 23–November 4, 1956. On November 1, there was still hope that the Soviet Union would let Hungary go its way, but starting at dawn on November 4, a brutal attack eliminated any such hope. The Nobel Prize ceremonies in December 1956 took place amid a gloomy international atmosphere. The inclusion of a Soviet laureate—though he could not be blamed personally for anything—gave added twist to the Nobel events. The organizers of the Nobel festivities introduced some modifications into the usual choreography of events. The banquet was moved from the traditional location in the grandiose City Hall to a more modest venue and the circle of participants was considerably curtailed. In a break with tradition, the diplomats of the laureates' countries—including the Soviet diplomats—were not invited. The ladies were supposed to dress in gray. Although Mrs. Semenov had brought with her an evening gown sewn in Moscow, she had to acquire a different one in Stockholm.

The Semenovs arrived in Stockholm without their children. The Soviet authorities had communicated that the Semenovs' thirty-one-year-old son, Yurii, would not come, but their twenty-eight-year-old daughter, Ludmilla, was supposed to be in their party. However, shortly before departure for Stockholm, Semenov was informed by the Soviet security organs that their daughter would not be allowed to go with them, either. They stipulated that the Semenovs were supposed to tell anybody who asked about her that she was preparing for her examinations (although she might have been a little too old to be a student). In the photographs taken at various functions, there is always an empty chair next to Mrs. Semenov for their missing daughter about whose no-show it was too late to inform the organizers. It was common practice in the Soviet Union not to let entire families go to the West lest they contemplate not returning. They would follow this practice for all Soviet scientists winning the Nobel Prize on subsequent occasions. The Semenovs, though, did not travel alone; they were accompanied by academician V. N. Kondratev, one of Semenov's first pupils in his Leningrad time, and also by one of his former bodyguards, now his "secretary."[46]

Semenov was a faithful friend, and his circle of friends was broad not only in terms of their number, but also in their age distribution. This was perhaps one of his ways of trying to put off aging. He indeed remained vigorous until very late in his life. He was seventy-five years old when he astonished his friends and family by divorcing his wife and marrying Lidia Shcherbakova (subsequently, Shcherbakova-Semenova), thirty years his junior. He announced this on his birthday, April 15, 1971. Semenov and Natalia Burtseva—Natalia Semenova—lived together for forty-seven years. Of their two children, Ludmilla married the future academician Vitalii Goldanskii and Yurii married the Kharitons' daughter, Tatyana Khariton. Natalia Semenova was devastated by Semenov's action, and many of Semenov's family and friends supported her. Kapitza and his wife—she was especially close to Natalia Semenova—stopped socializing with Semenov, and Semenova spent considerable time with the Kapitzas in their summer home. When she died, in 1996, she was buried in the Khariton grave in the Novodevichy Cemetery.

Lidia Shcherbakova worked in the area of scientific research, either as a researcher herself or involved with science administration. According to family members, she pushed Semonov to accept appointments in his late years and to stick to those appointments to the end of his life. As he was aging, she exercised increasing control over his activities. She made all the arrangements for Semenov's burial and grave, and the tombstone is Semenov's life-size statue. Those who felt close to Semenov and liked his joyful nature cannot recognize the man in the cold statue projecting authority and gloominess.[47]

The Akulov affair, discussed earlier, could have destroyed Semenov and his science. It did not, because Semenov was an excellent tactician and proved strong enough to withstand the attacks, but at a high cost. For some time Semenov lost the trust of the Soviet leadership. He had been instrumental in choosing Semipalatinsk as a proving ground for testing atomic bombs, and his Institute of Chemical Physics worked hard on the preparations for the first test. Nonetheless, when the time came in 1949, Semenov was not allowed to go and observe the explosion.[48] Khariton called Akulov "Semenov's Lysenko."[49] The analogy was more than superficial, because Akulov had made attempts to link his own theory of chain reactions with Lysenko's teaching of "theoretical inheritability."[50] Semenov probably felt that Khariton's characterization of the affair was apt, and appeared to be sensitive to the victims of Lysenko's oppression. When the geneticist I. A. Rapoport had no place to work, Semenov created for him a laboratory in his own institute for doing theoretical genetics and investigating the genetics of microorganisms. He also assisted the biophysicist L. A. Blumenfeld. Toward the end of the 1950s, Semenov actively supported the establishment of a biophysics department—with Blumenfeld in charge—at the Faculty of Physics of Moscow State University (more about this in chapter 10).

These were rare signs of courage in Semenov's career. In most other instances, Semenov chose to accommodate rather than resist. At the time when there was an intensive struggle against cybernetics and the quantum chemical concept of resonance (see in chapter 12), Semenov invited a group of leading associates of his institute to his office and declared that they must fight bourgeois idealism in chemical kinetics. For example, he considered Henry Eyring's theory of activated complex to be such an idealistic deviation, whereas in the international scientific literature it was a broadly accepted approach in discussing the mechanism of chemical reactions. Fortunately, there were hardly any consequences following Semenov's announcement, and soon he resigned himself to the notion that the theory of activated complex contained nothing harmful.[51]

Semenov was a product of his time. What he wrote at the age of twenty-six at the conclusion of a long letter to Kapitza could be taken as his *ars poetica*[52]: "I am not giving too much significance to our works; I consider us to be merely fertilizer for the next generation that will create true science in Russia, a living science with plenty of discoveries and inventions. The reason is that science—as everything else—is not moved by individuals, but by society, I'd even say, by the people. Individuals must be the pioneers; they have to facilitate reaching what the people will have the foundation to build on." It sounds pathetic to our ears today, but I tend to think that Semenov expressed his sentiments accurately at the time; possibly these remained his sentiments later as well.

In our 1965 conversation, I asked Semenov to prognosticate—something I seldom did in my next few hundred interviews because in my early attempts I had learned that my interviewees as a rule declined responding to such request. Semenov did not, and here are excerpts from what he said. His response, looking at it almost half a century later, indicates that he had a good grasp in 1965 of where science was heading:[53]

Statue of Nikolai Semenov over his grave in the Novodevichy Cemetery.

Source: Photograph by and courtesy of the author.

Nikolai Semenov on Russian postage stamp, 1996.

It is increasingly difficult to delineate the various sciences, to draw the border lines between physics and chemistry or between chemistry and biology, and who needs these borderlines anyway? I remember we used to say about the difference between physics and chemistry that physics deals with dirty materials but operates with clean techniques whereas chemistry deals with clean materials but operates with dirty techniques. This would not even stand as a joke today because physicists very often use super-pure substances, for example, in atomic energy research and in radio-techniques, whereas there is an ever broadening application of physical techniques in chemistry. Thus it is increasingly difficult to find a sharp dividing line between physics and chemistry, except, perhaps, in high-energy physics, but even there a periodic system of the elementary particles seems emerging not unlike the periodic system of the elements.

Let's consider another field, biology. It is mandatory to get to know in great detail all those substances that determine the various phenomena related to life. It is no longer possible to carry out serious research in biology without chemistry and without the participation of chemists. The living matter is highly developed not only from the point of view of its structure but also from the point of view of chemistry. The biological structure is a higher order chemical and physical structure. This is the area where we should be expecting the next chemical revolution. In the 1820s Wöhler showed the possibility of making organic substance from inorganic materials when he synthesized carbamide. Before Wöhler, the ability to create organic substances was attributed to some vital force. This discovery gave a tremendous impulse to the development of organic chemistry, leading to the synthesis of plastics and many other entirely new classes of compounds.

However, the investigation of the processes in the organism is a highly complex task. The chemical reactions in the living matter happen at room temperature whereas in industry often extreme conditions are needed to maintain them. These processes consist of large series of small steps each of which need large activation energies. The synthesis of proteins is a good example of how different the reactions in the living organism may be. Insulin has been produced in an extremely lengthy procedure in the laboratory but it takes only a few minutes in the organism. The high speed is made possible by the catalysts operating in the organism. They are called enzymes. We should learn about them and when we do, it will mean the beginning of another industrial revolution of unprecedented magnitude.

Let me conclude with an example. There is much effort going on in utilizing solar energy. However, the efficiency of the various solutions pales in comparison with the efficiency of the photosynthesis in the living organism in which solar energy also plays a decisive role. If we could only learn how nature does photosynthesis in the living organism, it would lead us to high-efficiency utilization of solar energy. Uncovering the secrets of biological structure would at least as much facilitate progress in other fields as in biology.

On April 15, 1996, there was a big commemorative celebration in Moscow on the occasion of the centenary of Semenov's birth. Those who graced the festivities wanted to demonstrate that they were not only a celebration of a scientist but of the country's glorious past. The president of Russia, Boris Yeltsin, was in the center of the presidium of the meeting; next to him was Prime Minister Viktor Chernomyrdin. There was also the mayor of Moscow, Yury Luzhkov, among many others. There were also dignitaries in the audience. Semenov's grandson, Alexey Semenov, was accompanying his other grandfather, Yulii Khariton, who was seated in the first row. He was ninety-two years old, and it was his last appearance in public; he died later in the same year. The era of the Semenovs and Kharitons had come to end.

Young Yulii B. Khariton.

Source: Courtesy of Alexey Semenov, Moscow.

Russian stamp honoring Yulii Khariton's centennial in 2004.

9

Yulii Khariton

DIRECTOR OF "LOS ARZAMAS"

Yulii B. Khariton (1904–1996) was for forty-six years in charge of the principal Soviet nuclear weapons laboratory, Arzamas-16. He was part the Soviet J. Robert Oppenheimer, the initial director of Los Alamos, and part Edward Teller, the life-long leader of the Lawrence Livermore National Laboratory in the United States. Khariton had a middle-class Jewish background with family connections abroad that made his position a miracle under Soviet circumstances.

He started his scientific career under Nikolai Semenov and early on made an experimental discovery that eventually led to the concept of branched chain reactions. He spent two years at the Cavendish Laboratory, where he earned his PhD degree. Following the discovery of nuclear fission, he became a leading member of the Soviet nuclear physics community. He was much decorated for the achievements of Arzamas-16, but stayed in the shadow of classified activities for much of his life.

Yulii Borisovich Khariton was born on February 27, 1904, in St. Petersburg and died on December 19, 1996, in Arzamas-16.* He is buried in the Novodevichy Cemetery in Moscow. He came from a middle-class Jewish family; his father was a journalist, and his mother an actress. This did not recommend him to the rulers of the dictatorship of the proletariat. A closer look at Khariton's family background reveals yet more serious problems from this point of view. In the autobiographies submitted to the authorities at different times, Khariton painstakingly detailed the fates of his parents, avoiding even the hint that he was hiding anything.[1]

His mother, Mirra Yakovlevna Burovskaya, had an unusual career for a provincial Jewish woman; she became a well-known actress in the famous Moscow Artistic Theater. She left Russia in 1910 to be treated for an illness at a European resort and never returned to Russia. She met a medical doctor by the name of Max Eitingon, a well-known Berlin psychiatrist and follower of Sigmund Freud. She divorced Khariton's father and married Dr. Eitingon.

* Arzamas-16 was the first secret nuclear laboratory in the Soviet Union; "Los Arzamas" is how its associates liked to joke about it, referring to the American Los Alamos.

When the Nazis came to power in Germany, the couple moved to Palestine and lived there to the end of their lives. She is buried in Jerusalem.

His father, Boris Osipovich Khariton, was more than a journalist; he was also an editor and publisher. He had a law degree from Kiev University from a time when it was rare that Jews would be admitted to law school. After the 1917 revolution he was the director of the House of Writers in St. Petersburg. He had clashes with the authorities. Following the departure of his wife, he hired a young woman from the Baltics to care for Yulii, whom she taught to speak perfect German. Boris Khariton was a well-known member of intellectual circles in Petrograd (which was the new name of St. Petersburg for a few years after the start of World War I). His activities were judged to be so alien to Soviet ideology and the Soviet state that in 1922, at the age of forty-six, he was exiled from the country together with a group of journalists and professors. When, in 1940, the Soviet Union annexed the Baltic States, Boris Khariton was arrested, tried, and sentenced to seven years in labor camp. The sixty-four year old Boris Khariton was sent to the Gulag, and died either on his way to or in one of the camps.

When Khariton reached school age, he started learning at home. He began attending school when he was eleven years old. He went to a trade school, which he completed at the age of fifteen. He could not go to college at once because of the prescribed minimum age of sixteen years. He started working in a workshop, where he learned how to operate various machines that proved useful in his future career.

At the age of sixteen, he became a student at the electrical-mechanical faculty of the Petrograd Institute of Technology. Abram Ioffe was the physics professor. Khariton found physics to be his most stimulating subject. Under Ioffe's influence, he moved to the physical-mechanical faculty. Aleksandr Friedman, another of Khariton's professors, published papers about the structure of the universe and corresponded with Einstein. Khariton was especially fascinated by another Ioffe disciple, Nikolai Semenov, whose work used the techniques of physics in chemistry; Semenov called his field of research "chemical physics" rather than "physical chemistry." The two are hard to distinguish; the new label stressed that it was basically physics, but the adjective "chemical" pointed to the objects of its inquiry.

Semenov recognized Khariton's talent early and let him develop his research with great independence. This trust soon paid off with the discovery of the branched chemical chain reactions, which had started with Khariton and Zinaida Valta's experiments. They wanted to investigate the oxidation of phosphorus vapors. The light-emitting ability of phosphorus—called "phosphorus luminescence"—was well known. It was also known that white phosphorus would glow at room temperature in air, but not in pure oxygen unless the oxygen pressure was *below* a certain value. This was a puzzle. Khariton and Valta decided to vary the experimental conditions for the oxidation of phosphorus, starting with applying very low oxygen pressures in the experiments. Khariton designed a glass apparatus; he asked the glass blowers to prepare the more difficult parts, but fused the different parts of the apparatus with his own hands.

In this experiment, a glass container with a piece of white phosphorus was pumped to sufficiently low pressure. Oxygen gas from another container was let into the principal container through a narrow capillary while the pressure was monitored by a sensitive pressure gauge. On the way from the principal container to the pressure gauge, there was a cold trap to collect the phosphorus vapors and prevent them from contaminating the pressure gauge. As the oxygen pressure increased, at one point light appeared. The light remained on while the oxygen supply stayed open, but subsided as soon as it was closed. There was thus a lowest pressure below which the reaction would not go. Later, Khariton and Valta found a higher pressure limit above which the reaction would not go either. Khariton and Semenov found the observations puzzling but did not have the impression that they would lead to a great discovery. Khariton and Valta reported their experimental observations both in Russian and in German.[2]

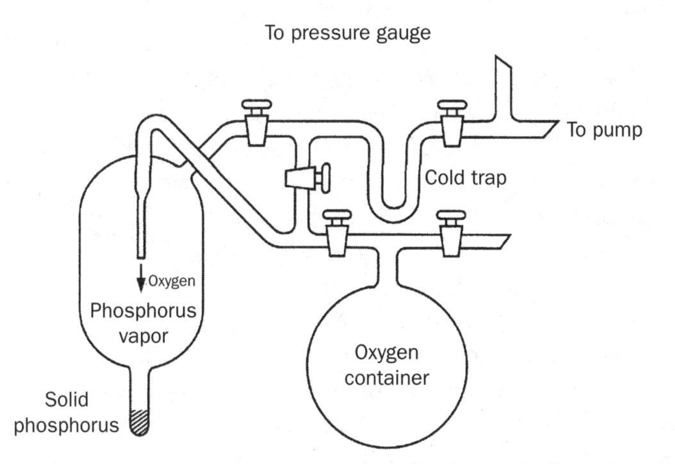

Sketch of the glass apparatus used by Yulii B. Khariton and Zinaida F. Valta for the investigation of the oxidation of phosphorus.

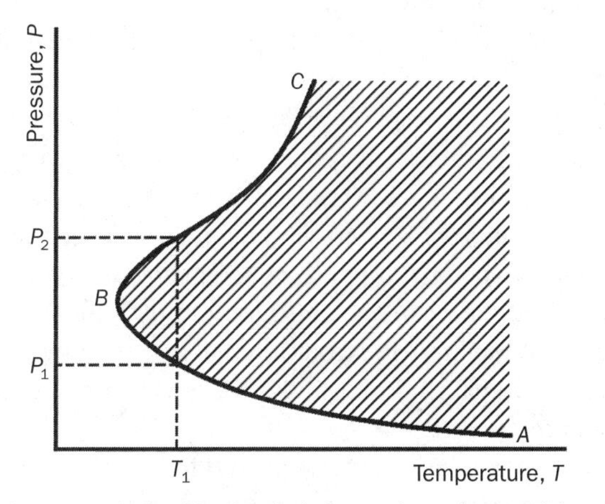

In the temperature/pressure relationship, the shaded area shows the experimental conditions at which the luminescence of phosphorus happened. For example, at a given temperature, T_1, there is a lower pressure, P_1, at which luminescence occurs, and with increasing pressure it stops at the higher pressure, P_2.

Valta graduated from Leningrad State University (Leningrad was the new name for Saint Petersburg under the Soviets) and in 1925 she became a coworker of the Institute of Physical Technology. She worked with Khariton as a postgraduate student. In 1927, she discontinued her graduate studies and moved to the State Geophysical Observatory in Leningrad, where she worked in atmospheric physics. She stayed there to the end of her life (the time of her death is unknown).

Khariton stopped working on the project because he was preparing for a foreign trip. He was to be a doctoral student at the Cavendish Laboratory in Cambridge, England, one of the most famous centers of physics research, under the directorship of Ernest Rutherford. Semenov's close friend, Petr Kapitza, had been Rutherford's esteemed associate since 1921 (see chapter 4). When in 1926 Kapitza returned to the Soviet Union for his first visit since leaving, Semenov introduced Khariton to Kapitza as a bright rising star of Soviet science. He asked Kapitza to help the young man to receive a fellowship at the Cavendish.

Khariton prepared for his studies in Cambridge with considerable anxiety, but he would prove equal to the challenge. He traveled to England by train in 1926, and on his way, he visited his mother in Berlin for a few days. He stopped again in Berlin in 1928, on his way back from England to Leningrad. On this second stopover, he was especially appalled by the Nazi propaganda in the media, which he found frightening already at this early stage. He returned to the Soviet Union with the impression that real danger was brewing in Germany.

In Cambridge, Kapitza helped Khariton get established and join Trinity College. Semenov did not let Khariton disappear from his radar during his stay there. He wanted to build connections with the chemists of Cambridge and Oxford using his protégé's stay in England as a bridge. Khariton worked primarily with James Chadwick investigating the sensitivity of the eye with respect to weak light impulses and alpha-radiation. In 1928 he earned his PhD degree from Cambridge University.

Khariton enjoyed the excellent facilities of the Cavendish Laboratory, including its splendid library. At the beginning of his stay, he noticed an intriguing paper in the German physics journal, the *Zeitshrift für Physik*, where he and Valta had published their report. The article was written by a German authority of chemical kinetics, Max Bodenstein, who declared Khariton and Valta's observations impossible and ascribed them to some unknown errors in their experiment. Khariton's initial fears soon subsided because he knew that they had performed their experiments with utmost care and that their observations were reproducible. He wrote a letter to Semenov, who took up the challenge. Semenov's painstaking studies led to the discovery of the branched chemical reactions (see chapter 8).

Branched chain reactions in time proved to be a widespread concept that would be used for the interpretation of many other chemical reactions, such as polymerization, cracking hydrocarbons, and combustion. Soon enough, there was yet another branched chain reaction discovered that had even higher significance—pertaining to the fate of humankind—the nuclear chain reactions, which are also branched chain reactions. Khariton recognized early the connection between branched chemical chain reactions and nuclear chain reactions. He became involved with the latter in 1939, right after the discovery of nuclear fission.

Upon Khariton's return to Leningrad from England, he continued working under Semenov. Khariton focused his attention on explosions and explosive materials and was building a research unit for these studies. In 1932, a whole series of discoveries happened in physics; among them, the discovery of the neutron (by Khariton's Cavendish mentor, James Chadwick) and the discovery of heavy water. Also, nuclear reactions were produced by means of artificially accelerated particles.

Although the Soviet scientists were not among the initiators of the new nuclear physics, they followed it closely. In December 1932, a group of nuclear physics was created in the Institute of Physical Technology, which soon expanded into a division. Igor Kurchatov, also one of Abram Ioffe's disciples, was in charge of it. Elsewhere in the Soviet Union, notably in Kharkov, in the Ukrainian Institute of Physical Technology, there was increased interest in nuclear physics. In the 1930s, nuclear physics did not yet hold the special position it would in the postwar period, when it would be exempted from the targets of Stalin's ideological crusades against the sciences.

In the Soviet Union, the first nationwide conference on nuclear physics was organized in September 1933 in Leningrad. At that time interactions with foreign scientists were still possible in the Soviet Union. The meeting included stellar foreign scientists, such as Victor Weisskopf, Paul Dirac, Frederic Joliot-Curie, and others. Note that in the United States, it would take a few more years for two immigrant physicists, George Gamow and Edward Teller, in Washington, DC, to organize similar meetings.

In 1933, a British friend of the Soviet Union, the engineer George Eltenton, arrived in Leningrad and was employed by Semenov's Institute of Chemical Physics, which had been established recently. Khariton's wife, Maria, worked for Eltenton as a laboratory assistant. Eltenton learned Russian fast and stayed at the institute until 1937, when his friends and acquaintances started disappearing and he returned to England. He continued working in his profession, but his friendly feelings toward the Soviet Union persisted. He surfaced in connection with the Manhattan Project as the one who tried to get Robert Oppenheimer to cooperate with the Soviet Union by supplying them with information about the American nuclear project. Khariton wrote in 1978–1979 that he had read recently in the *Bulletin of Atomic Scientists* about Eltenton's attempts.[3] The story must have been interesting for Khariton since he was at the receiving end of the intelligence about the American nuclear project at the time of the development of the first Soviet atomic bombs.

Khariton collected a superb group of researchers in Semenov's institute for the study of explosives; the most brilliant among them was the prodigy Yakov Zeldovich (see chapter 2). When in 1939 they learned about the discovery of nuclear fission, Khariton and Zeldovich immediately jumped into this area of research. They produced three important papers, of which two were published, one in 1940 and the other in 1941, in the Russian journal *Uspekhi Fizicheskikh Nauk* (Advances in Physics). World War II had already started, but initially the Soviet Union's involvement was limited. By the time the third paper should have appeared, Germany had attacked the Soviet Union, and for a couple of years nuclear research was put on hold. Only in 1983 would the third Zeldovich-Khariton paper be published (they always listed themselves in this order, in accordance with the Cyrillic alphabet). They investigated the fission of uranium nuclei; established the regularities in the process; pointed out the necessity for isotope enrichment, that is, to increase the relative abundance of the light, fissionable isotope of uranium; discussed the possibility of the production of slow neutrons by means of heavy water; and estimated critical mass.

In nuclear physics, Khariton and Zeldovich could only pick up in fundamental research what was left for them by the true pioneers. The similar roles of Robert Oppenheimer and Edward Teller come to mind. Where Khariton and Zeldovich—like Oppenheimer and Teller on the American side—could be at the top was in building weapons. Here, priority for individuals did not matter because the work would be classified. Even if Khariton and Zeldovich were second to the Americans, the value of their efforts would be equivalent to that of the Americans, provided that they acted quickly. Soviet nuclear research resumed during the war, and the man in charge of it, Igor Kurchatov, invited Khariton and Zeldovich to join. Intelligence reports about the Manhattan Project were a decisive factor in Stalin's decision to go in this direction and to proceed rapidly. This was a very different situation for Khariton and Zeldovich from the one before the war, when they had labored on the nuclear chain reactions only in their free time.

Khariton gave a talk at one of Kapitza's seminars on March 8, 1944, in the Institute of Physical Problems in front of an audience consisting of outstanding physicists. He spoke about some of the peculiarities of explosions. Khariton reported that, according to his experiments, if the time of flight of the pieces scattering from the detonating charge was shorter than the time needed for the completion of the chemical reaction, the explosion would die. Kapitza asked him why he was talking about this. Khariton explained: Imagine war preparations in the Lilliputian country. If they wanted to use grenades of Lilliputian size filled with trinitrotoluene, grenades larger than ten millimeters in diameter would not explode. If, instead, they used a more efficient explosive, for which a diameter less than one millimeter would suffice, the grenades would work beautifully. Eventually, the specialists would start calling these conditions of explosive geometry "the Khariton criteria."[4]

Khariton and Kapitza had become good friends during their interactions in Cambridge. This friendship was different from the one between Kapitza and Semenov that had formed in their early youth. In the Kapitza-Semenov interactions two very different personalities bonded. Khariton was more like Kapitza, delicate, although he eventually was placed in charge of many thousands people at the "Soviet Los Alamos," Arzamas-16. According to observers, both Kapitza and Khariton had to mask their sensitivity because they lived and operated in a world that demanded harshness, especially from people who were in charge of projects.[5]

In April 1945, as the Soviet war machine was taking over the eastern part of Germany, a group of Soviet physicists was dispatched there to find out about the German atomic bomb project. The group included such renowned physicists as Khariton, I. K. Kikoin, L. A. Artsimovich, and others. They held the temporary rank of colonels of the Interior Ministry and were under General A. P. Zavenyagin of the Interior Ministry (Narodnii Komissariat Vnutrennikh Del [NKVD]). The mission was smaller than but similar to the American "Alsos."[6] It could promise only limited success, especially in locating famous German scientists, because most of them had moved to areas that were to be occupied by Western Allies. They preferred to be captured by Americans and the British rather than the Soviets.

Yulii Khariton, Ludmilla Semenova (Nikolai Semenov's daughter), Evgenii Lifshits, and Lev Landau.

Source: Courtesy of Alexey Semenov, Moscow.

Yulii Khariton and Igor Kurchatov.
Source: Courtesy of Alexey Semenov, Moscow.

Khariton was not only a superb scientist; he spoke fluent German as well. He and his colleagues flew to Berlin and found that the German atomic bomb project had not progressed very far. Kikoin and Khariton decided to pursue another goal, to find possible reserves of uranium that the Germans might have accumulated for their project. At that time the Soviet Union did not have reserves of uranium. Kikoin and Khariton interviewed German physicists and discovered a building in Berlin, not far from Hitler's headquarters, where all the records were kept about the reserves of various materials. Getting hold of these reserves was a race between the Americans and the Soviets. The Soviets did not win, but they still collected about one hundred tons of uranium ore in the form of U_3O_8, which made it possible to begin plutonium production in a Soviet reactor one year earlier than it would have been otherwise.[7] This was a major boost to the Soviet project. Another one was the German scientific personnel taken back to the Soviet Union.

The first, and for a long time the only, nuclear weapons installation in the Soviet Union was developed in the Gorky Region—known before and again today as the Nizhnii Novgorod Region—near the Sarov Monastery. It was eventually called Arzamas-16 (today, the whole development that used to be Arzamas-16, is called Sarov). There is a town by the name of Arzamas, from which Arzamas-16 was situated at about a forty-five-mile distance. From the start, Khariton was in charge of the installation although he never had the title of director. His first title was chief constructor; later he gave this position to someone else and became the scientific leader. This was the position he found most fitting. Khariton had been so impressed by his experience at the Cavendish Laboratory that he tried to follow its working style at Arzamas-16. Of course, he never succeeded; the Soviet environment was so different from the British conditions, but it is telling that he tried.

The nuclear project did not only bring successes, but the manner in which the scientists referred to the outcome of tests showed that they related their work to their devotion to the Soviet Fatherland. If there were six successful tests and five did not work, they would say, Soviet Union, 6–Harry Truman, 5. Defeat was reported as, for example, Harry Truman, 2–Soviet Union, 1. The first Soviet atomic bomb explosion on August 29, 1949, broke the American monopoly on nuclear weapons. Subsequently, about a thousand people involved with the project were recognized with various awards and medals. The top were Kurchatov (who worked in Moscow) and Khariton. During the last months of the preparation for the test, a slogan spread: "Let's 'overkhariton' Oppenheimer."[8]

Initially, Khariton was appointed to his Arzamas-16 position, at Kurchatov's recommendation, by Lavrentii Beria, the Soviet leader charged with supervising the nuclear weapons program. Khariton knew that his family background carried a lot of uncertainty for his standing. On the other hand, Beria may have thought that Khariton's roots and family history would make him even better suited for the position. One cannot help recalling the suggestion in connection with the Manhattan Project that General Groves might have thought it easier to keep Robert Oppenheimer under control because Oppenheimer's leftist past made him more vulnerable before the security services.

The first Soviet atomic bombs were a copy of the American atomic bombs. The information came from very efficient intelligence in which one of the key persons was the refugee communist German physicist Klaus Fuchs. He immigrated to England from Germany, and then moved to the Manhattan Project as member of the British team. The design of the first Soviet bomb may have been the result of intelligence, but its production was stilla tremendous achievement. The Soviet Union was in ruins as a consequence of a devastating war when it embarked on this most ambitious project. Even if their atomic bomb was a copy, it required the highest level of technological production. The work began in 1943, when the war was still being waged on the territory of the Soviet Union. And there were other problems for the Soviet project, including self-inflicted ones. Stalin's anti-Semitism started in the postwar period and was coupled with ongoing anti-science measures that devastated cybernetics and biology, damaged chemistry, and nearly destroyed physics. The latter was saved at the last minute when Beria, and through him Stalin, understood that without modern physics and the physicists, there would be no Soviet atomic bomb.

At one point, the anti-Semitic campaign reached even the classified laboratory of the atomic bomb construction. Leading scientists, such as Veniamin Tsukerman, David Frank-Kamenetsky, and Lev Altshuler were accused of producing scientific results contradicting Marxist philosophy. Among their "crimes" was also that they held and disseminated views about music and biology that were in disagreement with the party line. Today such stories sound hilarious, but at that time they could have repercussions, such as slave labor and exile for many years. At one point, Khariton had to call Altshuler suggesting to him that he not to come to work for a while.[9] In the meantime, Khariton called the defending angel of the atomic project, Beria, who then took the necessary measures to ensure that Altshuler and the others would be left alone. Khariton greatly contributed to the normal atmosphere at Arzamas-16, and anti-Semitism would not become an important force there. Of course, being Jewish himself, Khariton would have found it difficult to do this alone, but other leading scientists, non-Jewish, took also a strong stand against any kind of discrimination and persecution. The most famous among them, Igor Tamm and Andrei Sakharov, were known to consider being free of anti-Semitism as a litmus test measuring a person's decency (see chapters 1 and 3).

Khariton's involvement was ubiquitous in Arzamas-16, including the research projects of scientists that fell outside the development of nuclear weaponry. But he never let his name appear as coauthor on publications by his colleagues. His attention to the minutest details in every facet of the work in developing and producing nuclear bombs was legendary. His approach might be called meticulous, exacting, captious, and scrupulous, but there was also a special expression coined to characterize it using his initials, Yu. and B.—"yubism." A conspicuous example how important minute details might be was the tragedy of the US. space shuttle Challenger in 1986. Richard Feynman was a member of the president's commission that investigated the disaster. He demonstrated in a dramatic presentation for a huge TV audience that a simple and inexpensive O-ring, used for insulation, was sensitive to temperature changes. The O-ring could fulfill its function due to its flexibility, but at cold temperatures it became rigid and thus could not perform its function of insulation. It was a minute detail; a typical example that closer attention to minute details should have uncovered a potential source for disaster.[10]

Khariton did not leave anything to blind trust; neither did he like generalities in discussions. He preferred facts and insisted on receiving them from those who gathered and produced them. His heads of divisions and subdivisions had to get used to his approach; the official service channel meant nothing to him, and he approached directly the engineers or others, even if they were of the lowest rank, to get the necessary factual information. He involved so many people at lower ranks that it would have been difficult for the people in charge to enact repercussions against their subordinates whom Khariton invited to participate in the discussions.

Except for the first periods of his career, Khariton was a scientific director rather than a researcher. He did not create new science, but encouraged others to do so and ensured the necessary conditions for their work. He often determined the tasks and problems to solve, but he seldom contributed to their solution. Of course, his relentless questioning, prodding, and constructive criticism of the scientists and engineers was part of the creative process.

Khariton was a most kind and considerate person whom friends and family could have found difficult to associate with the most rigorous and demanding leader of the secret nuclear installment. But the sensitive and considerate Khariton could become tough and ruthless once he was in the environment of Arzamas-16, where he could not have done his job without such qualities. He took pride in seeing his subordinates working day and night, literally. In this, he differed considerably from his Cavendish role model, Ernest Rutherford, who attributed great value to stopping actual work and renewing one's intellectual capabilities. Khariton would call his colleagues any time he wanted to discuss anything or had a question for them. He invited them to his office or to his home even on holidays. He may have been polite and apologetic, but nobody would have ever declined such an invitation.

However, having been a scientist, Khariton also knew that it was impossible to work without errors. If everything succeeds every time, it might be a sign of not taking any risks in innovations. Thus, demanding absolutely flawless performance of one's associates could backfire, prompting them always to take the safest route. He avoided being such a leader, and his was not a unique approach. For example, the first tests conducted by the second American weapons laboratory in Livermore, California, were unsuccessful. Ernest Lawrence had the magnanimity to see in these failures something positive. He told his associates that always succeeding in everything might speak to a lack of innovation and healthy risk taking.

Under the Soviet circumstances, however, Khariton's approach was not only reasonable, but, it was also daring, and showed that Khariton must have wielded considerable authority.

The defense projects were under tremendous pressure created not only by the enormity of the task but by outside interference. Orders came from above, and the scientists often did not have the freedom to determine the pace of their own activities. In the words of Zeldovich, "Sometimes the decision about next day's work had to be made overnight. They could follow only one single version that had to be realistic, simple, hopeful, and the most economical. Making mistakes was not permitted."[11] While Khariton tolerated mistakes, he expected and even demanded that his associates be well prepared for their tasks. One of his principles was that one has to know ten times more about a given problem than the minimum amount necessary to solve it. He was also willing to invest his time and efforts to enable him to follow his own maxim.

Khariton usually went around Arzamas-16 with a bodyguard, not that there was need for one, but bodyguards were assigned to him. On his travels, he was accompanied by two bodyguards. His office was on the second floor in the theoretical divisions building. When entering his locked office, he would personally examine the security stamping of the door prior to removing it. Outside Khariton's office was a reception room in which one of his bodyguards fulfilled the role of secretary when Khariton was in his office. The office was huge and long, with a large desk at one end and a small table in front of it with comfortable leather armchairs on both sides. In the middle of the office was a long table with many tall chairs around it. When one entered, a small, thin man stood up at the far end of the office and then came out up from behind his desk to greet the visitor with an outstretched hand. Some of his visitors noticed the conspicuously large slide rule on his desk always ready to be used.

There was a safe in Khariton's office. When it had to be moved, he let his "secretaries" handle its contents, except for a locked internal compartment. From the memoirs of one of his secretaries we know that the safe contained gold watches, uranium half-spheres, and many documents. Among them was a copy of the famous letter, from 1973, signed by forty academicians, including Khariton, condemning Andrei Sakharov's social activities. In 1996, the 92-year old Khariton, shortly before his death, finally let an assistant open the inside compartment of his safe and pulled out a crumpled, thin envelope from it. There were a few dollar bills in the envelope.[12]

The bodyguards were KGB officers. In contrast with some of his colleagues, Khariton did not mind having bodyguards around and found them useful. They helped the Kharitons with various household chores, made his travel arrangements, got the medicines for Khariton and his family, even if they were difficult to find, and so on. They were always with him during his travels. The arrangement of assigning bodyguards to top scientists of classified projects lasted until the fall of 1965 when the Soviet government ended such services.

For his safety, Khariton was not allowed to fly in the early period of the nuclear weapons project. Instead, he traveled by rail, and he had a comfortable, personalized carriage that was set up as a fully equipped office so he could work while riding the train. Between the town Arzamas and Moscow, in either direction, his train always departed in the evening to arrive in the morning, so he could use his time most efficiently. Often, a number of his associates traveled with him. An attendant was assigned to the carriage who prepared meals from the food Khariton and his associates brought with them. There was a forty-five-mile car ride to the Arzamas railway station from the institute or his home, which took an hour and a half. Usually two cars were dispatched for his trips in case one broke down. The train ride between Arzamas and Moscow was far from optimal. The rest of the train (apart from Khariton's personal carriage) was in neglected condition and the town of Arzamas could do nothing to change the situation. When the town leaders turned to Khariton for help, his intervention resulted in an improved service.

Khariton was a seasoned, even shrewd bureaucrat. One of his principles was to avoid at all costs getting rejection from the superiors above Arzamas-16—the Ministry of Medium Machine Building, which was in fact the ministry of nuclear matters. He never turned to the ministry with requests that had not been most carefully prepared. He thought that it would take only one rejection to open the way to further rejections, and he did not find that acceptable. His situation was delicate because if the ministry rejected a request, he was still in a position to turn directly to the supreme leadership of the country. He knew, however, that such a step would alienate the ministry, which already viewed him with suspicion because of his very special position, and he could not be turning to the leaders of the country with every one of the needs of Arzamas-16. On the other hand, years of experience taught the ministry to take Khariton seriously. Once he requested something they knew they had better pay full attention to what he wanted.

It was during the most successful period of Arzamas-16 that Khariton decided to join the Communist Party. It happened shortly following the 20th Congress of the Party in February 1956, when Nikita Khrushchev unmasked many of Stalin's crimes in a secret speech. This may have prompted Khariton to become a party member. His involvement in "political" activities was not limited to party membership. For decades, he was "elected" to be member of the Supreme Soviet—the Soviet parliament. It was always to represent a certain district, and Khariton took the interests of his constituency seriously. He did not seem to spare time or effort going after their various problems, and in this, his "secretaries" helped him a great deal. He seemed to have an affinity for this public service. It augmented his classified world, which was interrupted only by annual vacationing with friends and family.

When Khariton took his vacations, he knew that sometimes—probably out of ignorance—somebody would inquire about his work. Khariton would tell the would-be intruder into his other life that he never discussed work while on vacation. He took secrecy very seriously, and the whole regime in which he operated was based on secrecy. There is an amusing human story in this connection. The drivers of automobiles carrying the leading scientists involved in the classified nuclear weapons program were not only forbidden to discuss anything they had overheard of their passengers' conversations; they were even forbidden to repeat the individual words they may have overheard. It was noticed that after a while the drivers stopped swearing and using "four-letter" words.[13]

The first Soviet nuclear device was tested successfully on August 29, 1949, near Semipalatinsk, in Eastern Kazakhstan. It was a copy of the American plutonium bomb, and it meant a great victory for Khariton and Arzamas-16. It was also a great victory for the Soviet Union because it broke the American monopoly of nuclear weaponry, and Stalin now possessed what he badly wanted. Accordingly, the Soviet State expressed its gratitude to those who helped create this bomb. Large amounts of money were given out as bonus, but money did not mean much in Soviet society as there was not much to buy. So the top participants, including Khariton, in addition to the title of "Hero of Socialist Labor," received an automobile and a week-end house (dacha). Their children were entitled to receive education in any institution of higher education in the country. The awardees, their wives, and children received the right to free, unlimited travel within the country, by trains, ship, and airplanes for as long as they lived (this privilege was later withdrawn by Khrushchev). Numerous other participants in the project received progressively lesser benefits. According to some sources—whether it is true or not, we don't know—the order of awardees was determined by a simple scheme devised by Beria. Those who would have been shot had the test failed, became Heroes of Socialist Labor; those who would have been sentenced to the longest prison terms received the Order of Lenin, and so on.

The Soviet officialdom for decades denied that intelligence played any role in creating the first Soviet atomic bomb. After the political changes, however, Khariton admitted that its success was the result of espionage. It seemed that the Russians were now assuring the world that they were not holding back anything about the atomic bomb. Perhaps this was to make their story about the Soviet hydrogen bomb more credible—they have always maintained that it was created exclusively by Soviet efforts.

The first Soviet hydrogen bomb test of August 12, 1953, utilized lithium(6) deuteride as solid fuel for the thermonuclear reaction. Its power was 0.4 megaton TNT-equivalent, and it was not a bona fide hydrogen bomb; rather, it was a boosted atomic bomb containing a thermonuclear reaction component. The application of lithium(6) deuteride was the so-called second idea. The layered arrangement of thermonuclear fuel and uranium was the *sloika,* or so-called first idea. By this time the Americans had already introduced the Teller-Ulam approach: radiation implosion. It was utilized in the "Mike" test on November 1, 1952, the test of a thermonuclear device, huge and heavy, which produced a 10.4 megaton TNT-equivalent explosion.

Radiation implosion would eventually become the so-called third idea in the Soviet program. There were further labels of the various solutions in the Arzamas-16 project. Thus, the utilization of the layered arrangement of fuel was also called "sakharization" a derivative of Sakharov's name (sakhar means sugar), referring as if to the caramelization of the fusion fuel. There was yet another test using the *sloika* arrangement, on November 6, 1955. Very soon after this test, came the test on November 22, 1955, in which the compression of the thermonuclear fuel was achieved using the third idea, that is, by radiation implosion. The numbering of the ideas rather than revealing what they really were was a way of describing them from the time when all this was still classified information.

The desire to prove that the Soviets were on a par with the Americans—and at times might have even performed better or earlier—seems to have been present from the beginning to the end in Khariton's activities. A. K. Chernyshev was twenty-three years old, a fresh physicist graduate, when he joined Khariton in 1969; he stayed with him for twenty years. He noted that in the 1980s they were still looking to American projects for guidance. They worked hard to emulate the American approach in Project Plowshare, which was about peaceful applications of atomic and thermonuclear explosions, and did not prove a success in the United States. When there were discussions of various issues in this connection, and the participants had run out of arguments, Khariton would ask, "How about the Americans? Do they have it? What can we know about this?"[14] Chernyshev thinks that even the great successes of the Sputnik and Gagarin's first manned flight did not suffice to create true Russian self-confidence.

In 1983, on the occasion of his recognition by the Soviet Academy of Sciences with its highest distinction, the Lomonosov Gold Medal, Khariton gave a prescient address. He talked about the future of energy consumption by humankind. He mentioned the sharp increase of carbon dioxide in the atmosphere and about global warming. He also warned about acid rain and about radioactive contamination. His conclusion was that there was a real danger of ecological catastrophe, and that the solution was safe atomic energy. Furthermore, he called for continued efforts for developing controlled thermonuclear synthesis. In conclusion, he brought up a third example of branched chain reactions, that of the exploding growth of the Earth's population. This showed his interest in the most diverse problems of society, about which, during much of his life, he could not voice his opinion because he was living in the shadow of his classified activities.

In his old age, Khariton gradually lost his vision and the ability to write. His grandson, Alexey Semenov, helped him with his correspondence. Khariton's last printed contributions were prepared with the help of his associates, especially V. N. Mokhov. Mokhov was compiling a volume of reviews of the scientific contributions of Arzamas-16 and thought it would be nice if Khariton would introduce it. Khariton dictated his thoughts but could only make changes when the text was read to him; so the work progressed slowly. It was not a long piece, but it took about one week of hard work to produce. Obviously, Khariton's meticulous attention to detail, to every word and phrase persisted. The product was titled "Appeal to the Readers."[15]

In it, Khariton stated that he did not regret having worked on nuclear weapons, in part because it was interesting physics (as if echoing, for example, what Enrico Fermi had expressed). It was also a decisive contribution to maintaining peace because they made the world more stable. He warned that even without further development of nuclear weapons, the existing arsenal required maintenance and the involvement of highly qualified personnel. It was difficult to create such a collective, but it could be destroyed easily. It was important to stop testing in order to terminate the arms race. The associates of Arzamas-16 had started studying the related questions long before they became popular. At the same time, it is necessary to ensure the safety and security of existing nuclear weapons. Neglecting these questions would be hazardous. As long as nuclear weapons exist anywhere in the world, they must be serviced by highly qualified scientists and engineers. Keeping the necessary collective together would be impossible without cultivating research projects in fundamental science. Even during the most intensive period of weapons development, the associates of Arzamas-16 conducted fundamental research, thereby turning their knowledge and experience to peaceful use. They can be proud of what they achieved in advancing fundamental science, in addition to the means of defense they created.

The last test explosion of the Soviet Union was conducted on October 24, 1990. In his late eighties when asked whether or not he thought about stepping down from his position, Khariton insisted that his involvement was needed to maintain the position of the Arzamas-16 institute. Nuclear weaponry was no longer considered as important as before, and the institute was making efforts to extend its activities to maintain its integrity. Khariton deeply worried about the future of Arzamas-16. In a speech in 1993 about the fundamental research at Arzamas-16, he mentioned that twenty-five thousand people worked at the institute and a whole town had grown up around it. He called for new challenges and new tasks under the changing conditions.

There have been attempts to draw parallels between Khariton and Oppenheimer and between Khariton and Teller. The Oppenheimer parallel is especially attractive, even starting with their first names Yulii (Khariton) and Julius (Robert Oppenheimer, though Oppenheimer did not use Julius), their same year of birth (1904), and other similarities.

However, Oppenheimer was the director of Los Alamos only for a short period. Regarding the length of their service as leaders of weapons laboratories, Teller is closer to Khariton. An important difference in the comparison with either American is that Khariton did not conduct any activities outside his immediate realm (his activities as a member of the Supreme Soviet do not count for anything substantial).

The very quiet Khariton, though, joined a few protests over the years. In 1952, he was in the group of physicists who turned to Beria protesting the attacks against the theory of relativity and quantum mechanics. In 1955, Khariton joined his colleagues by signing a letter to Khrushchev protesting the antiscience activities of Trofim Lysenko. In 1966, Khariton was in a small group of scientists (the others were N. N. Semenov and A. P. Aleksandrov) who asked Leonid Brezhnev to prevent attempts to exonerate Stalin. Alas, there was then the letter in 1973 condemning Andrei Sakharov, which Khariton signed and then regretted to the end of his life. It may be argued that the signers of the former three letters were not threatened by repercussions, whereas Khariton must have been afraid that the refusal to sign the letter against Sakharov might have made the authorities remove him from his position at Arzamas-16.

Considering Khariton's roots and family background, it was an ironic quirk that he occupied the position he did and occupied it for such a long time. A person with his background was most unlikely to have such a position in the Soviet Union. People in much lesser positions could not keep their jobs, sometimes not even their lives, under Stalin. It is then a sad irony that after the collapse of the Soviet Union, Khariton's roots seem to have hindered him from receiving the honors others who had been in similar positions had already been given. In Russia, there are many institutes that carry the names of their founders, such as the Ioffe Institute, Semenov Institute, Kapitza Institute, Kurchatov Institute, Zababakhin Institute (Zababakhin was Khariton's pupil), and so on, but Arzamas-16 has not been yet named after anybody, and there is no Khariton Institute. Khariton has been accused of causing harm to the Sarov Monastery. The accusation may be linked to the reluctance to name Arzamas-16 after him. This reasoning seems doubtful, however, when it is suggested that had he converted to Christianity, the naming would have had no opposition. Thus, it is not Khariton's deeds; rather, anti-Semitism has been the cause of unwillingness to name an institute after him.

In 1977, Khariton's wife died; he never married again. He was then already seventy-three years old; he could have returned to Moscow and lived there, enjoying his extended family. He could have found interesting occupations, could have even received a prestigious assignment. However, he opted to stay at Arzamas-16 and continue as its supreme leader. He was helped by subordinates and family members who came and stayed with him for extended periods. Yet he must have felt lonely. His wife died in Arzamas, but due to her husband's position was buried in Moscow at the Novodevichy Cemetery, with the expectation that one day Khariton would join her in the Novodevichy grave.

Alexey Semenov, a professor of chemical biology at Moscow State University is the grandson of two of the most famous Soviet scientists: Tatyana Khariton, the Kharitons' only child, was his mother and the Semenovs' son, Yurii Semenov was his father. Both his parents are buried in the Khariton grave at the Novodevichy. That also became the burial place of Semenov's second wife, the mother of the Semenovs' children, Alexey's paternal grandmother.

Yulii Khariton, widowed, in Arzamas in front of the house where the Kharitons' apartment was.

Source: Courtesy of Alexey Semenov, Moscow.

Khariton loved to travel and during his busy doctoral studies at the Cavendish he found time to travel to southern France. Then for six long decades it did not even occur to him to visit any place in the world outside the Soviet Union; there was, though, an exception in the early 1960s when he traveled to Prague on business. By the time the borders opened for him, he was close to ninety years old. In 1991, for the first time since 1928, he embarked on a trip to the West. He had a severe case of glaucoma, and on his trip to the United States he was accompanied by his grandson and by his doctor. His right eye had been operated on by Soviet doctors—unsuccessfully. The American doctors did not recommend surgery for his left eye. Khariton became completely blind by 1994, when he was ninety years old.

In 1993, Khariton and one of his associates, Yurii Smirnov, published an article in the American *Bulletin of the Atomic Scientists* about the Soviet nuclear program.[16] In it there is this passage: "Of course there was little joy in watching the columns of prisoners who built the installations initially. But all that receded into the background, and people had little regard for the difficulties of everyday life—they were trying to achieve success in the best and quickest way."[17,**] This passage provides food for a lot of thought. There was nothing

** This is a laconic reference to the fact that the laboratories of Arzamas-16 were built by slave labor, by prisoners of the Gulag who existed under inhuman conditions.

more about this in the long article, and one wonders whether or not Khariton might have had more emotions about those prisoners who built the installations than the passage revealed. Had his father not been killed in 1941, he might have been among those prisoners, and we do not even know whether or not Khariton knew that his father had perished. In the passage quoted above there is "the end justifies the means" approach. For fairness, however, I must add that according to Khariton's grandson, this joint paper was written when Khariton could no longer see, and this incriminating passage may have slipped in without his knowing about it. When Alexey questioned him about it, Khariton wanted it removed, but it was too late, the paper had already been printed.[18]

The passage quoted earlier was not the only one that a reader of that article might find stunning. It states summarily that "today, many people realize that it was Soviet physicists who first developed thermonuclear weapons."[19] This sentence introduces the section entitled "The First Hydrogen Bomb." The section describes a *version* of the history of the development of the hydrogen bomb in which emphasis is given to Soviet preeminence. When scientists claim authorship of the hydrogen bomb and give the impression of exaggerating their roles at the expense of others, John von Neumann's words come to mind: "Some people confess guilt to claim credit for the sin."[20] Khariton and Smirnov recognize that the first Soviet atomic bomb was a faithful copy of the American plutonium bomb (dropped over Nagasaki). In contrast, they go out of their way to ascertain that intelligence played no role or that its role was negligible in developing the Soviet hydrogen bomb. Our purpose here is not to analyze the history of nuclear weapons, but the title of Khariton and Smirnov's article "The Khariton Version" is an inevitable reminder of another "version." Werner Heisenberg and Carl Friedrich von Weizsäcker concocted a "version" of history about the failed German attempts to build an atomic bomb. It will be difficult to offset decades of secret operations and deception, and understandable suspicion and doubt linger about statements from Russian sources about espionage and priorities.

Edward Teller had great respect for Khariton, and proposed him for the prestigious American distinction, the Fermi Award. This award was originally established by the US Atomic Energy Commission as its highest recognition for achievements in nuclear science and technology. The Fermi Award is presented to its recipient by the president of the United States. In 1954, Enrico Fermi was its first recipient, followed by John von Neumann (1956), Ernest Lawrence (1957), Eugene Wigner (1958), Glenn Seaborg (1959), Hans Bethe (1961), Edward Teller (1962), and Robert Oppenheimer (1963), to mention just the first few awardees. Although it is an American distinction, there have been a few international recipients in its history. For example, in 1966, Otto Hahn, Lise Meitner, and Fritz Strassmann, the discoverers of nuclear fission, received it.

The Khariton grave at the Novodevichy Cemetery.

Buried here are Yulii Borisovich Khariton; Yurii N. Semenov, Khariton's son-in-law; Natalia N. Semenova, Yurii's mother (Nikolai Semenov's second wife); Maria Nikolaevna Khariton, Khariton's wife; and Tatyana Yulevna Khariton, Khariton's daughter.

Source: Photograph by and courtesy of the author.

Teller wrote a long letter of nomination dated January 17, 1995, in which he carefully enumerated Khariton's merits.[21] It was a rather bizarre proposition. In my Teller biography I wrote: "Had the cold war and its arms race been a sporting event, this [the Fermi Award to Khariton] might have been an expression of fair sportsmanship, but the cold war could have hardly been considered just another sporting event."[22] Teller's proposal was declined, and it was never repeated because Khariton died in 1996, and the Fermi Award cannot be given posthumously. A few years later, Teller stated something about Khariton—he meant it as praise, but for most other people it would have read as condemnation; either way, it lacked foundation. He wrote in a letter of February 9, 1999, to Siegfried S. Hecker, a former director of the Los Alamos National Laboratory: "As I hear the story of Russian workers on nuclear weapons, I notice a remarkable difference between us and them. In our case, there was strong controversy whether or not to proceed on nuclear weapons. In the Soviet case, there was *no controversy* and this may have been in part due to reasonable *administrative work* done by people like Khariton (emphasis added)."[23] Khariton was indeed part of the regime, but the "administrative work" that was meant to avoid any "controversy" was not done by Khariton but by Stalin, Beria, and their men.

One wonders how much and how often Khariton had to bow to expediency as he did in 1973 when he signed the letter condemning Sakharov. He had an introvert personality; for all his exceptional decorations, he seems to have been in his own private internal exile within the Soviet system. He must have had tremendous nostalgia for the brief period of his time in Cambridge. In March 1964, there was a Labor Party science delegation in Moscow, and the Soviet Academy of Sciences held a reception for the delegation. One of the British delegates, Tam Dalyell, described a strange encounter there: "As we dispersed, an ascetic, gaunt, dapper man of some 60 years with piercing yet kindly eyes, who had not opened his mouth, sidled up to me and said very quietly: 'I see you were a student at King's College, Cambridge—how is Edward Shire?' E. S. Shire was the physics tutor at King's and a distinguished member of the Cavendish. I told my questioner about Shire and felt that he had a genuine affection for his friends in Cambridge from 40 years ago. But he avoided my best efforts to find out who he was."[24] Eventually, Dalyell learned that he had met Yulii Khariton, a conspicuously rare event for a Westerner.

Khariton's privileges, including his private railway carriage, resemble a golden cage; he could not have felt himself truly free, ever. It is also telling that in the innermost compartment of his safe he of all people, for decades, kept a few dollar bills—for what purpose? In 1990, Khariton, who was always careful with his words, greeted the first American visitors at Arzamas-16 by saying, "I was waiting for this day for forty years..."[25] For forty years Khariton lived under a variety of tension; the burden of his roots, the enormity of responsibility of his position, and the possibility of confrontation utilizing the weapons he built. Forty years is a long time to wait, and waiting so that he could finally share this feeling of relief with someone, might have been almost unbearable.

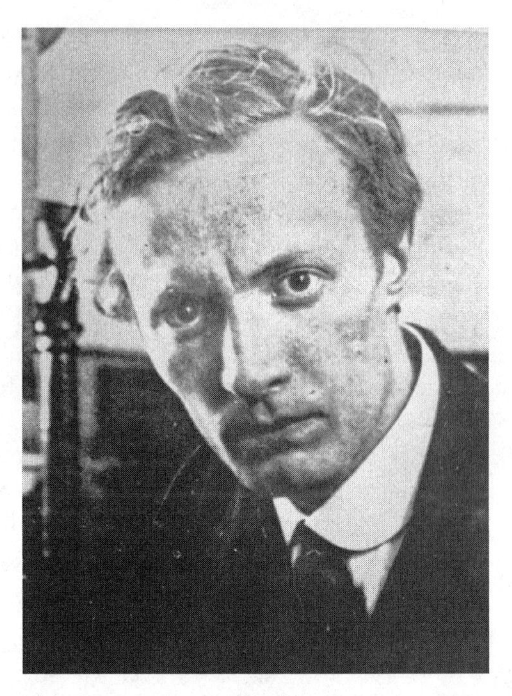

Young Boris Belousov

Anatol Zhabotinsky lecturing in Pushchino

Source: Both images courtesy of Simon Shnol, Pushchino, Russia.

10

Boris Belousov and Anatol Zhabotinsky

"IMPOSSIBLE REACTION"

Boris P. Belousov (1893–1970) and Anatol M. Zhabotinsky (1938–2008) gave their names to the Belousov-Zabotinsky oscillating reactions, which the Nobel laureate Ilya Prigogine considered to be one of the most important discoveries in the twentieth century. The military and medicinal chemist Belousov was the original discoverer, and the biophysicist Zhabotinsky worked out its theory and made the reactions known. The two lived in Moscow, a few kilometers apart, and talked over the phone but never met in person.

In 1980, they shared the prestigious Lenin Prize with three other researchers; for Belousov, it was a posthumous award. Zhabotinsky moved to the United States in 1991 and spent the rest of his life in Waltham, Massachusetts, working in an untenured position at Brandeis University. After the Lenin Prize, he never won any significant award, but he is considered to be the father of a whole area of modern science, called *nonlinear chemical dynamics*.

Oscillating reactions* belong to the larger group of phenomena described as "nonlinear chemical dynamics." In oscillating reactions, there are periodic changes in the concentrations of the reaction components. In those where concentration changes are manifested as color changes, the phenomenon is visually stunning. "Nonlinear chemical dynamics" may sound formidable, but deconstructing the expression shows that it is not so at all. It is about chemical reactions in which the changes of various properties may not have a simple proportional relationship (accordingly, their relationship is "nonlinear"), and the event is examined as it evolves in time (the word "dynamics" refers to this feature).

Boris Pavlovich Belousov was an obscure Soviet military and medicinal chemist and Anatol Markovich Zhabotinsky was a graduate of Moscow

* "Oscillating" reactions is how they are known, but strictly speaking, they are "oscillatory" reactions: it is not the reaction that oscillates; rather, the concentrations of the participants in the reaction oscillate—that is, change periodically.

State University who then worked in various research institutes in and near the Soviet capital. Belousov died in Moscow and Zhabotinsky in Waltham, Massachusetts. Except for the prestigious Soviet award, the Lenin Prize, which they shared with three others, no other recognition came their way. Belousov did not live to enjoy the Lenin Prize; by the time it was announced, he had been dead for ten years. Neither Belousov nor Zhabotinsky was elected to the Soviet Academy of Sciences. Their fame is due to the fame of the reactions named after them, the Belousov-Zhabotinsky reactions. Neither seems to have taken a prominent place in the annals of science. This is why I am referring to the late Ilya Prigogine, the Russian-born Belgian Nobel laureate, to place Belousov's discovery and Zhabotinsky's follow-up work in proper perspective.

Prigogine was born in Moscow just a few months before the communist revolution and died in Brussels as a Belgian viscount. His father was a factory owner and chemical engineer, and his mother, a former student of the Moscow Conservatory of Music. The family left the Soviet Union in 1921 and settled in Belgium. Prigogine had a distinguished career and became one of the most decorated scientists of all time. I met him for the first time in 1969, in Austin, Texas. There, at the University of Texas, a center was named after him and he spent a part of his time there annually. We were both visiting the chairman of the physics department at his lake house for a weekend. I was a research associate at the department. Prigogine was not yet a Nobel laureate but he was already *Prigogine*—his name already had an aura about it of one who was making great contributions to science.

He received his unshared Nobel distinction in chemistry in 1977 for his contributions to nonequilibrium thermodynamics.[1] Thermodynamics is about the movement of heat, and since all systems move toward equilibrium, the state of equilibrium used to be almost exclusively the focus of attention of researchers. Prigogine was among those pioneers who recognized the special importance of learning about nonequilibrium states. To illustrate their importance: living organisms are highly organized systems that are not in equilibrium; they are only moving toward equilibrium through irreversible changes.

When Zhabotinsky died, his colleague at Brandeis University Irving R. Epstein published his obituary in the prestigious British magazine *Nature*. He stated that Ilya Prigogine "regarded the BZ [Belousov-Zhabotinsky] reaction as the most important scientific discovery of the twentieth century, surpassing quantum theory and relativity."[2] This sounded like an exaggeration and there was no source given for this statement, but Prigogine seems to have represented this view consistently. Years before, in 1995, I recorded a conversation with Prigogine, and he being a great authority in the field, I asked him, "How important was the discovery of the Belousov-Zhabotinsky reactions?" His response was less extreme than the one referred to in the obituary, but the claim was still quite substantial:[3]

> I think it was one of the most important discoveries of the century. It was as important as the discovery of quarks or the introduction of black holes.

The significance of the Belousov-Zhabotinsky reactions is in demonstrating a completely new type of coherence. It shows that in non-equilibrium, coherence may extend over macroscopic distances in agreement with our theoretical results I mentioned. Again, at equilibrium coherence extends over molecular distances while in the Belousov-Zhabotinsky reactions this coherence extends over macroscopic distances, of the order of centimeters. This is a striking example of non-equilibrium structures.

Belousov's story goes back to czarist times, and we know about it from the Soviet-Russian biochemist Simon Shnol, who narrated it in a double capacity. Shnol was personally involved in the story of the Belousov-Zhabotinsky reaction. Lately, he has become a science historian who has taken it upon himself to preserve and disseminate information about—as he labeled them—the heroes, villains, and conformists of Russian science.[4]

Ilya Prigogine in 1998 in Austin, Texas.

Source: Photograph by and courtesy of Vladimir Mastryukov, Austin, Texas.

The Belousov-Zhabotinsky reactions signify a whole class of chemical events; they are oscillating reactions in that their properties change periodically, and when the changes are accompanied by color changes, they present visually profound phenomena. For years, the reviewers and editors at Soviet journals declined to accept Belousov's discovery.

Of course, what Belousov found did not happen without antecedents. As early as the end of the seventeenth century, the famous Robert Boyle observed flashes of luminescence when studying the oxidation of phosphorus. In the nineteenth century, Boyle's observations were confirmed. The great time gap shows how slowly things moved back then; but there were no incentives to enhance the interest in such studies. The oxidation of phosphorus was a heterogeneous process; it involved solid phosphorus and gaseous oxygen. It happens that this very reaction is discussed in two other chapters in this book (see chapters 8 and 9).

In the beginning of the twentieth century, further heterogeneous reactions were described.[5] Attempts also began to understand these oscillations and to describe their mechanism. However, they were still too complicated for scientists to grasp their essence. One of the main findings of the thermodynamic research of these processes was a negative one, that is, that it is impossible to have oscillations in the vicinity of the thermodynamic equilibrium state.

The first description of an oscillating reaction in the liquid phase—that is, in a homogeneous reaction—was described in the early 1920s. However, rather than welcoming it and launching further investigations, the chemistry community found it suspect. It was suggested that the oscillations were caused by some heterogeneous impurities. This remained the attitude of chemists toward such phenomena through the mid-1960s. The German physical chemist Karl F. Bonhoeffer tried to convince his chemistry colleagues about the possibility of oscillating reactions in solution, but in vain. A heterogeneous system is indeed easier to imagine yielding oscillating properties, whereas the causes of oscillations may remain hidden in a homogeneous system, such as a solution. In the extreme, mechanical oscillations, like a swinging pendulum, are the easiest to understand.

A periodic color change of a solution is not accompanied by any visible movement in space, and this makes it hard to accept even when one sees it with the naked eye. In addition, there were thermodynamic arguments for rejecting the idea of oscillations in solutions. There was a substantial crack in this conservative edifice when in the mid-1950s Prigogine and one of his associates published in an obscure journal two papers about oscillations far from thermodynamic equilibrium.[6]

Under normal circumstances, after the Soviet chemistry journals rejected Belousov's manuscripts about his discovery of an oscillating reaction in solution, he might have turned to other publications. For example, he could have sent his manuscript to physics periodicals, which he did not consider. Yet more straightforward might have been to send out his manuscript to international journals, but this was not a possibility for Belousov at that time in the Soviet Union. It was certainly not for his lack of ability to produce his manuscript in a foreign language. He spoke very good German and French. In any case, his possibilities were severely limited because he made his discovery during one of the darkest periods of Soviet life, the last years of Stalin's reign, in the late 1940s and early 1950s.

The sad fate of Belousov's discovery during his lifetime tragically befitted his reclusive persona. He was born in czarist Russia into a family of a bank clerk; the family had six sons. The fate of the family was to a great extent determined by the revolutionary activities of some of the sons, and part of the family, including Boris, found themselves in exile in Switzerland. There was at that time a community of Russian exiles and political immigrants waiting for the opportunity to return to a revolutionary Russia. Legend has it that in Zurich the young Belousov met the most famous Russian exile, Vladimir Lenin, and played chess with him.

Belousov received most of his education in Switzerland at the Swiss Federal Institute of Technology in Zurich (Eidgenossische Technische Hochschule, ETH, Zürich). He became a chemical engineer, alas without an official certificate. He returned to the Soviet Union, and for a long time the lack of such papers did not matter. Eventually, he was appointed to be in charge of a big laboratory. Later, due to his inability to prove his qualifications, his position was degraded without diminishing his responsibilities. He was engaged in industrial chemistry and, increasingly, in defense-related military research. Another legend says that when he was removed from his high position, but continued the same work, Iosif Stalin personally ordered that he receive remuneration corresponding to his previous position. There is realistic foundation to such a story in that Stalin was known to make personal decisions and to micromanage segments of Soviet life, down to the minutest details.

The areas of Belousov's scientific activities encompassed mostly gaseous chemistry and analytical chemistry; he also investigated poisonous materials. When in the mid-1930s he left his military job, he found employment in a secret medicinal institute where he dealt with substances that could be used for protection from radiation. Throughout his professional career, all his activities were classified; hence there were no publications on which his name could be found.

Belousov survived the purges of Stalin's terror in 1937–1938, but many of his friends and colleagues did not. When two decades later, first Shnol and then Zhabotinsky wanted to involve him in their work and meet with him, he consistently declined. He said that he had lost his friends and did not want to make new friends. Belousov asked them not to bother him again about personal meetings or about including him in joint publications, but he offered his assistance and indeed proved to be very helpful. He also revealed that the idea for his oscillating reaction came out from his efforts to build some cyclic reactions, in a way analogous to the Krebs cycle of biochemical reactions, but much simpler.[7] The Krebs cycle involves citric acid and Belousov's recipe also involved citric acid. It was a simple recipe and was modified over the years, but it essentially remained as he had first communicated it. For example, Belousov used flasks and test tubes, whereas later, it was more convenient to use a Petri dish for conducting the reaction.

Here is the list of ingredients in Belousov's recipe: a solution of $KBrO_3$, citric acid, and sulfuric acid; and the solution contained cerium ions. The reaction showed an astonishing effect. First the liquid was yellow, then colorless, then yellow again, continuously alternating. When Belousov first observed the bands of changing color in another, similar reaction, he called his flask a "zebra."[8] This was Belousov's discovery. He duly described his observation in a manuscript and submitted it to serious chemistry journals, such as *Zhurnal obshchei khimii* (Journal of General Chemistry) and *Kinetika i kataliz* (Kinetics and Catalysis). He made two series of attempts, the first in 1951 and the second in 1955. On both occasions, he received unpleasant responses declining publication and expressing disbelief. According to the teachings of thermodynamics, such a system in equilibrium could not have shown the effect he observed and described. It could have been considered to be a certain kind of *perpetuum mobile* that—we all know—cannot exist. Based on equilibrium thermodynamics, indeed, such a system was impossible. What he did not know, and what the editors and reviewers could not fathom at that time, was that the system displaying this peculiar behavior was *far from equilibrium*.

The journals should have checked whether or not Belousov's observations were valid and then worried about the explanation, but nobody bothered. Later, Zhabotinsky joked to me that it was a pity that the chemists knew their thermodynamics well and so were unable to accept the report about the oscillating reaction. Later, Zhabotinsky found no difficulty in getting his papers about oscillating reactions published in biology journals, because the biologists were not versed in thermodynamics. The truth is that the chemists who had to decide about Belousov's manuscript were limited in their knowledge of thermodynamics. What they had learned about thermodynamics at that time did not include the characterization of systems far from equilibrium. Nonetheless, they should have been willing at least to *look* at Belousov's reaction; of course, this is easy to say in hindsight.

Belousov did not attempt to prove his claim, even though he could have just taken his dishes and chemicals to the editorial office and demonstrated his experiment. The editors and reviewers might have not believed their own eyes but at least should have felt uncomfortable and puzzled. Instead, Belousov just left things alone. When Shnol later urged him to get at least something about his reaction printed, he wrote up a brief manuscript and had it included in a volume of conference abstracts compiled in the Institute of Medicinal Radiation where he worked at that time. It became his only printed communication ever.[9]

Had Belousov been left alone, his name and discovery might have disappeared into oblivion. However, there were changes in Soviet science that impacted him and his reaction. After Stalin's death, during the second half of the 1950s, Lysenko and his unscientific terror of biology continued. Now, instead of Stalin, Lysenko found another great protector for his unscientific views in the new Soviet leader, Nikita Khrushchev, who badly wanted to see Soviet agricultural production enhanced and wanted it to happen quickly. He was willing to believe in the miracles Lysenko promised rather than make the necessary changes in the system. Leading Soviet physicists, such as Igor Tamm, and others, were getting increasingly tired of Lysenko's reign and damaging activities. They would have found it difficult to directly interfere in what was happening in Soviet biology, but they could introduce some initiatives in their own field concerning biology. They organized seminars on modern biology in physical research institutes, and, in 1958, they decided to create a chair of biophysics at the Faculty of Physics of Moscow State University.

The rector of the university at the time, the internationally renowned mathematician I. G. Petrovskii, supported the initiative and appointed Lev Blumenfeld to be the head of biophysics. Blumenfeld was a physicist interested in biochemistry. His father was killed in the 1937–1938 purges of Stalin's terror. Blumenfeld was fired from his jobs more than once, the last time as a "cosmopolite" in the anti-Semitic actions during Stalin's last years. Nikolai Semenov invited him to be in charge of a laboratory in his Institute of Chemical Physics. This was followed by the university appointment, where he made a lasting contribution by developing his department into a significant scientific center.

Blumenfeld invited Shnol to give a course in biochemistry, and Shnol now had the opportunity to involve students, including doctoral students, in conducting research in projects that had previously been neglected. Shnol received Belousov's recipe through Belousov's grand-nephew; then, together with his students, reproduced Belousov's reaction. In 1961, Tamm came for a visit, and he was much taken by the demonstration of Belousov's oscillating reaction, which two of Shnol's students had put together. Tamm opined that it would take a long time and a lot of effort to understand the phenomenon. This was where one of the first graduates of the new department, Anatol Zhabotinsky, excelled. But it did not happen right away, only following a detour in his career.

Anatol Markovich Zhabotinsky was born in 1938 in Moscow into a Jewish family of intellectuals. Both his parents were physicists, graduates of the Faculty of Physics, Moscow State University. His father, Mark Zhabotinsky, had studied under Mikhail Leontovich, another of the great Soviet physicists, who also participated in the movement to free Soviet science from Lysenko's charlatanism. Mark Zhabotinsky worked at the Physical Institute of the Academy of Sciences (FIAN). Anatol's mother, Anna Livanova, specialized in the history of physics and authored books about it.

Anatol's first love in his youth was biology, but he wanted to avoid getting into a science where Lysenko reigned. The new biophysics at Moscow State University was a godsend. It did not exist in 1955, when he started his university studies, but was already being created by 1958, when he had to choose his specialization. Several of his fellow students were the children of influential members of the Soviet Academy of Sciences or of Soviet politicians, including the son of Georgii Malenkov, who had been the prime minister between 1953 and 1955 and who, even after being demoted from that position, remained a member of the supreme leadership of the country—of the Politburo—until 1957. By then, the idea of biophysics had taken root at Moscow State University. Besides, Malenkov's son was not the only student working to further modernization. It was a rare phenomenon in Soviet society that demands from below—in this case from students—had any impact on what was happening at all.

When Zhabotinsky, still an undergraduate in 1958, joined the biophysics department, he became Shnol's student. Zhabotinsky was to be one of the first students to graduate from this newly organized department. Thus, his presence in Shnol's group was equally memorable for both. According to Shnol, young Zhabotinsky was a "typical product of the [Soviet] intellectual world."[10] In this world, "the children from very early age were taught to deliberate about their environment. They loved mathematics, and during family meals they were engaged in solving puzzles and paradoxes." Shnol saw great value in these children but found that "they could be quite unbearable. When they find themselves in normal society and experience how ignorant their peers can be, they tend to consider themselves geniuses. They think that their superior knowledge was of their own making, whereas it stemmed from their families and the favorable circumstances of their childhood." Shnol found that this description fit his new student, Anatol Zhabotinsky, well.[11]

At the Faculty of Physics of Moscow State University, Zhabotinsky had a great opportunity to acquire experience in scientific research. Another great attraction was the summer field trips, including one to the field laboratory of the famous biologist Nikolai Timofeev-Resovsky. There, the students could attend brilliant lectures on genetics and theoretical biology, still taboo in most of contemporary Soviet biology. Timofeev-Resovsky was a world-class scientist, knew many famous biologists, and was a colorful personality and terrific storyteller.

Zhabotinsky's choice for Diploma work (master's thesis) was unusual. The biophysics department wanted to build its own electron-paramagnetic-resonance (EPR) apparatus to study the electronic structure of substances. Blumenfeld intended to form a team for the task when Zhabotinsky volunteered to do the job alone. He was warned that it was an impossible task for a Diploma work, but he was too enthusiastic to consider the friendly advice. It did not work out the way he had hoped; in fact, it did not work out at all. Luckily, he was able to fulfill his Diploma work requirement, but only barely.

He had to give up his hope of staying on at the university on a postgraduate fellowship. Instead, he was "distributed" to a radiology institute. Soviet graduates did not have the freedom to choose their jobs. The term "distributed" was the official Soviet technical term for assigning jobs to graduates. Zhabotinsky started his career in the radiology department of a medical institute specializing in cancer treatment. He spent most of his time in the library because in the department he found hardly anybody from whom he could learn anything.[12]

From time to time he returned for consultation with his former professor Simon Shnol in the biophysics department. Eventually, Zhabotinsky decided that he would formulate his own research project. Thus, the backwater character of his first workplace had one advantage in that it indirectly encouraged him to go his independent way. By this time, Shnol and his students had become interested in Belousov's oscillating reaction, received Belousov's recipe, and reproduced it in their experiments. Under Shnol's influence, Zhabotinsky developed a research idea involving oscillatory phenomena in biochemical processes, such as photosynthesis. Shnol found Zhabotinsky's goal interesting but not practical in that the necessary materials and instrumentation were not readily available to acquire for the grandiose project he had in mind. At this point Shnol raised the possibility that Zhabotinsky could pursue his postgraduate studies about Belousov's reaction. Shnol had no doubt about the validity of Belousov's observation; he had observed it with his own eyes, but he also realized that the real challenge was to understand the mechanism of the reaction.

A simplified description should help gain an understanding about oscillating reactions.[13] For start, consider the reaction A + B → C + D in the presence of a catalyst X. Now, suppose that the mechanism could be reduced to two consecutive steps:

A + X → C + Y (reaction 1)
B + Y → D + X (reaction 2)

A + B → C + D (the sum result)

Here, during the reaction, the catalyst X transforms into Y, which is a different state of X, and X and Y display two distinctly different colors; the changes in their relative concentrations will appear as color changes. Initially, the reaction mixture displays the color of X. In reaction 1, in the presence of X, part of A transforms into C while X converts into Y. In reaction 2, in the presence of Y, part of B transforms into D, and Y coverts back into X. With the advancement of reaction 1, it stops at a certain point, because there is a finite amount of X, and when it disappears, the reaction cannot continue. Since Y is now present, the reaction mixture will display the color of Y; thus there is a color change. Reaction 2 now continues and as a consequence, X is being produced, which leads eventually to a color switch. But the presence of X makes the renewal of reaction 1 possible and the process continues as long as both A and B are present in sufficient amounts. The color changes appear as oscillations. They can be sustained if reactants A and B are being continuously fed into the system.

Oscillations can be illustrated with another example. Consider a system of rabbits (R), foxes (F), and grass (G), and consider the following events: The rabbits eat the grass and multiply (here let us assume that the grass grows continuously, so no matter how much the rabbits consume, no shortage of grass develops). The fast growth of the rabbit population, however, is adversely affected by the foxes because they eat the rabbits. As a consequence, the number of foxes grows—there are plenty of rabbits around so the foxes eat and multiply. As they eat the rabbits, the rabbit population diminishes, and at a certain point, the fox population stops growing, lacking sufficient food, and starts diminishing. The diminishing fox population favors the growth of the rabbit population, and so on.

To express this story like the reactions above, first, the growth of the rabbit population is described as $R + G \rightarrow 2R$. Then, the growth of the fox population at the expense of the rabbit population is described as $R + F \rightarrow 2F$. Before the total of R would become extinct, the F population starts diminishing for the lack of sufficient R, letting the R population grow again, and so on. The result will be a periodic change; starting with a certain rabbit population and fox population; a growth of the rabbit population, then a decline of the rabbit population, and the simultaneous growth of the fox population, followed by a decline of the fox population and growth of the rabbit population, and so on.[14]

Zhabotinsky did not hesitate to accept Shnol's suggestion to make Belousov's reaction his postgraduate project, but he did not want a conflict with the undergraduates who were already studying it. They were two female students; one was a fourth-year student, Anna Bukatina. Shnol was a popular student adviser; students were taken by his enthusiasm and the fact that he could suggest exciting and doable research projects even to undergraduates. Zhabotinsky thought that duplication of work on Belousov's reaction would be superfluous, but if the students dropped the project, he would take it up. For the two students it was not difficult to drop this project because there were other intriguing problems that Shnol could suggest to them. Bukatina's interest had already cooled anyway because she felt that the Belousov reaction had become "the subject of rather intense local interest and a high-profile topic, and she did not want to be involved with something like that."[15] She switched to a biological project, and her involvement with Belousov's reaction did not fully stop, it was in a different aspect: she became Zhabotinsky's wife. When their son, Mikhail, was born in 1964, his parents decided that he should have his mother's Russian surname rather than his father's Jewish surname. When the Bukatins moved to the United States, Mikhail became Michael.

As Zhabotinsky accepted Shnol's suggestion, he started working on the reaction right away. At the same time, his previous idleness in his work place was changed into exciting activities, because in the meantime he had also changed his job. It was about a couple of months into his new job when Shnol gave him Belousov's recipe on a piece of paper and he gave him also small amounts of the necessary reagents.

Zhabotinsky's second position was again in a cancer research institute, but it was very different from the previous one. His boss was Leon Shabad, a world-renowned expert in cancer hygiene who maintained a strict order in the work place. This did not prevent him from developing good personal relations with all the members of his department, including the low-ranking ones, which was foreign to the Soviet medical community. Zhabotinsky's task was organizing the analysis of cyclic hydrocarbons in car exhausts, and he was given independence to pursue this project. He purchased equipment, assembled the necessary instrumentation, and increased the sensitivity of the measurements by three orders of magnitude in comparison with the method that was used before. Once Zhabotinsky had organized the improved performance of his unit, he could return to his own research project.

Belousov's recipe was simple and old-fashioned; Zhabotinsky had to weigh the reactants, dissolve them in moderately diluted sulfuric acid, and add water to reach the final volume. Once he had done all this, he could observe the oscillations in the color of the solution. The mixing and dissolving the ingredients produced a lot of heat, and this accelerated the reaction. The period of oscillations was rather short. Zhabotinsky devised a setup to record the oscillations, and repeated the experiment many times by using different amounts of reactants. Thus, he could see how the oscillations depended on the concentration of the solution. This was the easy part of the work.

Then came the more difficult part, understanding the mechanism of the reaction. Zhabotinsky and Shnol consulted Belousov (on the phone), and he had some ideas. Still, Zhabotinsky felt that he lacked experience and turned to Lev Blumenfeld for advice. Blumenfeld had acquired good training in quantum chemistry in the early 1950s, when quantum chemistry had become anathema for official Soviet science. Although this was a different area of research, Blumenfeld still gave Zhabotinsky useful pointers. He even wrote down a possible scheme for the reaction; it eventually turned out to be wrong, but it was a start. During this time, Zhabotinsky, in addition to his job, was a doctoral student of the university under Shnol's supervision. Two years of uninterrupted research followed.

First, Zhabotinsky reproduced Belousov's results. Then, he modified Belousov's system and made it more convenient to study. Further, he succeeded in determining the mechanism of several steps in the oscillating reaction although he did not yet have a complete understanding of the mechanism of the entire process. Nonetheless, what he already had was deemed sufficient for his first paper. When Zhabotinsky was preparing his first manuscript for publication, he faced the problem of coauthorship. He wanted Shnol as coauthor, and he included Belousov's name in the subtitle of the paper. He also wanted to show his draft to Belousov. Shnol declined coauthorship; for him, it sufficed to see the work progressing. He volunteered to send the draft to Belousov using the same means as he had received the recipe from Belousov.

The manuscript came back from Belousov after two weeks with a nice note saying that Belousov was happy that someone was continuing his work, but declined to be a coauthor. In 1962, Zhabotinsky sent off the manuscript to the Soviet journal *Biofizika* (Biophysics).[16] The article duly appeared two years later. The long wait for publication was quite normal at that time in Soviet scientific literature.

After having written and submitted his first paper, Zhabotinsky continued his research as Shnol's doctoral student. His main contribution was in setting up the mathematical description of the various steps of the reactions. Mathematical modeling was his forte. His model helped him understand the mechanism of the missing steps of the oscillating reaction. Once he did, it became easy to devise new oscillating reactions that were similar to Belousov's original reaction. What varied were the ingredients and the catalysts.

At this point, Zhabotinsky felt that he had completed his goal and embarked on preparing his second manuscript. He wanted it to appear fast. This could be possible if a full member of the Science Academy were to present his manuscript to the special periodical of the Academy called *Doklady Akademii Nauk* (Proceedings of the Academy of Sciences). He approached Aleksandr Frumkin, the renowned physical chemist and director of the Institute of Electrochemistry in Moscow (today, the Frumkin Institute), who himself had published two papers on electrochemical oscillations. It was the beginning of 1964, and it took two months for Zhabotinsky to make an appointment with Frumkin. The academician was known to read everything before he would sign anything. The wait was worthwhile, because once Frumkin gave his approval, it took only three more months for the paper to appear.[17] However, this was a one-time opportunity only; Frumkin asked Zhabotinsky not to bring him any more manuscripts, pointing to a high stack of manuscripts on his desk, submitted by members of his own institute.

With his two papers and being the sole author for both—a highly unusual feat for a doctoral student, especially in experimental work—Zhabotinsky wrote his dissertation. It was on the topic of periodic chemical reactions in the liquid phase, and he defended it without any difficulty. Having thus earned his PhD-equivalent Candidate of Science degree, he was appointed as a junior research associate in the Institute of Biophysics. This institute had, in the meantime, moved from Moscow to Pushchino, a small town about sixty-five miles due south of Moscow. Zhabotinsky did not mind because he had a young family, and in Pushchino he had a much better chance of receiving an apartment than in Moscow; and this is what happened.

In Pushchino, Zhabotinsky's perhaps most fruitful research period followed. An active group formed around him, and they had plenty of results. They generated considerable interest, first within the Soviet Union, but soon internationally as well. In 1966, they organized the first ever Symposium on Oscillatory Processes in Biological and Chemical Systems.[18] At this symposium,

Zhabotinsky was glorified, and the name Belousov-Zhabotinsky started being used. Belousov was still alive but did not attend; one can only hope that he felt satisfaction when he learned about the brewing triumph of his discovery. One of the most active participants in the meeting was D. A. Frank-Kamenetsky, the renowned nuclear physicist who had been one of the leading scientists at the secret nuclear laboratory Arzamas-16.[19] Zhabotinsky was moving ahead with his academic career and in 1971, he defended his Doctor of Science dissertation.[20] The culmination of his academic career during this period was the publication of his Russian-language monograph about the reactions with oscillating concentrations.[21]

Soon, Zhabotinsky's interest turned to the application of oscillating reactions for modeling analogous biological processes. In some ways he was now reaching back to the original idea that his mentor, most realistically, did not find feasible for his doctoral studies. It is these biological relevancies that show the outstanding importance of oscillating reactions. For Zhabotinsky, the most attractive topic was the propagation of excitation in the heart. The normal operation of the heart is controlled by very long waves of excitation. He was familiar with the theory that attributed the most dangerous cardiac arrhythmia to the emergence of short spiral waves of excitation in the myocardium. But in talking with specialists, he found that very few of them believed in this theory. He and his associates—because he was now developing a strong group at the Biophysics Institute—started to study chemical waves in thin layers of solutions containing the oscillating reaction and found that they formed a wonderful target with spiral patterns, which were relevant to those in the myocardium. Their very first paper on chemical waves in two-dimensional media appeared in 1970 in *Nature*.[22]

Zhabotinsky experienced some difficulties in his personal life at this point that caused him to leave the Pushchino institute, and return to Moscow. Divorce and remarriage followed. Albina Krinskaya, the former wife of one of his closest colleagues, became his second wife. Zhabotinsky was looking for a place where he could continue his research, but it proved difficult. None of the research institutes of the Academy of Sciences in fields that would have suited his work had an opening for him. Finally, he was employed by the Institute of Biological Tests of Chemical Compounds, a new institution set up by the Ministry of Medical Industry. He was appointed head of the laboratory of mathematical modeling. The laboratory was staffed by only a small group, but it started growing as soon as Zhabotinsky joined it. Its coworkers were graduates of the biophysics departments of Moscow State University and the Moscow Institute of Physical Technology, both strong institutions. Soon, the group of about a dozen members was doing interesting work on the biochemical regulation in red blood cells.

At this time, Zhabotinsky's interest shifted again, and he wanted to apply his knowledge to the improvement of cancer chemotherapy. He enjoyed having the independence to choose the research projects for his group, and he could make them appear attractive to his associates. They discovered a resonance response of dividing cells to periodic administration of highly intensive anti-cancer drugs. The essence of Zhabotinsky's idea was based on the observation that chemotherapy has side effects, which can become lethal with increasing doses. The side effect manifests itself in the killing of healthy cells in addition to the cancerous ones. The solution may be to administer chemotherapy that would take into account the time scale of the cell division of healthy cells and thus avoid damaging them. Of course, there are many different kinds of the healthy cells, and this approach would first have to consider the most vulnerable cells. Once this was solved, the target for cell protection would be the next most vulnerable cells, and so on. This approach was promising for the development of more efficient chemotherapies for cancer because it would make it possible to increase doses without risking most of the healthy cells. They hoped that the same approach might be possible to apply in the struggle against AIDS as well. They had limited resources, but their expertise and dedication helped them a great deal in their efforts.[23]

This was a happy period for Zhabotinsky and this was also the only period in his life when outside recognition encouraged him in his activities. The idea of awarding a Lenin Prize to Zhabotinsky and three of his closest associates was born. It seemed increasingly realistic, and it was a big deal, because it was the highest scientific award in the country. A closer look, however, revealed a great deficiency in that Belousov's name was missing from the initial list of awardees. The Lenin Prize can be bestowed posthumously, and it would have been utterly unfair not to include Belousov, even though he had been by then dead for years. It was through Simon Shnol's organizational effort that, at the last moment, Belousov's name was added to the roster of awardees. The winners received the Lenin Prize in 1980.

International fame, the Lenin Prize, and his exceptional creativity in target-oriented research did not, however, suffice to protect Zhabotinsky's family from discrimination. Around 1980, upon graduation from high school, Michael Bukatin applied to the Faculty of Physics, Moscow State University, to be admitted as a student. This was the institution where his grandparents and his parents had studied, and Michael was a gifted and motivated student whose preparation included much from his family background that is not available in the school curriculum. Yet he was flunked in his physics entrance examination.[24] Michael made another attempt two years later but to no avail.[25] It was common knowledge that tacit instructions existed about limiting the number of Jewish students at institutions of higher education. This could not be given as reason for declining acceptance; so the university entrance committees administered the entrance examination in such a way as to make the applicant earn a failing grade. (I am sensitive to such injustice, having gone through a similar experience in my chemistry entrance examination due to "unfavorable" social origin, one generation before at Budapest University.[26])

The 1980s were not a happy period in Zhabotinsky's life. He experienced various bureaucratic difficulties in the management of the institute, which made his situation increasingly unbearable. Once again, he decided to leave. It took a while before he found another place. Only in 1989 was he able to join the National Scientific Center of Hematology, where he planned to apply the results on directing cell division to the treatment of leukemia. He had high hopes, but was disappointed when he was unable to get out of endless processes of reorganization. It was even worse when he was finally appointed to be in charge of a big and diverse department without any hope of doing reasonable research work any time soon. This happened when there was a rapid deterioration of science during the last years of existence of the Soviet Union.

Given this situation, it was a lucky development when in 1991 Irving Epstein offered Zhabotinsky a one-year visiting position at the chemistry department of Brandeis University. He joined the department and stayed there to the end of his life. Zhabotinsky was fifty-three years old; it might not have been too late to build up an independent career, but for him, apparently it was. His American period started well; he went for a lecture tour and visited a number of cities and universities. Back at Brandeis, he participated actively in discussions at seminars and in the research of his colleagues. Initially, he wanted his desk to be placed in a laboratory to be in the midst of the life of the group which he joined. Eventually he acquired his own office, but he still preferred spending his time in the lab.[27]

Lenin Prize winners, Albert Zaikin, Genrikh Ivanitskii, Anatol Zhabotinsky, Valentin Krinskii, immediately following the award ceremony at the Kremlin in Moscow.

Source: Courtesy of Michael Bukatin, Waltham, Massachusetts.

Anatol Zhabotinsky in 1995 at Brandeis University.

Source: Photograph by and courtesy of the author.

Memories of his experience during his years at Brandeis differ among his colleagues. The Hungarian visiting professor Miklós Orbán spent every summer for thirty years at Brandeis and overlapped with Zhabotinsky for seventeen summers. They had known each other since 1978. With Orbán, Zhabotinsky liked to talk about his father and about his own youth, but he avoided two topics: politics and science. Orbán's impression was that Zhabotinsky did not feel at ease. On the other hand, he visibly enjoyed it when he could get involved in laboratory work, preparing solutions and participating in the experiments with his hands. Orbán and Zhabotinsky coauthored half a dozen publications. Zhabotinsky had a good number of joint papers with other associates of Epstein's group, too, because he liked to involve himself in projects and make useful contributions. Alas, he hardly initiated new research. Orbán's general impression was that somehow Zhabotinsky was left in the shadow during his American years.[28]

He was never appointed to a tenured position, and he held the title of adjunct professor for his last few years at Brandeis. This arrangement had its merits in that he did not have to teach much or to worry about proposals of his own; he could be engaged in what he did best—research. Yet he was in many respects less independent than he had been in his Soviet life, especially at the Institute of Biological Tests of Chemical Compounds. According to Shnol, the joyful and friendly atmosphere in which Zhabotinsky worked in Moscow and Pushchino contributed to his productivity. Shnol wondered whether Zhabotinsky's lack of noteworthy results in his American period might be ascribed to his changed environment.[29] However, even during his Soviet life, there were two sharply different periods. Through the mid-1970s, he was at his most productive; whereas after that, up to his departure, he was underperforming compared with his prior achievements. In his American period, he had great productivity, except for the very last few years when an illness was gradually taking over. He never tried to capitalize on the authority of his name stemming from the fame of the Belousov-Zhabotinsky reactions.

The Belousov-Zhabotinsky reactions have much broader applications than what Zhabotinsky, let alone Belousov, might have envisioned when they started their respective work on them. To mention just one aspect of very general and fundamental significance, the reaction can be used to understand symmetry breaking—a literally vital characteristic of many life processes. Prigogine emphasized the importance of symmetry breaking in the most diverse processes. He stated:[30]

> For me the most interesting thing is that far from equilibrium you automatically break symmetries. For example, in thermal diffusion, you create a situation where there are different concentrations in the "hot" part than in the "cold" part. Therefore, space symmetry is broken. Another example is the changing role of time. In the Belousov-Zhabotinsky reaction, for example, two instances of time are no longer playing the same role. At one point you have blue molecules, then yellow molecules, then blue molecules again, and so on. This is a time-symmetry breaking, but there are many other possibilities for symmetry breaking. The non-equilibrium structures have opened an entirely new chapter in symmetry breaking, and this may be not so familiar to some physicists.

The significance of Belousov-Zhabotinsky reactions is still being enhanced. Belousov did not live to see appreciation of his great discovery. However, Zhabotinsky saw real hope for continuation of their work. His full page obituary in *Nature*, alas, was—like Belousov's Lenin Prize—a posthumous recognition. Zhabotinsky died in Boston, but his ashes are buried in Pushchino, near Moscow. His tombstone displays a scheme symbolizing the oscillating reactions, and the text carved into the stone reads: "founder of nonlinear chemical dynamics."

Erlen Fedin facing Aleksandr Kitaigorodskii with two other associates of the laboratory, Natalia Gorskaya and Ilya Amiton.

Source: Courtesy of Jan Kandror, Wiesbaden, Germany.

11

Aleksandr Kitaigorodskii

SOVIET MAVERICK

Aleksandr I. Kitaigorodskii (1914–1985) was a rare original thinker in science and had a high standing in the Soviet and the international science community, yet he was never elected to the Soviet Academy of Sciences. He is buried in the exclusive Novodevichy Cemetery on his father's plot, who himself was a major authority in chemical technology.

Kitaigorodskii uncovered the governing principles of how molecules build up crystal structures, and initiated a new direction in crystallography. He correctly predicted the distribution frequencies of molecular crystals among the possible 320 classes using rudimentary considerations.

Today, his pupils and the pupils of their pupils continue in leading positions at universities and research laboratories all over the world. He was a great intellect and a free spirit in an environment that frowned upon independent thinking.

One of Aleksandr I. Kitaigorodskii's former pupils, the noted spectroscopist Erlen Fedin, said of his first meeting with his future mentor: "I for the first time met a free man."[1] Coming from Fedin this was a heavy statement, because he had been in the Gulag,* and he had a critical mind. In 1995 he left Russia and immigrated to Germany.

Aleksandr Isaakovich Kitaigorodskii was born on February 16, 1914, in Moscow. His father was a well-known professor, Isaak Kitaigorodskii, a specialist in the chemistry and chemical technology of glass, ceramics, and silicates. Before the 1917 revolution, he was also the co-proprietor of an industrial plant. Young Kitaigorodskii enjoyed the advantages of a comfortable and enlightened home, where he received education, culture, and the knowledge of

* GULAG: *Glavnoe upravlenie ispravitelno-trudovikh lagerei* (Main Administration of Correctional-Labor Camps).

European languages. From his youth, he spoke German and French. English, he would learn later. He became a good pianist and an excellent dancer.

The Kitaigorodskii family was part of the Jewish community in czarist Russia at a time when many in this community were already trying to overcome the confining framework of Jewish life. The regime was restrictive for much of the population, but there were additional restrictions for Jews. The two most conspicuous restrictions were geographical confinement and the numerical limitations in public education. The designated territory where Jews were supposed to live was sometimes called the Pale and it extended to considerable regions of Poland, Lithuania, Belarus, Bessarabia (today, mostly, Moldova), Kurlandia (today, part of Latvia), and large areas of the Ukraine. The confinement was introduced in 1791 and was abolished by the February 1917 so-called bourgeois revolution.

Even in the regions to which the Jews were confined, the educational institutions, both gymnasiums and universities, had to limit the ratio of Jewish students to 10 percent. It was worse outside the Jewish areas, where the limit was 5 percent, and in Moscow and St. Petersburg, where it was 3 percent. These restrictions were communicated by the minister of education of the czarist government in an *unpublished* circular letter to the officials in charge of the educational districts. The czarist measures remained in effect until 1916. The czarist minister's unpublicized mode of operation was remarkable in that he must have known how negatively such orders could be viewed by enlightened people, organizations, and nations. But it was even more remarkable that decades later, under the Soviet system, *unpublished* instructions would once again be circulated limiting the number of Jewish students to be accepted by some institutions of higher education; in certain institutions, it was next to zero.

Kitaigorodskii's father did not attend high school; he was educated at home. He passed examinations to gain admission to university. From 1906, he attended the Kiev Institute of Technology. His interest in glass began during his student years. He completed his studies in 1911. Kitaigorodskii's mother, Asya Sindelevich graduated from a very good institution of higher learning for women in St. Petersburg. Incidentally, the numerical restrictions for Jews in education mentioned above did not apply to women.

Although Isaak Kitaigorodskii belonged to the bourgeois class, the revolution did not change his life in a substantial way. He was no longer a proprietor of his plant, but he continued directing it, and became a recognized expert in the glass-ceramics industry. He was even sent abroad to broaden his experience in his field. He spent almost half a year, in fall and winter 1925–1926, in Western Europe and the United States. In 1929, he was given the title of professor; and in 1935, he received the of Doctor of Science degree without a dissertation. The smooth career of the father was in contrast with what his son would experience a generation later. The father was well respected by outstanding scientists and other representatives of the intelligentsia. When in 1939, the doyen of Soviet physics, Abram Ioffe, needed special glasses for experiments in his institute, he turned to Isaak Kitaigorodskii.

Aleksandr Kitaigorodskii graduated from high school at the age of sixteen. Maybe seeking adventures or maybe to escape his bourgeois background, he left home, and went as far as the Ural Mountains. There, he found employment as a factory worker and augmented his earnings with additional part-time jobs. For a while, he was satisfied with taking life as it came, but soon he started studying on his own, as if preparing for more demanding occupations. At the age of eighteen he returned to Moscow, got married at the age of nineteen, and became the father of twins—a girl and a boy. He also learned that the Faculty of Physics of Moscow State University was seeking outstanding third-year students, who had completed their first two years of study elsewhere. Kitaigorodskii applied, but then had to admit to the dean of physics, the world-renowned physicist Sergei Vavilov, that he had not completed the prerequisite first two courses. When the dean understood that Kitaigorodskii lacked the prerequisites for third year studies, he advised the young man to start his attendance the following fall as a freshman. Kitaigorodskii pointed out that the Faculty of Physics must have had a reason to advertise for third-year studies, and offered to be tested. Fortunately for Kitaigorodskii, Vavilov was not afraid of unorthodox approaches; Kitaigorodskii had impressed him favorably, and he decided to examine the aspiring student.

The result was that Kitaigorodskii was allowed to join the third year at once but had to promise to make up for the missing exams of the first two years. He fulfilled his promise. Once he had completed the required courses, he did not just catch up with the rest of his class, he started doing scientific research parallel to his regular studies. The area happened to be X-ray crystallography, the structure determination of solids by X-ray diffraction. This early choice determined his career in science for the rest of his life. He graduated from the Faculty of Physics in 1935, and immediately started doing postgraduate work. He defended his dissertation for the Candidate of Physical-Mathematical Sciences degree (PhD equivalent) in 1939. The topic of his dissertation was the X-ray study of amino acids. It was at this very time that the world-renowned Linus Pauling at the California Institute of Technology (Caltech) expected every one of his numerous doctoral students to use X-ray crystallography for the determination of an amino acid structure.[2]

Kitaigorodskii did not stay at the university after graduation; he started working at the biophysics division of the Institute of Experimental Medicine. After the German attack on the Soviet Union on June 22, 1941, he moved again, and worked as head of the physics department of an armaments plant in Ufa, in the region of the Ural Mountains. Kitaigorodskii labored there diligently, but he understood that routine activities were not for him. He thought to change fields as far as the subject of his work was concerned, but not his principal technique. He recognized the value in that he had already earned a reputation in the X-ray analysis of materials.

As he looked around for possibilities, he was invited to head a research laboratory at the Institute of Organic Chemistry of the Soviet Academy of Sciences.

The director of this institute, Aleksandr Nesmeyanov (see chapter 12), was a rising star in Soviet science. Although Nesmeyanov was primarily a synthetic organic chemist whose interest was in making new compounds rather than in determining their structures, he had broad views. During the institute's wartime evacuation to Kazan, Nesmeyanov used some of his time—having been left without the means of experimental work—to get acquainted with techniques he had been unfamiliar with, including X-ray crystallography. By accepting Nesmeyanov's invitation, the physicist Kitaigorodskii found himself in a chemistry research institute. He would become a pioneer in Soviet chemistry, where before his work, scarce attention was paid to the application of X-ray analysis of organic molecules.

Kitaigorodskii carried out innovative work, not only relative to Soviet science, but also on the international scene. The war was still being waged when he was already formulating his research strategy. His interest was in finding regularities among the crystal structures built by various organic molecules. He had observed that in a most diverse collection of organic crystals, the distances between molecules were very similar. This observation led him to formulate the concept of molecular shape, and once there was a shape, the next question he asked was about the mode of the *packing* of molecules in a crystal structure.

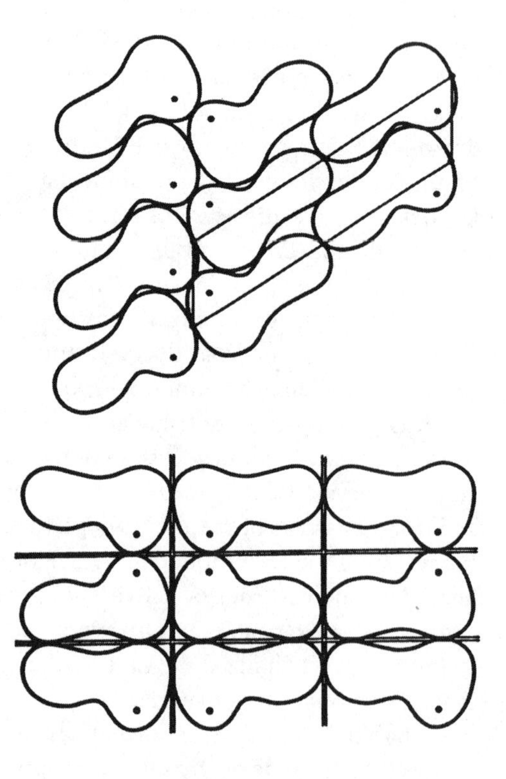

Two modes of packing in Kitaigorodskii's representation. The upper arrangement represents complementarity. In the lower arrangement the individual motifs are arranged in a face-to-face—mirror-symmetrical—manner. The complementary arrangement yields more efficient space utilization than the arrangement of higher symmetry; hence, the three-dimensional analog of the former is more frequently found in crystals than the latter.

Once again, unwittingly, he brushed elbows with the famous Linus Pauling. A notable German physicist Pascual Jordan had published a series of papers suggesting that quantum mechanically stabilizing interactions operated preferentially between *identical* or nearly identical molecules or parts of molecules. In response, Pauling and the German-American Max Delbrück wrote a short article in *Science*, "The Nature of the Intermolecular Forces Operative in Biological Processes."[3] The polemics was focused on the process of biological molecular synthesis that leads to replicas of the molecules present in the cell. This was a forward-looking consideration at a time when not much more than speculating could be done about it. The implications of the arguments were of great general importance. Pauling and Delbrück suggested precedence for interactions between *complementary* parts rather than between identical parts. According to them, all the possible interactions between molecules, including weak van der Waals attractions and repulsions, strong electrostatic interactions, or intermediate hydrogen bonding, would provide stability to a system of two complementary rather than two identical parts in juxtaposition. This description of intermolecular forces by Pauling and Delbrück seemed directly applicable to the packing of molecules in crystals. It is unlikely that Kitaigorodskii had access to the relevant literature during the war and thus could have been influenced by the ongoing debate.

Thus, in all probability, Kitaigorodskii entirely independently declared his views on the preeminence of the interactions between complementary shapes. Further, he not only pronounced his ideas about the preference of complementary packing in molecular crystals but also carried out painstaking analyses to test his idea. He was confident enough to communicate his research strategy in 1945 in a brief paper entitled "The Close-Packing of Molecules in Crystals of Organic Compounds," published in the English-language Soviet *Journal of Physics (USSR)*.[4] In this very first paper on the subject, Kitaigorodskii stated "the mutual location of molecules is determined by the requirements of the most close-packing." Incidentally, this English-language Soviet periodical soon ceased to exist in the Soviet Union as the result of the country's closing up before ostensibly disturbing foreign influence.

In 1946, Kitaigorodskii defended his Doctor of Science dissertation, "Arrangement of Molecules in Crystals of Organic Compounds," at the Institute of Physical Problems (Petr Kapitza's institute). His degree was in the physical-mathematical sciences. The structure of organic crystals remained his leitmotif throughout his career, and it did not change when in 1954 he moved from the organic chemistry institute to the newly established Institute of Element-Organic Compounds (Institut elementoorganicheskikh soedinenii, INEOS) of the Soviet Academy of Sciences. Kitaigorodskii became the head of the Laboratory of X-ray Structure Analysis. "Element-organic" means that in addition to the usual composition of organic compounds, atoms of other elements may be present in forming these substances. In Western literature "hetero-organic" is the accepted label.

Nesmeyanov orchestrated the spinoff of the new institute from the Institute of Organic Chemistry whose director he had been for many years, and now he became the director of INEOS. Kitaigorodskii moved with Nesmeyanov to the new institute. This showed both his allegiance to the outstanding organic chemist and Nesmeyanov's recognition of the importance of Kitaigorodskii's structural work. This research is equally necessary in organic chemistry and in element-organic chemistry. However, since in element-organic chemistry, new compounds of virtually all elements may figure, the structural diversity is considerably greater than in traditional organic chemistry. Thus Kitaigorodskii found a greater challenge in the activities of the new institute than in those of the old one. However, he never became a performer of routine structure determinations. He typically found it more exciting to look for answers to the question, why? than to the question, what? In time, a separate laboratory for routine structure determination spun off from his unit, headed by his former pupil, Yurii Struchkov.

When Kitaigorodskii started his investigation of how molecules build up crystal structures, he initially assumed a general three-dimensional ellipsoidal shape for all organic molecules. This was a useful simplification. Before him, for example, Lord Kelvin—more famous for his temperature scale and other inventions—had also described all molecules by one general shape.[5] Kitaigorodskii, however, soon realized that he needed to go one step further if he wanted to find regularities in describing the packing of molecules in organic crystals. He decided that he would look for such regularities for *arbitrary* shapes—the arbitrary shape being the least restrictive in his considerations. For the classification of packing, he invoked a very general characteristic of crystal structures: symmetry.

Crystals are often looked upon as symbols of symmetry. Usually, the symmetry of the external shape of the crystal is what is meant by this. However, even the symmetrical external shape and many other properties of crystals are determined by the symmetry of the arrangement of the building blocks in their internal structure. This is the most important characteristic of a crystal. Conversely, almost any solid, even a piece of glass may be cut into a beautiful symmetrical shape; yet this does give it the special properties of a crystal.

At the end of the nineteenth century, three scientists, independently of each other, determined that 230 different symmetries are possible for the internal structure of crystals. So the task Kitaigorodskii set before him was to determine the relative probabilities of the occurrence of all 230 different internal structures. This probability could be guessed by finding out how well the various arrangements—all 230 of them—might foster the most economical and energetically most advantageous "closest packing." Kitaigorodskii had noted that such packing was a crucial requirement for forming crystal structures.

Today, the task Kitaigorodskii set himself to solve would be simple. Hundreds of thousands of crystal structures have been determined, and the data are stored in data banks. The task would reduce to making statistics of the distribution of those hundreds of thousands of structures among the 230 possibilities. When Kitaigorodskii embarked on this study, however, only a handful of structures had been determined, and that was certainly insufficient for establishing any meaningful statistics. Kitaigorodskii liked models and considered a good model a crucial tool in scientific research. He devised a most rudimentary mechanical instrument, which he called "structure finder." It was merely a stand to which he fastened a wooden molecular model, again, very rudimentary because it was of an arbitrary shape, and brought in identical models to surround the first model.

His art was in making the arrangements of the models mimic every one of the 230 possibilities of packing alluded to earlier. It was a painstaking work in which, for each of the 230 cases, he determined the density of the packing arrangement. This was admittedly a very rudimentary approach and was soon replaced by more sophisticated ones. But the principle Kitaigorodskii had proposed remained valid in even the most sophisticated investigations. The goal remained the same, to establish the order of the various arrangements from the point of view of density of packing, in other words, from the point of view of space utilization. It is obvious that the better the utilization of available space, the higher the density of the arrangement. Looking back, it is not surprising that he found complementarity of molecular shapes to be the key concept in building crystal structures. Thus, if we take a molecule of an arbitrary shape and add to it molecules of the same arbitrary shape, the collection of such molecules will better utilize the available space by taking complementary positions relative to each—for example, lining up head to toe—than by facing each other. This is easily seen in the arrangements of arbitrary but equivalent shapes in the plane, that is, in two-dimensional arrangements (see, p. 254).

Incidentally, there has been a long tradition of illustrating and teaching the packing principles in crystallography by invoking two-dimensional patterns.[6] The most famous representative of this approach has been the Dutch graphic artist M. C. Escher whose intriguing patterns were developed and utilized by the Dutch crystallographer C. H. MacGillavry.[7] In the Soviet Union, the Azerbaijani crystallographer Khudu Mamedov and his colleagues developed a plethora of periodic drawings illustrating crystallographic packing in the plane.[8] These drawings were quite unique because their periodic patterns represented motifs of their cultural heritage.

Kitaigorodskii was original in applying the complementarity principle—indeed, discovering its applicability here—to the structure of molecular crystals. But the importance of geometrical complementarity had been recognized ever since science has been cultivated. Two thousand years ago, the natural philosopher Lucretius, in his *De rerum natura* (*The Nature of Things*), proclaimed complementarity as the fundamental principle of best packing arrangements[9]:

> Things whose fabrics show opposites that match,
> one concave where the other is convex,
> and *vice versa*, will form the closest union.

The "structure finder" proved to be an excellent aid in taking the first steps in the analysis of the arrangements of molecules in crystals. It embodied the principal advantage of modeling, namely, that it overemphasized the importance of one feature and ignored everything else. Usually, there are a plethora of effects to be considered and a good model selects one or a few to take into account. This is also why there are serious limitations of the findings when the results are based on rudimentary models. And this is also why eventually scientists embark on refining their models. Kitaigorodskii's model also needed refining, because many lesser features remained hidden in the results obtained by his initial approach. However, he achieved spectacular success because he was able to predict the relative frequency of occurrence of each of the 230 possibilities of arrangements as molecules of arbitrary shape build up crystal structures.

Today, with the knowledge of hundreds of thousands of crystal structures, it can be said that his findings, by and large, have withstood the test of time. This is all the more significant if we consider that it is still difficult, if possible at all, to predict the crystal structure of a new substance merely on the basis of its composition. This limitation once again underlines the importance and originality of Kitaigorodskii's principal ideas. One might expect in today's enormity of structural information a more straightforward approach. When a new substance is produced and its composition determined, it should be possible to deduce its crystal structure simply on the basis of the hundreds of thousands of structures already determined. Alternatively, on the basis of all our knowledge about the most diverse interactions within molecules and between molecules, it should also be possible to predict the most probable crystal structure for the new substance. Yet a more forceful tool should be a utilization of the combination of the two approaches. Alas, this is not the case; it is still impossible to predict the crystal structures for most new substances. It is still true that only doing the necessary experimental determination and/or painstaking computational work can lead to such knowledge, at least in most cases.

During the first half of Kitaigorodskii's career, even potentially dangerous situations turned out to be harmless for him. At the time of the ideological attacks on the theory of resonance in the Soviet Union (see chapter 12), Kitaigorodskii was among those under attack. However, his attackers soon forgot about him. They were after bigger names, in particular the director of the Institute of Organic Chemistry, the recently appointed president of the Soviet Academy of Sciences, Aleksandr Nesmeyanov. As it happened, Nesmeyanov turned out to be too big a fish for them. What Kitaigorodskii had been accused of was using the theory of resonance in interpreting some of his results in structure analyses. The compounds in question were mercury-organic compounds—Nesmeyanov's favorite field of research. Fortunately for Kitaigorodskii, Nesmeyanov managed to protect those under him from the attacks of the "anti-resonance" people.[10]

The first period of Kitaigorodskii's professional life was all success, augmented by his great popularity among his colleagues. He became friends with the legendary physicist Lev Landau (see chapter 5), and they coauthored the bestselling popular-science series, *Physics for Everyone*. Given Landau's grapho-phobia, it is safe to assume that they probably talked about the contents of these books, but Kitaigorodskii actually wrote them. Also, it remained to Kitaigorodskii to revise subsequent editions of the books after Landau was incapacitated by an automobile accident in 1962 and after his death in 1968. The English translations of these popular books have been published repeatedly. Kitaigorodskii continued the series on his own.

Kitaigorodskii loved all kinds of adventure, downhill skiing as well as water skiing, driving automobiles, dressing according to the latest fashion, playing the piano and singing; he could speak well and he could give exciting talks at a moment's notice, not only in Russian, but also in foreign languages. He enjoyed life and his enjoyment was visible. He loved having girlfriends, and even if he was not faithful to them, they did not complain. His behavior was conspicuously different from the generally accepted "Soviet" norm, but he got away with it; at least for a time.

He was the pioneer of the field he created, organic crystal chemistry, and his position at the Science Academy was strengthened by the fact that the director of his institute was also the president of the institution. Kitaigorodskii did not appear to work hard though he certainly did; but he liked being thought of as spending little time and effort laboring. He cultivated the image that most of the time he just lectured about popular topics, wrote books, and—something that came later in his career—traveled abroad. In reality, he did his homework meticulously and diligently. For example, while others took long vacations during the hot summer months in Moscow, he stayed mostly in town and focused on his popular-science books. He co-chaired a weekly seminar at the institute with another physicist, the academician I. V. Obreimov. It was a free-spirited gathering where no topic was off the table.

Kitaigorodskii even encouraged his associates to study the philosopher Spinoza and others. His coworkers and the participants of his seminar kept abreast of developments in other fields beside organic crystal chemistry. They invited the physicist Igor Tamm and the chemist I. L. Knunyants to speak about the recently discovered genetic code. When computers were becoming available in his institute, he quickly adjusted himself and his group to their enhanced possibilities for research. He saw in them great potential in solving the problem he considered fundamental in his specialty, namely, the prediction of crystal structures on the basis of the composition of newly synthesized substances. This goal still has not been reached, but the stakes are high, because one day this may lead to the creation of new substances of any desired structures and properties!

Kitaigorodskii invariably impressed people at their first meeting with him. One described him as "clear-thinking, broadly educated, and full with interesting ideas."[11] According to another, "he was thin, tall, and his external appearance was nothing short of extravagant, he appeared carelessly dressed."[12] He was in his mid-thirties at the time, and his lectures radiated his enthusiasm for his subject. He showered his students with terms and concepts unknown to them. The students found his presentation very stimulating.

As a young mentor, he took several students for Diploma work (master's degree equivalent). He was always available to his students to discuss ideas and to tutor, but he did not suggest thesis topics: his students had to find them by talking with chemists. To the students he represented great authority. They were looking for Academician Kitaigorodskii, and when they found him, he told them, "Kitaigorodskii—this is I, though I am not *yet* academician."[13] This was in 1948. In 1949, his first doctoral student, Sarra S. Kabalkina, defended her dissertation brilliantly but had to take a job far away—this was ascribed to the anti-Semitism that was developing parallel to the Cold War. Many years later, Kitaigorodskii and Yurii Struchkov asked for the help of other authorities in science to let Kabalkina return to scientific research.[14] Struchkov was his assistant, to whom he gladly delegated most administrative work in the laboratory. Struchkov became a world-renowned crystallographer and one of the world's most prolific research scientists.[15] In addition to his science, Struchkov loved doing administrative work and was good at it, so this was a fortunate match with Kitaigorodskii's aloofness.

It was not always easy for Kitaigorodskii, working in an essentially chemistry—moreover, an organic chemistry—institute, to convince others that his structural science, including bond lengths and bond angles, could have meaning for chemists and chemistry. But he was tireless in telling his colleagues, including his boss, the director of institute and virtually the boss of Soviet science at that time, that one day they would also understand the significance of the metric aspects of molecular structure.[16]

Kitaigorodskii organized a course about the packing of organic crystals for doctoral students of the Faculty of Chemistry at Moscow State University. He lectured in a large auditorium, and not only doctoral students but undergraduates as well as instructors attended his presentations in droves. A future crystallographer of international renown, Petr Zorky, met Kitaigorodskii for the first time while attending this course. Zorky was at once infected by the lecturer's enthusiasm and decided to follow him in his doctoral studies. His own mentor warned him to stay with more or less predictable structure determinations for his dissertation, but Zorky was set on understanding the packing in crystal structures. In his dreams, he was shaking a bag of three-dimensional models of molecules and wanted to see whether or not after sufficient shaking they would assume the packing similar to what they determined in the structures of molecular crystals. The results were new findings, which he could publish, but his dissertation work took him six years rather than the usual three. Zorky, though, never regretted taking the road less traveled.[17]

Zorky later headed the crystal chemistry laboratory of Moscow State University. Kitaigorodskii by 1958 had the *title* of professor, which he had received as early as 1947, but he was not a professor of Moscow State University. Zorky noted about him in a conversation in 1996:[18]

> He was not a professor at that time, although he should have been. When the Laboratory of Crystal Chemistry of Moscow State University was created in 1955, of which I am in charge today, it was originally meant for Kitaigorodskii. However, as it turned out, he did not get this position. He was never elected to the Academy of Sciences either. He always made the elite of the Academy uncomfortable. He irritated them. He stressed his independence and his superiority, and they didn't like it at all. Some were even trying to belittle his contributions and, besides, there were some unpleasant episodes too. One was an incident at the French exhibition some time in the late 1950s. He was accused of trying to steal a couple of French books there. There was a big fuss over it and there was a French correspondent present who made a big story out of it too. Everything got out of proportion. His enemies then capitalized a great deal over this incident.

I was intrigued by Kitaigorodskii's failure to get a university professorship and appointment as head of laboratory at Moscow State University, and I pursued this in my conversation with Zorky:[19]

> IH: Why didn't Kitaigorodskii get the position?
> PZ: There was a set of criteria and he failed one of them.
> IH: Which one?
> PZ: We had a questionnaire and he failed No. 5 on it.
> IH: What was No. 5 about?
> PZ: It was about nationality classification.
> IH: What was his problem?
> PZ: He was Jewish.

The question may be asked, why could Kitaigorodskii's father have a career without the barrier of anti-Semitism, whereas in Kitaigorodskii's case it played a role? Kitaigorodskii Senior's career developed before the war, in a period of Soviet history free of anti-Semitism, at least free of the state-sponsored kind. Soon after World War II, during Stalin's last years, very strong anti-Semitism characterized the Soviet Union. Although the harshest measures disappeared after Stalin's death, some survived, if not in overt legislation, certainly in covert instructions. They concerned, for example, severe restrictions in admitting Jewish students to certain institutions of higher education and in hiring Jewish associates in such institutions and within the research establishment. Soviet personal documents—identity cards—clearly identified Jews as a nationality.

In the late 1960s and the 1970s, I had several meetings with Kitaigorodskii, mostly at conferences, and in 1967, I had a face-to-face conversation with him. That meeting impacted me. When I was building up my experiments in Budapest during my doctoral studies, I wanted to conduct some experiments in Moscow at INEOS, and I used the exchange program between the Hungarian and Soviet Academies of Sciences to arrange a visit. The electron diffraction group whose apparatus I intended to use was part of Kitaigorodskii's laboratory, but I did not anticipate meeting with him since he was the head of the whole laboratory, which was a much larger entity than the group I was to visit. As it turned out, a few days before my arrival, there was a fire in the institute, not a large one and far from Kitaigorodskii's laboratory, yet it was unthinkable to let a foreigner wander about the institute in the aftermath of the fire. I should have been notified and told to cancel my trip but was not, and I duly arrived at the institute at the appointed time. I was told that there would be no experiments, but I would be received by the head of the laboratory, the famous Professor Kitaigorodskii. This sounded like a consolation prize, and I was not happy about it; but in hindsight, it was better than any experiment I could have carried out at that time.

Kitaigorodskii received me in his office, which he shared with the administrative assistant of his laboratory. There was a stream of visitors to her during the time I was there, but Kitaigorodskii appeared oblivious to the constant commotion; he could concentrate on what he was doing. He asked me about my aborted plans for experiments and quickly came to the conclusion that instead of the proposed series, I ought to devise a different one. My first thought was that he just wanted to ease my frustration over the missed experiments; but soon enough I understood that he was right, and I acted accordingly. In any case, he immersed himself in my research project and in no time came up with a useful suggestion. The rest of the time—and it was not at all a hurried conversation—we talked about many things, that is, mainly he did, and I find it hard to remember anything concrete. What does remain is the impression that I left his office satisfied and happy, with a feeling that I had gained a lot, but what exactly, I found difficult to put into words.

Kitaigorodskii's principal interest was in creating a global understanding of organic crystal structures, and he was less interested in the precise details. He was also capable of admitting defeat in the lively arguments he loved to engage in, and also of changing his views. For example, he used to maintain that a molecule has one structure regardless of its environment. This meant that he ignored structural differences between various states, such as a gas and a solid. In a gas, the molecule is so far from other molecules that it can be considered isolated from everything else. By contrast, every molecule in a solid is closely surrounded by other molecules. Interactions with those surrounding molecules are inevitable, and they cause structural changes. Early on, the information about structures was so approximate that such differences could be ignored. Later, more precise data pointed to the importance of such changes. It was not easy for Kitaigorodskii to admit that his views were oversimplified, but he had the strength to do so.

Aleksandr I. Kitaigorodskii lecturing.

Source: Courtesy of Irena Akhrem, Moscow.

Aleksandr N. Nesmeyanov and Aleksandr I. Kitaigorodskii before an institute meeting.

Source: Courtesy of Irena Akhrem, Moscow.

Kitaigorodskii's discoveries and teachings were important enough to be published in English, and this happened well before he could travel abroad and meet with his colleagues in the West. One of them gave this description of their first encounter:[20]

> By about 1973, Kitaigorodskii began to appear in Western countries. I met him for the first time at the 1st European Crystallography Meeting in Bordeaux [France] in the summer that year. I am not at all sure how I had imagined him as a person, perhaps as some deadly serious bookish type of scholar, perhaps as some Dr. Zhivago-like figure—after all, I had never been to the Soviet Union and had met very few Russians; in fact, my knowledge of Russians was mainly confined to impressions from Russian literature, Tolstoy, Chekhov, Turgenev, Pasternak. In any case, he was quite different from my expectations: a vivacious, high-spirited personality with a healthy appetite for the good things that life had to offer.

Kitaigorodskii's discoveries in organic crystal chemistry were so fundamental that his contributions became part of general knowledge, and his teachings were often used in other people's papers and books without attribution. This, however, did not enhance his reputation back home as a great scientist. In addition, his popular lectures, in which he behaved as a true performer, further contributed—though this should not have been the case—to the erosion of his authority as researcher. His habit of pretending that he avoided hard work and that things came easily to him did not enhance sympathy toward him. He was irreverent toward authority, especially the authority of some of his pompous colleagues, among them full members of the Academy of Sciences, and they had a very low threshold of tolerance for his clever jokes.

There was something inherently sad about Kitaigorodskii's fate. His whole existence was a resistance to the existing Soviet regime, but he seldom did anything that could have resulted in punishment. Besides, he was so well known internationally that any measures that might have been taken against him would have made waves. He did not receive awards, medals, or positions, but at one time he was given a green passport as one of the Soviet representatives for UNESCO, which made his travels more flexible. This was a special distinction in the Soviet Union, where not only foreign travel but even the denial of foreign travel carried some aura of recognition. Kitaigorodskii could not contain himself and wanted to use his green passport as if he were a free citizen. For example, rather than return to Moscow following a UNESCO assignment in India, he flew to Paris. He saved some money of his travel allotment, which he was supposed to "donate" to the Soviet authorities. Instead, he bought himself a few days in Paris. The Soviet authorities reprimanded him and threatened him with appropriate measures, but he declined even to apologize.

Kitaigorodskii was an excellent speaker who could entertain his audience with scientific explanations as well as he could with anecdotes and stories on the most diverse topics. He had the ability to express his thoughts concisely so that they would stay in the minds of his listeners and reverberate in them. It would be difficult to give exact quotations of most of his sayings, because they have appeared in a variety of ways and have become part of the folklore. He said, for instance, that a first-rate theory predicts, a second-rate theory forbids, and a third-rate theory explains things after the fact. Of course, when something is well understood, it makes it possible to predict phenomena before they are observed experimentally. Rules pronouncing the impossibility of certain features or events also reflect considerable knowledge. But even deductions made from already available data have importance.

When he wanted to describe the sizes of molecules in crystal structures, he simply stated that he dressed the molecules in the fur coat of van der Waals spheres. This meant that when he established the shapes and sizes of molecules, he took the rather weak but not negligible van der Waals forces into account. When asked, shortly before his death, which of his achievements in science he considered the most important, he said, "I've shown that the molecule is a body. One can take it, one can hit it—it has mass, volume, form, hardness. I followed the ideas of Democritus."[21]

The next example is my favorite; he stated that when molecules are free, that is marvelous; but molecules may get deformed by interactions with other molecules. As explained earlier, the meaning is that the structure of a molecule may undergo changes: in the gas phase the molecules are so far apart that they can be considered free. In the crystal, they are closely packed, and their structure, especially if it is not very rigid, will undergo changes. In certain political situations, this statement could convey some hidden but unambiguous message. I once quoted this statement in a talk I gave in East Germany, and for years afterward, my East German colleagues made references to it in private conversations.

Petr Zorky continued in his own research, faithfully following Kitaigorodskii's teachings. Zorky noted that although there have been great advances in organic crystal chemistry, "the significance of close packing theory remains in force...this theory is an important part of the *mentality* of crystal chemists"[22] (emphasis added).

In time, Kitaigorodskii felt increasingly hurt by being ignored by the authorities at the Academy of Sciences. He watched with growing irritation as others, often less deserving, were elected, and he was not. Those who recognized his scientific value and were not bothered by his flamboyance, nominated him. In the 1970s, citation data were becoming available that indicated his strong impact in world science.[23] It was to no avail, however, because he was anathema to many members of the Academy of Sciences. He continued his activities, obviously pretending that he was not affected by the lack of recognition. He must have suffered from the view that he was gifted at popularizing science but was not quite taken seriously by fellow scientists. He published substantial articles in *Literaturnaya Gazeta* (Literary Newspaper), and he was even accepted as member of the Union of Soviet Journalists. His critical views of Soviet scientific life had relevance to broader aspects of Soviet life and generated lively debates, but he was no political dissident. His broad cultural activities enhanced respect for him by some, but not by the Soviet scientific establishment. Gradually, he developed a drinking problem and parallel to this, his health was slowly deteriorating. At some point he was no longer permitted to drive. His close friends and former pupils were devastated observing the disappearance of not only his physique but—even more painfully—his free spirit.[24]

Soviet postage stamp with Aleksandr Nesmeyanov's portrait, 1980.

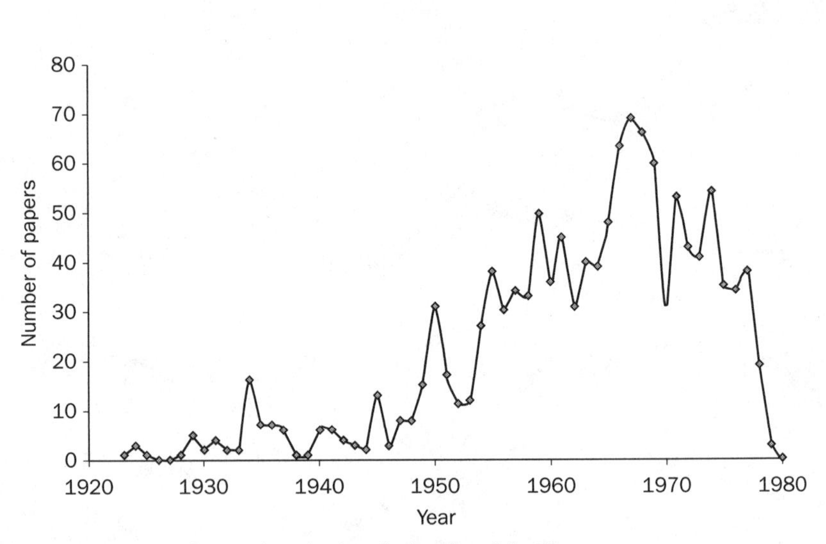

Aleksandr Nesmeyanov's annual production of scientific publications.

12

Aleksandr Nesmeyanov

BRILLIANT ADMINISTRATOR AND SOVIET COURTIER

Aleksandr N. Nesmeyanov (1899–1980) was exceptionally prolific both as a chemist and as an administrator of Soviet science. He created new organic compounds and a new field in chemistry—he called it element-organic chemistry—known in the rest of the world as hetero-organic chemistry. He tried to improve Soviet life by his own means. When agricultural production proved inadequate to feed the country, he initiated a new branch of chemistry to produce artificial food.

As Rector of Moscow State University, his achievements included the establishment of a new campus. As the President of the Soviet Academy of Sciences for a decade, he created a plethora of research institutes, including one for his own field and another for scientific information. He served the Soviet regime as courtier, first under Stalin, then under Khrushchev. He did not like independence in people, but he tried to guard the independence of the Soviet Academy of Sciences. He held an unprecedented number of public positions. His actions were limited by the restrictive nature of the Soviet regime whose boundaries he did not attempt to overstep.

In fall 1961, I attended Aleksandr Nesmeyanov's course in organic chemistry at Moscow State University. Although organic chemistry was not my favorite area of chemistry, I sat close to the front in the huge auditorium because once I was there I wanted to follow his lecture. I did not find Nesmeyanov to be a great orator, but my experience differed from others who described his lecturing style as colorful and exciting. Maybe that characterized his younger years. To me, he seemed to monotonously recite the subject. However, his tremendous knowledge shone through his dull style, and he did not use any notes to help him remember what he wanted to say. I was a master's degree student; later, it turned out that I did not have to pass his examination: My course credit from Budapest sufficed. Nonetheless, attending his lectures turned to be a useful exercise in getting to know this notable contributor to Soviet scientific life.

When I attended his course, Nesmeyanov had just left the most important science administrator's position in the country, the presidency of the Soviet Academy of Sciences. However, he retained positions that each would have sufficed to keep anybody busy full-time. He was director of a large research institute, chair of a large university department, and held many other appointments. He gave the impression of being rather aloof. He did not interact with his students in the lecture theater; he came, lectured, and left. Even during the breaks, he did not mix with us. This was, again, different from how others described their experience with him as a professor. In 1961, he was in his early sixties, seemed more sleepy than alert, and dressed shabbily. He appeared assiduous, talking to the last second of the lecture time, though at that point he quit, and he never missed a lecture.

Aleksandr Nikolaevich Nesmeyanov was born on September 9, 1899, in Moscow, thus his life began under the czarist regime. His father held a law degree from Moscow State University and served as the principal of an orphanage. His mother had artistic ability, and Nesmeyanov inherited her flair for the arts. Even while in his highest positions, he painted. Between 1908 and 1917 he attended secondary school at a private gymnasium. A book turned his interest toward chemistry: when he was thirteen years old, he came across an inorganic chemistry textbook and found it more exciting than the books of Jules Verne and H. G. Wells. A year later, he acquired an organic chemistry text, which fascinated him even more, and he started a home laboratory. He fell in love with materials, their smells, colors, and shapes.

His home laboratory became the young Nesmeyanov's sanctuary. In youth, he was not strongly built, was shy and introverted, and found it difficult to enter a room if there were other people there. He was afraid of public speaking and refrained from it throughout his university years. He was religious but gave it up by the time he started reading books about science. He decided early on that he would be a professor and achieve something big in science.

He became a vegetarian very early. When he was about nine or ten years old, he decided not to eat meat, and a few years later he decided not to eat fish either. This was not due to the influence of his immediate surroundings: he was alone in this in his family, and his mother tried to talk him out of it. She told him that the animal world was organized in such a way as to provide the foodstuff of others, adding that this was the law of nature. Nesmeyanov responded that if that were true, then people should not attempt to fly because it would contradict another law of nature. He went through an especially trying period during the famine of 1919–1920, but it did not shake his resolve. He stressed that his vegetarianism was not a protest against anything; rather, he was appalled by the hopeless situation of animals selected for slaughter.

In 1917, he started to study at the Physics and Mathematics Faculty of Moscow State University, which included chemistry. In one or another capacity he remained affiliated with this school for the rest of his life. He completed his studies in 1922 and continued as Nikolai Zelinsky's doctoral student. Zelinsky was one of the great Russian organic chemists. Nesmeyanov, being his protégé, was assured a bright career.

Zelinsky was not only a superb chemist; he was also a man of social conscience. In 1911, he and his colleagues walked out of the university in protest against the educational policies of the czarist government and returned only in 1917. At that time Zelinsky was appointed department head of organic and analytical chemistry. This is when Nesmeyanov entered university, and his eventual involvement with Zelinsky coincided with one of Zelinsky's creative periods. Nesmeyanov spent six years with Zelinsky, 1922–1928. The professor had a great impact on his disciple, but not by giving him ideas for research projects or instructions for using the literature; rather, by encouraging him to be independent. The two never published any paper jointly. In his later life, Nesmeyanov increasingly appreciated what he learned from Zelinsky's example and felt "infinitely grateful to him."[1]

After completing his studies, Nesmeyanov first worked in a pesticide laboratory. It was his entrance into the world of organic compounds incorporating a metal, first of all mercury. This turned into an interest that remained with him throughout his career. He created new routes to synthesis and new mercury-organic compounds, and nothing in his scientific career would surpass these early successes. Many years later he jokingly lamented that his organomercury chemistry overshadowed everything else he did. He sympathized with Sir Arthur Conan Doyle who said: "But I have written more than just Sherlock Holmes."[2] His innovative approach was soon extended to the synthesis of large classes of other organic compounds. They included derivatives of a diverse set of metals and nonmetals. This gave the idea to Nesmeyanov to coin a new name, element-organic chemistry, to replace the previously popular, but more restrictive metal-organic chemistry. Through the preparation of a large number of new substances, he was involved with a variety of chemical concepts and structures. He did not invent new concepts; rather, he extended the application of existing ones with great success.

Nesmeyanov's organizational talents were already manifested in his first work place. His fledgling unit started growing, and in 1930 he established the Laboratory of Organic Chemistry at the Institute of Fertilizers and Insectfungicides. In 1934, he organized a separate Laboratory of Metal-Organic Compounds. This process of expanding would be a pattern he would repeat. In 1935, he moved to the Soviet Academy of Sciences, with which he remained affiliated for the rest of his life. The expressions "rest of his life" and "rest of his career" can be used interchangeably for him, because his career and his life were one. Science was his life.

In 1939, he became director of the Institute of Organic Chemistry of the Science Academy (today, the Zelinsky Institute). But he never abandoned his direct involvement in research, even if not with his own hands then through his trusted associates. In the institute, he remained in charge of one of the research laboratories, not surprisingly, the Laboratory of Element-Organic Compounds, which in 1954 had been spun off from the institute. It became the basis of a new institute, the Institute of Element-Organic Compounds (Institut elementoorganicheskikh soedinenii, INEOS), with Nesmeyanov as its director. Today, it is the Nesmeyanov Institute. In the meantime, his university career continued to advance. He was appointed a professor of organic chemistry, then chair of the Department of Organic Chemistry; and from 1948 to 1951, he was the rector (president) of Moscow State University.

His name became associated with the establishment of the conspicuous new university campus on the Lenin Hills in Moscow, known before and again today as Sparrow Hills. It is a beautiful area of the city, overlooking the Moscow River and downtown Moscow, but far enough away to enjoy broad spaces, plush vegetation, and fresh air. The decision about the new campus was not part of a broad plan for the development of the university. Rather, when Stalin decided to build several high-rises in Moscow, he started looking for uses for the buildings. One of the party leaders told Nesmeyanov to request one of the planned high-rises for the university. Nesmeyanov wrote a letter to Stalin to this effect, and the dictator approved it right away.

Nesmeyanov's tenure as rector was cut short because in 1951, following the sudden death of the president of the Soviet Academy of Sciences, physicist Sergey Vavilov, Nesmeyanov was made the new president. He first heard about this possibility from his driver; then, Georgii Malenkov, one of Stalin's closest aides, invited him for a meeting and told him that his presidency had already been decided. Nesmeyanov oversaw an extraordinarily intensive period of development in Soviet science, comparable only to the development in the 1920s. But this was at a different level. In the 1920s, the goal was to resuscitate science in the Soviet Union; in the 1950s, it was to achieve world preeminence.

Three generation of Russian organic chemists: Sergey Nametkin, Nikolai Zelinsky, and Aleksandr Nesmeyanov.

Source: Courtesy of the late Lev Vilkov, Moscow.

The Soviets succeeded spectacularly if judged by their impact in shaking up American science and science education. But in reality, only selected areas of science progressed, primarily those related to defense, and especially to the development of nuclear weaponry and rocketry. For a while, the establishment of atomic and hydrogen bombs took precedence over everything else, but the nuclear program was not under the jurisdiction of the Science Academy. It was under the Ministry of Medium Machine Building—a camouflage; in practice it was under the close control of the security organs. These areas of science made fast progress with outstanding government support. Other fields suffered fatal blows, such as genetics, and generally, biology, and also areas that fall under the umbrella of "cybernetics," including computer technology. The consequences of the destruction of modern biology were manifested in the disastrous performance of agriculture. The fate of cybernetics had similarly long-ranging consequences for the entire Soviet infrastructure, rendering it obsolete. On the other hand, Nesmeyanov's presidency of the Academy saw the establishment of about twenty-five new research institutes, including one for scientific information.

The development of science in the 1950s was unfavorable with respect to international interactions when compared with the 1920s. In the 1950s, young researchers had few opportunities to gain international experience. This was not for want of trying on Nesmeyanov's part. He tried and achieved some success, but the meagerness of his success tells us how hopeless the situation was. He made efforts to send young researchers abroad, but this he did as a private initiative rather than in any institutional way, which would have had a broader impact. The British chemist Alexander Todd of Cambridge University told the story that a Soviet deputy prime minister helped to put together an official agreement whereby Nesmeyanov could send two of his young associates to Todd.[3] This was all he could achieve despite his high office. Years later, one of the two, N. K. Kochetkov, succeeded Nesmeyanov as director of the Institute of Organic Chemistry.

Nesmeyanov was one of the few Soviet scientists who had direct contact, though infrequent, with the supreme leaders of the land. He was in charge of the committee for the Stalin Prize between 1947 and 1961. The committee did not make decisions under Stalin, it only made recommendations, even though a member of the highest party organ, the Politburo, was always present during its meetings. The final word was Stalin's at a Politburo meeting to which Nesmeyanov's summon came always at a moment's notice. The meeting usually started late in the evening and lasted till 2 a.m. Everything about these meetings was determined by Stalin. In case of the recommendations concerning scientists he asked only questions of clarification. In the fields of inventions and constructions, and especially weaponry, he participated much more. All those present sat at a table while Stalin walked around, asking questions, obviously in charge on the issue of who should and should not receive the prize named after him. Under Khrushchev, the procedure was simplified, and Nesmeyanov's committee was given the right to make decisions rather than only recommendations.

Nesmeyanov's dealings with Khrushchev were broader than with Stalin and went beyond who should be getting what prize.[4] Once Nesmeyanov and the atom czar Igor Kurchatov initiated a conversation with Khrushchev about the impossible situation in biology where, following Stalin's death, Lysenko's damaging influence continued. Lysenko managed to convince Khrushchev that only he could lift Soviet agriculture out of a crisis characterized by terribly low productivity. The meeting with Khrushchev started ominously when Kurchatov told the Soviet premier about the success of hybrid corn in the United States and the damage to Soviet science caused by the denial of genetics. Khrushchev became visibly agitated and waved a couple of long ears of corn from his desk at his visitors. He advised in no uncertain terms Kurchatov and Nesmeyanov to stay with their physics and chemistry and leave biology and Lysenko alone.

Nesmeyanov appeared discouraged after this experience and shared his gloom with his associates at the department of organic chemistry of Moscow State University. However, he did not give up and began formulating an efficient approach to make changes. If the country's leadership did not support his action, he decided to conduct it in a clandestine manner.[5] He had a couple of courses organized at the chemistry faculty that would be concerned with the new directions in biology. It would have been impossible to organize such courses at the Faculty of Biology, but the Faculty of Chemistry was not under Lysenko's influence; on the contrary, there Nesmeyanov held great authority. What happened was more like a conspiracy than the normal way of making changes in the curriculum and in the organization of scientific research.

One evening in early 1958, two leading biologists were invited to meet with Nesmeyanov, not at his Academy office, but at the university. He invited five young associates from his department to attend the meeting. Nesmeyanov told his visitors that the Academy was preparing a resolution about the development of chemistry and that it would be possible to sneak into that resolution a decision to develop a center for biological research, using a camouflage label like the "physical chemistry of biology." Nesmeyanov appeared very practical, because he had in mind at once two biological institutes rather than only one, and he had already selected his two visitors, Vladimir Engelhardt and Mikhail Shemyakin, for the positions of the two directors.

Together they decided that the two biological institutes should have names that did not "smell" of biology. One would be called the Institute of Radiation and Physical-Chemical Biology, to be directed by Engelhardt. The other would be the Institute of Natural Products Chemistry, to be directed by Shemyakin. Nesmeyanov called their agreement "our little biological conspiracy."[6] The operation was not without danger. Just to give a sense of the risk they were taking, it was also toward the end of the 1950s that several articles critical of Lysenko appeared in the Soviet *Botanical Journal*. In response, Lysenko published an authoritative article in the central newspaper of the Communist Party, *Pravda*, and as a consequence all members of the editorial office of the *Botanical Journal* were fired and had to be replaced. However, Nesmeyanov's plan worked; the Academy's chemistry resolution was soon accepted, and nobody noticed or called attention to the insertion concerning biology.

Nesmeyanov knew that he had to act quickly. It appeared that one of the Academy institutes was moving out of its building to take up residence elsewhere. Nesmeyanov directed his two co-conspirators to occupy the building right away. It was a big place, and they divided it between themselves, assigning one half of it to one new institute and the other half to the other. Much of the action took place overnight. Eventually, both institutes received the names their founders had wanted from the start; Engelhardt's institute became the Institute of Molecular Biology and Shemyakin's, the Institute of Bio-organic Chemistry. Today each of these two institutes carries the name of its founder.

Incidentally, in spite of Khrushchev's protecting Lysenko and his unscientific domination of Soviet biology, the situation was slowly progressing. The international community had decided that in 1961 the Fifth International Biochemistry Congress was to take place in the Soviet capital. The organizers agreed to have this big meeting in Moscow provided they had the freedom to choose the speakers and topics. This was granted by the Soviet authorities, but not the freedom to select the titles of the sessions. Lysenko was still the director of the Institute of Genetics in Moscow, and the term "molecular biology" was still anathema to the authorities. When Engelhardt wanted to name one of the sessions "molecular biology," he could not get it through, and they settled on a compromise name, "biological functions at the molecular level."[7]

Ever since Nesmeyanov's encounter with Khrushchev about Lysenko, he felt Khrushchev's scorn toward him and, what was worse, toward the Soviet Academy of Sciences. Khrushchev intended to "improve" the activities of the Academy, and, referring to the Academy, he declared publicly: "For the watch to go, you have to shake it."[8] The airing of his wrath culminated at the end of 1960, when Khrushchev, again publicly, criticized the Science Academy for wasting its resources on "little flies." This was another thinly disguised attack on the Academy and on genetics since the "little flies" were the famous *Drosophila melanogaster*, the favorite fruit fly in genetic research. To everybody's—Politburo members and others—astonishment, Nesmeyanov just as publicly explained the importance of such investigations.

On another occasion, when Khrushchev threatened to abolish the Science Academy, Nesmeyanov remarked, "So this is the way it is; Petr Alexeevich [the czar, Peter the Great] created the Science Academy, and Nikita Sergeevich [Khrushchev] is going to destroy it."[9] But Nesmeyanov offered to step aside as president if Khrushchev so desired. Nesmeyanov was soon replaced. In a few years time, Khruschev was removed from office and his Politburo colleagues ascribed the need to replace him to Khrushchev's damaging activities; as an example, they mentioned that he might have abolished the Academy of Sciences.

Nesmeyanov's removal from the office of the Academy presidency did not cause any loss of his authority among his peers. Soon, they elected him to be head of the chemistry division of the Academy of Sciences. For the next dozen or so years he was responsible for the entire science of chemistry in the Soviet Union. He continued as director of his institute INEOS and as chair of organic chemistry at Moscow State University.

Nesmeyanov's scientific output did not suffer just because he held high administrative positions; the numbers even show the opposite. For his entire career, from his first publication in 1923 to his last in 1979, a total of fifty-six years, he figured as author or coauthor on 1,133 publications. Of those fifty-six years, he occupied high offices for a total of twenty-seven years, during which time his name appeared on 1,040 publications. So the average annual number of publications he produced while holding various official positions was thirty-eight papers per year, which means that he needed barely ten days to do a paper on average, continuously, for a twenty-seven-year period that included his being rector of Moscow State University and then president of the Soviet Academy of Sciences, along with other positions. This was a rare though not unprecedented accomplishment. A little later, during the period 1981–1990, a crystallographer in Nesmeyanov's institute, Yurii Struchkov, had a higher frequency of publications, averaging a paper a little less than every four days.[10] Unlike Nesmeyanov though, he held no administrative positions. Struchkov was a workaholic, and his X-ray diffractometer yielded structure data on crystals in an already near-publishable form. But Struchkov was known to rigorously scrutinize every manuscript before it left his desk.

Nesmeyanov had so many publications that their sheer number would have made it impossible for him to scrutinize them all, let alone to participate in all creatively. But he had great ideas and contributed to initiating a lot of new work. Also, Soviet publication habits were rather peculiar. Papers were often written and published that should have been considered only parts of a larger paper. Fragmentation of publications was very much in vogue, and not only to lengthen the publication lists of the big bosses—institute directors, for example. Very often, doctoral students needed those publications to get confirmed for their next year of study, and, ultimately, for the defense of their dissertation. Also, if there was a scientific discussion, the arguments flew back and forth, resulting in additional publications, and associates of INEOS often participated in such scientific debates. If both sides of the arguments involved INEOS people, it could happen that Nesmeyanov would appear coauthor of papers on both sides of the debate. It happened on such an occasion that he received two manuscripts for his signature at the same time, representing two sides of a debate, and he asked plaintively, "In which am I right?"

Nesmeyanov was a classical chemist who had experience in "wet" chemistry: he used to work in the laboratory with sizeable amounts of substances rather than with physical instrumentation that often required the use of only miniscule amounts of substances and no manipulation of the materials. He supported modern techniques and was interested in them, but could not always hide his unhappiness when they yielded results that differed from his expectations. But usually, sometimes after some persuasion, he accepted results that did not match what he had hoped for.[11]

His younger associates did not know that Nesmeyanov had helped his colleagues to acquire knowledge of these physical techniques. When, during the first period of the war in 1941, the research institutes in Moscow had to be evacuated, the Institute of Organic Chemistry moved to Kazan. The whole institute received two rooms at the local university, and the scientists were sent there without any equipment or other means to do their work except a few books. Nesmeyanov decided that the time should be used for learning. He divided the collective into six groups. Each group had to attend a course by a specialist in a given field. One of the specialists was a crystal chemist, later of international renown, Georgii Bokii. When Bokii started his course of crystal chemistry—the first course he ever gave in his life—he noticed among his "students" the director of the institute, Nesmeyanov.[12]

The year 1941 was sad for Nesmeyanov, not only because of the war in general, but for family reasons as well. His brother, Vasilii Nesmeyanov, fell victim to Stalin's Terror and was executed in 1941. This tragedy may have helped Nesmeyanov develop sympathy for the persecuted. Years later, there was, for example, the Razuvaev case. The chemist Grigorii Razuvaev had been declared "an enemy of the people" and was working in a labor camp in the faraway Komi Autonomous Republic (northeast of Moscow, west of the Ural Mountains). He was involved with the extraction of radium from water. Once he had to transport the product of his work to Moscow, and he used the occasion to visit his old acquaintance Nesmeyanov. He at once decided that Razuvaev should be given the opportunity to acquire his Candidate of Science (PhD equivalent) degree on the basis of his earlier publications, whose reprints had miraculously survived. The defense was successful and Razuvaev received his degree. Soon, he was set free from the camp, and it was a mere few months after the granting of his first degree that Nesmeyanov organized for him his defense of the higher doctorate, again, on the basis of old works published ten years before. During the defense, Nesmeyanov made a statement that Razuvaev's work had not become obsolete; on the contrary, it had withstood the test of the most objective judge—time. He called on everybody to vote "for" giving Razuvaev the degree. This happened under Stalin.[13]

Nesmeyanov was especially protective of his immediate coworkers although some of his actions may appear strange in hindsight. Thus, when in 1949, his close associate R. Kh. Freidlina's husband was arrested on trumped-up charges, Nesmeyanov immediately demoted her from the position of laboratory head. Nesmeyanov's action could be interpreted in different ways, but Freidlina was convinced that it saved her by making her "invisible."[14]

Another remarkable episode happened in 1952 at Moscow State University. It was the time of the "fight" against the "doctors-plotters"—the mostly Jewish doctors accused of planning the assassination of Stalin and other Soviet leaders. There was a big meeting of the associates of the department of organic chemistry. It was a large collective, among them was a member of the protein laboratory by the name of Maria Botvinnik—long-time scientific secretary of the department. This fifty-year-old lady was popular among her colleagues both for her professional and human qualities. The only exception was her immediate boss Nikolai Gavrilov. Part of the reason for his belligerence was that her experimental results did not agree with his theory. Once, as Gavrilov was entering the crowded auditorium, he said in a loud voice: "Among us, there is a Zionist agent," and pointed to Dr. Botvinnik. The accusation was the heaviest possible under the circumstances; the auditorium froze, and the silence was profound. It was broken by Nesmeyanov's quiet voice. "What you are saying, Nikolai Ivanovich is beyond what a healthy brain could produce," he said. He thus crushed at the outset a possible anti-Semitic campaign in the chemistry department.[15]

These incidents, though, were isolated cases. Nesmeyanov could not have occupied some of the highest positions in the Soviet hierarchy without showing harmony with Stalin's and then Khrushchev's regimes and without being their devoted servant. He did not tolerate political dissent in his INEOS. He did not like it when anybody deviated even to the minutest degree "from the only correct line," the party line. When a scientist (not a Nesmeyanov subordinate) published a highly critical paper about Lysenko and his "teachings" abroad, one of the associates of Nesmeyanov's institute, Natalya Gambaryan, organized a seminar about Lysenko, and only the strongest intervention of her boss, academician I. L. Knunyants, saved her from being fired.

Aleksandr Nesmeyanov, third from the right, with his closest associates still at the Institute of Organic Chemistry. Seven of the seventeen scientists in this picture became corresponding or full members of the Soviet Academy of Sciences: From the left, 2-Kuzma Andrianov, 5-Mark Volpin, 11-Dmitrii Kursanov, 13-Martin Kabachnik, 15-Rakhil Freidlina, 16-Nesmeyanov, and 17-Vasilii Korshak.

Source: Courtesy of Jan Kandror, Wiesbaden, Germany.

When in 1968, Czechoslovakia was occupied and the Prague Spring was brutally suppressed, there was a meeting at INEOS to express support for this despicable action. Yet when a vote took place, a few did vote against and a few others abstained. They could not be penalized directly just for a vote, but one of them, Yurii Aronov, was made into a scapegoat, and his life was made miserable.[16] He lost his job and his right of residing in Moscow (he was from a different place and living in Moscow was a special privilege). For a long time this excellent scientist was unable to find employment. Another unfortunate case involved Yulia Zaks, who made a statement in favor of the human rights activists Petr Yakir and Viktor Krasin, who in 1972 were being persecuted. Zaks was Jewish, which did not bode well for her because Nesmeyanov had been repeatedly accused of employing too many Jews in his Institute. She was summarily fired from INEOS.

When Sakharov was being denounced by some of his fellow members of the Science Academy for his dissident activities, Nesmeyanov asked the leading scientists of INEOS to sign a letter of condemnation. He was a true Soviet-style courtier. The readers of Nesmeyanov's memoirs, published two decades after his death and one decade after the collapse of the Soviet Union, were astonished by the great deal of nonsense about the greatness of Stalin in them.[17] It was impossible for Nesmeyanov to be oblivious to the conditions in Stalin's Soviet Union. He must have been well informed in general, but also through the fate of his own brother. It was a puzzle why those memoirs were released in 1999. He was compared to the churchman who did not believe in God, but served the Church for such a long time that he was unable not only to speak, but also to think differently anymore.[18]

Soon after Nesmeyanov was made President of the Soviet Academy of Sciences in 1951, his standing in organic chemistry was severely tested. There had already been attacks against science within the framework of Stalin's anti-science campaign. Now it was chemistry's turn. A meeting was called to discuss the structure theories of organic chemistry, and the irony in Nesmeyanov's position could not be missed. Unfortunately, the chemists, similarly to the physicists, could not avert an ideological attack. It is true, though, that the attack on chemistry resulted in much less severe consequences than what happened to biology. The chemistry meeting lasted four days, June 11–14, 1951, in Moscow. Four hundred and fifty chemists, physicists, and philosophers attended, representing major centers of scientific research and higher education from all over the Soviet Union. There was a report, "The Status of Chemical Structure Theory in Organic Chemistry," compiled by a special commission of the Chemistry Division of the Academy of Sciences, followed by forty-three oral contributions. An additional twelve contributions were submitted in writing.

The conference adopted a resolution and sent a letter to I. V. Stalin. This letter expressed self-criticism for past deficiencies in appreciating the role of theory and theoretical generalizations in chemical research. It stated that the foreign concept of "resonance" had spread among some of the Soviet scientists. It called this concept an attempt at the liquidation of the materialistic foundations of structure theory. The letter further mentioned that the Soviet chemists had already started their struggle against the ideological concepts of bourgeois science. The letter concluded that the falseness of the so-called theory of resonance had already been unmasked and that all efforts would be made to cleanse Soviet chemical sciences of the remnants of this concept. Looking back, this was obvious nonsense, but at the time it had serious consequences. It was ironic that at the meeting, which in itself was the manifestation of suppression of the freedom of research, there were repeated references to Stalin's teachings on the importance of the struggle between differing opinions and of the freedom of criticism. In reality, Stalin's regime did not tolerate any dissenting voice from the communist ideology or what was assumed to be the communist ideology. The minutes of the meeting have appeared in a hardbound volume of 440 densely printed pages.[19]

The theory of resonance in chemistry described the structure of a molecule by a set of "resonating" forms. The approach proved useful when the structure of a molecule was difficult to describe in a unique way. For example, the benzene molecule, C_6H_6, has a structure intermediate between two extremes that were shown as if resonating between the two. In reality there is only one structure of benzene, but it was found convenient to represent it by this "resonating" model.

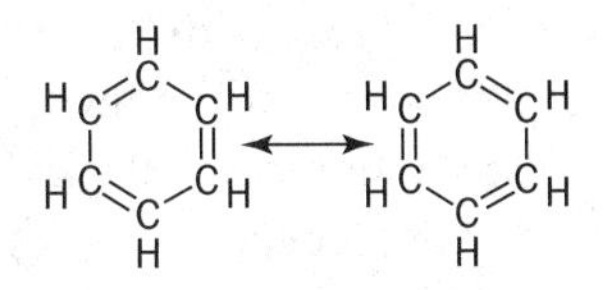

The average of these two structures was found to be in perfect agreement with the experimental data. Linus Pauling, one of the giants of twentieth-century chemistry, was the most visible among the creators and propagators of the resonance approach. He used it extensively in his classical monograph *The Nature of the Chemical Bond.*[20] Another American chemist, George Wheland, published a monograph about the theory and its applications to organic chemistry. In it Wheland stressed that "resonance is a man-made concept... it does not correspond to any intrinsic property of the molecule itself, but instead it is only a mathematical device, deliberately invented by the physicist or chemist for his own convenience."[21] Reviewing the Soviet chemical literature of the time reveals that an astonishing war was waged against this innocent approach, which, by the way, in the West had its proponents and opponents, without any ideological coloring.

The report of the Chemistry Division of the Science Academy was submitted to the meeting by a special commission whose members included a number of academicians and future academicians.[22] Some of them, M. I. Kabachnik and R. Kh. Freidlina, were Nesmeyanov's closest associates. One section of the report discussed the mistakes of Soviet chemists. Here, we learn that G. V. Chelintsev actively criticized the concept of "resonance" in the open press. It was to a great extent attributed to him that Soviet scientific society had turned to this question. The basis of Chelintsev's criticism was his own "new structure theory" which, however, completely contradicted the modern theory of chemical structure and was contrary to the experimental facts and theoretical foundations of quantum physics. Ya. K. Syrkin and M. E. Dyatkina were named as the main culprits in disseminating the theory of resonance in the Soviet Union. They were accused of having further advanced the erroneous concepts of Pauling and Wheland, ignoring the works of Soviet and Russian scientists, idolizing foreign authorities, and quoting works of secondary importance by American and English authors.

Others were also criticized, among them the organic chemists Nesmeyanov and Freidlina. They had interpreted the diverse reactivity of compounds of mercury using the resonance theory. These lesser sinners, however, along with many others, eventually repented, and have become critics of the application of resonance. This must have been humiliating for Nesmeyanov, but it was a lesser evil, and in the final account, it lifted him from a defensive position. In this situation, he certainly acted as a seasoned politician.

He and the other genuine scientists were lucky in that it was only a small, though very vocal, group that blindly attacked the theory of resonance and, especially viciously, the alleged proponents of the theory. Quantum chemistry and all of the science of the West were also under attack. Return to historical Russian achievements was advocated. Another lucky development was that the attackers of the resonance theory proposed their own theories for general acceptance. These theories were shown to be worthless by many. Nonetheless, all participants painstakingly dissociated themselves from the theory of resonance. At times, the self-criticism of some excellent scientists was humiliating in the extreme.

The philosopher B. M. Kedrov declared Erwin Schrödinger to be a representative of modern "physical" idealism, which put him in the same category as Pauling. Furthermore, he stated that Paul Dirac's superposition principle was as idealistic as Werner Heisenberg's complementarity principle and even more idealistic than Pauling's theory of resonance. Kedrov was obviously joining in the attack and applying no self-restraint. He was a philosopher of natural sciences who had been trained in chemistry at Moscow State University and subsequently had become a philosopher. He had started a promising career, but suffered a setback when his father, an old-time communist was arrested during the 1937–1938 Terror. After the war, Kedrov served for a time as a relatively liberal editor of the new journal *Voprosi Filosofii* (Problems in Philosophy). In 1948, he was demoted from this position in the course of ideological struggles and was soon fired from the Institute of Philosophy, where he had risen to the post of deputy director. In subsequent years, Kedrov became a fierce fighter against alien ideologies, considerably improving his career prospects. His performance at the chemistry meeting must have pleased the most hardline party officials. His outbursts against modern physics and similar statements by a few others give the impression that previously prepared materials against the physicists were being used.

It was characteristic of the times when the writer V. E. Lvov, who was absolutely ignorant of chemistry, criticized the commission report for a serious political error. He lamented that the protagonists of the theory of resonance were still being considered to be the greatest Soviet scientists. These protagonists had been unmasked as spokesmen of the Anglo-American bourgeois pseudoscience by the press and by Soviet society. According to Lvov, the report was vague about the main thrust of the ideological struggle taking place in theoretical chemistry. He also quoted, as a positive example, the criticism of Mendel by T. D. Lysenko, who "proved" that Mendel's work had nothing to do with the science of biology. Furthermore, Lvov fiercely attacked the theories of Heisenberg as well as those of Heitler and London—all universally recognized authorities in science. He protested the report's view that quantum mechanics was a development of the classical Russian scientist Butlerov's teachings. The most important political task of Soviet chemistry, he declared, was the isolation and capitulation of the group of unrepenting proponents of the ideology of resonance.

The last entry in the minutes of the meeting is a dissenting opinion in the form of a short letter by E. A. Shilov, member of the Ukrainian Academy of Sciences. He is critical of the report and the resolution of the meeting for looking backward rather than forward. He suggested concentrating on new results and new teachings instead of conducting scholastic debates about meaningless questions. Shilov added that the result of ending such debates would be that the efforts and time of Soviet organic chemists could be devoted to valid and productive work. Professor Shilov's contribution was not delivered as an oral presentation during the meeting. Considering the prevailing atmosphere, it was a uniquely brave contribution.

The significance of the Moscow meeting can be evaluated on a number of levels. At one level, the records speak for themselves and provide an excellent demonstration of Orwellian doublespeak; but even George Orwell's famous book *1984* pales by comparison. They also show the fear in the Soviet political system of everything coming from the West, even if it was only a chemical theory. The inferences of the theory of resonance appear exaggerated beyond any reasonable limit, and the statements reflect the atmosphere of a staged trial rather than a scientific discussion. There are chemists who do not like the description of molecular structure as a series of resonance structures. What is mind-boggling is that such a dislike was made into an official dogma with philosophical justification. The story of resonance, however, should not be viewed in isolation from the rest of Soviet life in the early 1950s, the last years of Stalin's reign. To me, the question that is most telling is the one asked of M. E. Dyatkina: "How do you explain that you are so conspicuously familiar with the teachings of foreign scientists? May it be that you, along with Professor Syrkin, are intentionally bowing to foreign scientists?"

According to a different evaluation, the Moscow meeting had a distinctly positive significance. It was governed by a healthy mechanism of self-defense by the higher echelons of the Soviet chemical establishment. Rather than letting harsher outside interference crush it, as had happened to some other fields of science, this establishment organized itself a milder purge. By the same stroke, it ridiculed and marginalized the pseudoscientific extremists in its own ranks, whose most vocal representative was Professor Chelintsev. It is true that jobs in chemistry were lost due to ideological controversy, but lives were not lost due to political interference as occurred in other areas.

There were long-term negative consequences of the resonance controversy for Soviet chemistry. Many of the brightest young scientists in the Soviet Union stayed away from theoretical chemistry, if not from chemistry, for decades. Theoretical chemistry was just not the field to be associated with. Vladimir Tatevskii, one of the most influential chemistry professors at Moscow State University, had participated in the attacks against the theory of resonance. He exerted influence at the Faculty of Chemistry for decades. His statements in 1951–1952 were as vitriolic against Pauling as against Syrkin and Dyatkina.[23] He was a meticulous specialist and a demanding one, so he was both revered and feared for a long time. Even today, six decades later, at the Faculty of Chemistry of Moscow State University, there is a great deal of ambiguity in judging him and his role in the resonance controversy.

One of the surprising features of the debate was the fierce attacks on Linus Pauling, who at about the same time was having also difficulties in the United States because of his leftist views. His passport was taken away to prevent him from attending conferences in Europe. When in 1993, I asked him about the Soviet resonance controversy, he put all Soviet chemists into the same group and opined that the Soviet chemists needed some time to properly appreciate the resonance theory.[24] Pauling's response appeared as ignorant as it was unfair. There were chemists at the Moscow meeting who not only understood the resonance concept but had applied it extensively. Syrkin and Dyatkina were the best known among them. Nesmeyanov had also utilized the theory, even if he subsequently dissociated himself from it. After Pauling's leftist politics became known, he was never again attacked in the Soviet Union; he was considered to be a great friend of the socialist state.

In the academic year 1964/65, Dyatkina gave a course on inorganic structures. It was not for credit, and it was not given at the University but at one of the research institutes in Moscow. I traveled every week to this institute to attend her lectures. The large audience came from all over the city. The theory of resonance was no longer an issue of ideology, but she freely used another concept, electronegativity. It also came primarily from Pauling's teachings, and it was also a useful though not very exact concept about the affinity of the atoms of different elements for electrons. At the University, there was a group of influential professors who considered the concept of electronegativity ideologically unacceptable. This was fifteen years after the resonance controversy. The opposition to the concept of electronegativity was, however, much less fierce than the opposition to resonance had been, and ideology had a diminished influence on science at the time, compared with the Stalin years; yet it seemed as if history was repeating itself, though on a mini-scale.

Aleksandr N. Nesmeyanov may have been humiliated at the 1951 meeting, but he was not defeated, and those who claim that he saved Soviet chemistry from graver damages may be right. Only a control experiment could have provided a more exact answer to this question. Nesmeyanov was a pragmatic person, and after this meeting he continued to serve as president of the Academy for almost a decade. In this position he survived Stalin, but did not survive Khrushchev. This was a paradox of his career.

It was the period following his Academy presidency that brought considerable publicity for Nesmeyanov on account of a special direction in chemical research he was developing, aimed at creating synthetic food. There could be two reasons he embarked on this path. His dedicated vegetarianism prompted him to ease the food situation in case animal sources ceased to be used for human consumption—this was an unlikely scenario. The other was more realistic, the continuing poor state of Soviet agriculture that was unable to feed the people in spite of the addition of tremendous virgin areas to arable land. Nesmeyanov's institute devoted a great deal of effort to working out the techniques for producing synthetic food. They produced black caviar, which was a popular item and proved to be a publicity success—I remember the excitement it caused in Moscow. In addition to the chemical syntheses, many other problems of technology would have to be solved before consumers could be offered the synthetic black caviar. Nesmeyanov held quite utopian views about the role of his synthetic food, and he declared, in part (italics added):[25]

> The laborious and *poorly productive agriculture is a thing of the past*.... Those industries will also be gone that produced machines, fuel fertilizers, and herbicides for agriculture. *34% of the population who worked in agriculture* is now available for more creative work....
>
> There are no more years of crop failure or places of failure. There are no more the vast food losses owing to inclement changes in the weather, natural disasters, pests, spoiling, rotting, freezing, etc., which now destroy the greater part of the harvest. The villages will become towns and the towns will be transformed to garden towns. The professional cooks and waiters associated with domestic food preparation will have completely disappeared. The housewife will be free since the food is ready, packed and tinned but complete with vitamins. It needs at most heating....
>
> The ploughed area gradually reduces while parks and forests expand. River drying and shoaling ceases and the great problem of water pollution is solved along with food excess.

When Nesmeyanov spoke about the poor productivity of agriculture and the large proportion of the population, one-third of the total, engaged in agriculture, he was obviously referring to the Soviet conditions. He let his imagination go, but it was a typical Soviet approach to get carried away by the potential benefits of a project while forgetting about the potential drawbacks.

Nesmeyanov was very good in thinking up projects on a smaller scale as well, and in particular in designing new molecules. One example is mentioned here. Among the element-organic substances, Nesmeyanov was thinking of carbon cages which would envelope one or more hetero-atoms within their cages. These were the endohedral polyhedral clusters, labeled $M@C_nH_n$ where C_nH_n means a hydrocarbon skeleton; M, the heteroatom, a metal, for example; and @ meant that the first part of the formula was enveloped by the second part. It was decided that first the feasibility of such structures would be tested by computations. The Quantum Chemistry Laboratory of Nesmeyanov's institute was assigned to the task. These structures were too large for the available computational possibilities at the time, so the task was simplified. First they wanted to examine the stability of cages without anything inside them, and ignoring the hydrogen atoms bound to the respective carbon atoms of the cage.

A young associate of the laboratory, Elena Galpern, was charged with the work under the supervision of the laboratory head D. A. Bochvar. She started with the carbon cage consisting of twenty carbon atoms, C_{20}, called "carbododecahedron." The name referred to the exclusively carbon cage having the shape of the dodecahedron. However, the cage of the C_{20} molecule would have been too small to house anything. The next target, C_{60}, was large enough to envision a metal atom inside. First, however, the shape of the cage had to be determined.

As she was computing various shapes, her mathematician colleague Ivan Stankevich, who often played soccer, brought his soccer ball into the laboratory one day. He told her that "twenty-two healthy men are kicking this ball for hours and it is not destroyed. A molecule of its shape must also be very sturdy."[26] The recently introduced soccer ball was sewn together from pentagonal and hexagonal patches; the ball resembled a truncated icosahedron, and the truncated icosahedron has sixty apexes. The subsequent computations showed this shape to be stable, and this result constituted a discovery whose significance Galpern, Bochvar, and Stankevich could not appreciate at the time. Nonetheless, they published a report about it in the prestigious Soviet periodical of the Science Academy, *Doklady Akademii Nauk* (Proceedings of the Science Academy).[27] Nesmeyanov acted as the academician sponsoring the manuscript for publication.

Elena Galpern, whose calculations predicted the molecule that later became known as buckminsterfullerene.

Source: Courtesy of Elena Galpern, Moscow.

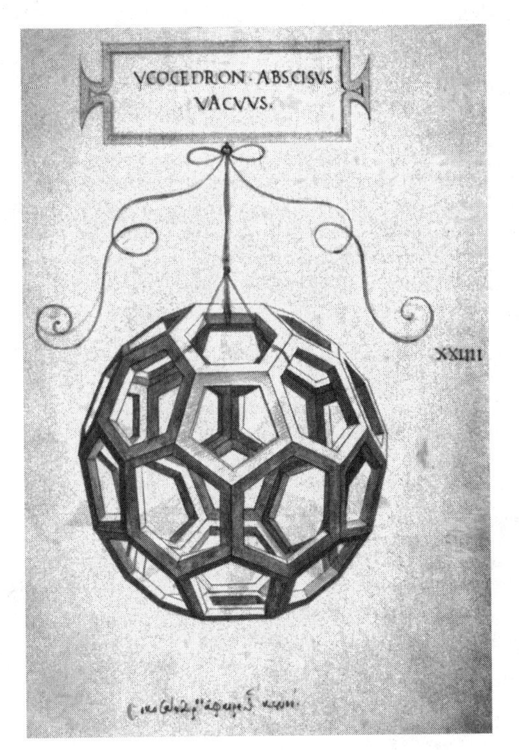

A truncated icosahedron—the shape of the buckminsterfullere molecule, represented here by Leonardo da Vinci's drawing for Luca Pacioli's *De divina proportione*.

Unknown to Galpern and the others, by the time they did this work, there had been a suggestion for an all-carbon truncated icosahedral molecular shape by Eiji Osawa in Japan. He based his suggestion on qualitative geometrical considerations. A dozen years later, the actual discovery of C_{60} happened, eventually leading to a chemistry Nobel Prize for the leading members of the British-American team.[28] In that work, the molecule was given the name "buckminsterfullerene," after the American designer R. Buckminster Fuller whose exhibition hall at the 1967 World Exposition in Montreal was constructed from pentagons and hexagons. The process culminated in yet another work in which for the first time C_{60} was produced in the laboratory by a German-American group.[29] This was when all those interested in the C_{60} story wanted to learn about the antecedents of the discovery, and Galpern's study entered the limelight. When asked about her computations, she always gave credit to the assistance she received from her colleagues and pointed to the originator of the idea, Aleksandr Nesmeyanov.

Epilogue

There are fourteen heroes in the twelve chapters of this book, and though I met personally only half of them, I no longer consider the other half to be strangers, either. I hope the reader will feel similarly. A few questions have puzzled me and might also have puzzled my readers. First, how could these extraordinarily gifted men stay loyal to an oppressive system for so long? Second, did conditions in the Soviet Union facilitate or hinder their creativity and productivity? Third, will the extraordinary performance of these Soviet scientists be continued by scientists in Russia and the other successor states?

The first question brings to mind the tragedies that the families of almost all the scientists suffered, particularly at the hands of Stalin and his henchmen. Tamm's brother was murdered and so was Nesmeyanov's; Khariton's father perished in the Gulag; Landau was incarcerated for a year; and Sakharov spent seven years in exile. All the scientists in this book were treated as prisoners in their own country until the end of existence of the Soviet Union.

Stalin's Terror did not appear out of the blue. The transition following the enthusiastic first years after the victory of the Socialist Revolution was gradual, in spite of the dictatorship of Lenin and his comrades from the beginning of their regime. From the very start, the Bolshevik terror led to tragedies for millions of human beings. Yet it is also true that in the 1920s tremendous creative forces were liberated in the Soviet Union. Suffice it to remember the peaks of creativity in the suprematist paintings of Kazimir Malevich and other avant-garde artists and in the discovery of the branched chain reactions in chemistry by Nikolai Semenov and his associates.

Initially, the Soviet State encouraged contact with Western scientists. This facilitated organized exchanges and the welcoming of foreign scholars. In the early 1930s the situation started to change. Parallel to buildup of Stalin's isolationism and totalitarianism, a unique socialist nationalism developed. It was conveniently reinforced by the life-or-death necessity of defending the Soviet Fatherland from the Nazi invaders in a war that was labeled in the Soviet Union the Great Patriotic War. This blended Stalin, communism, and patriotism in the minds of many Soviet citizens. During the war, quite a few Soviet scientists who had stayed out of the party now joined it.

World War II was barely over when the Soviets perceived a new foreign danger, one that had bombs of unprecedented power. Both sides were afraid of each other. Stalin grabbed whatever territory he could in Eastern Europe in the wake of the victory over Nazism, and erected an Iron Curtain over his empire.[1] This made Western Europe especially uneasy. And in the United States, there was talk of a preventive strike against the Soviet Union before it could fully develop its own nuclear capabilities. The great mathematician and contributor to the American nuclear power John von Neumann wanted a first nuclear strike on Moscow. In 1950, he declared, "If you say why not bomb them tomorrow, I say why not today? If you say today at five o'clock, I say why not one o'clock?"[2] The input of scientists was essential for the Soviet Union's ability to defend itself. The scientists participating in developing Soviet nuclear might—almost without exception—felt that it was their patriotic duty to work toward this goal.

Much of the letter-writing activities and other opinions expressed by scientists that were directed to the Soviet leaders also stemmed from the feeling of duty. Kapitza displayed extraordinary bravery in his sending his letters of protest to Stalin. Although the dictator seemingly ignored Kapitza's letters, almost never answering them, he was hurt when Kapitza stopped writing them. Khrushchev may have lectured and humiliated Sakharov for his stand on nuclear testing, yet the Soviet leader did not feel immune to the criticism of his decisions by this scientist giant. The picture forms that the Soviet leaders were apprehensive about what the intellectuals might do or say. Otherwise, why would they have blackmailed Boris Pasternak into declining the Nobel Prize? The situation of intellectuals in the Soviet Union and in the United States was different. The American intellectuals could freely voice their opinion about, including strong opposition against, administration policies without risking anything, but then would the American political leadership be influenced by their criticism?

There was a universal feeling of relief—at least among the Soviet intelligentsia—when the thaw started in the Soviet Union. It culminated in Nikita Khrushchev's secret speech in February 1956, on the closing day of the 20th Congress of the Communist Party, in which he unmasked Stalin and his crimes. But that this was done by one of Stalin's closest aids in itself meant that there would be a limit to the liberating changes that were taking place. The restrictions of Soviet life eased greatly in comparison to conditions under Stalin, but the totalitarian and dictatorial character of the Soviet regime continued.

The post-Stalin Soviet Union never truly confronted its history under Stalin. Even though physical extermination of their opponents was no longer practiced, Stalin's successors carried on his regime in other ways. In its actions, the post-Stalin regime recognized itself as the inheritor of Stalin's times. Rather than getting rid of past crimes and their perpetrators, they were excused and absolved and their crimes hushed over. Trofim Lysenko continued his reign over biology under Khrushchev, and when both he and Khrushchev were finally removed, the past continued to haunt the field.

Nikolai Vavilov, the world-renowned plant botanist, was one of the best-known victims of the Stalin-Lysenko terror. His principal "crime" was that he held fast to his scientific views and did not succumb to Lysenko's unscientific "theories."

In 1943, he starved to death in prison in Saratov. Many years later, after he was exonerated from all his "crimes," his pupils and former colleagues wanted to erect a memorial to him in the Saratov cemetery, where his remains rested. For years, Soviet officialdom stayed silent when petitioned for assistance. The money was raised from private contributions, and then the statue was done by a recognized sculptor. The official Soviet organizations now felt it incumbent upon themselves to participate in the planned unveiling ceremony, but they did their best to minimize the significance of the event. They changed its date so that people who had long planned to attend could not. When the statue was finally unveiled, the few hundred mourners who were there were stunned because the face bore no resemblance to that of Nikolai Vavilov. A representative of the Soviet cultural authority, without whose permission the statue could not be erected, had forced the sculptor to alter the features of the face. Wrinkles had to be removed lest they suggest the harsh conditions of Vavilov's prison stay, and a smile had to be carved on it, radiating happiness rather than an expression that might imply an unhappy past.[3] The Soviet authorities had been willing to allow the monument but preferred to mask what really happened rather than to acknowledge it, even though it had happened three decades earlier. This was in the 1970s when the Soviets were still in power.

Even in today's Russia, there is a tendency to gloss over the darker sides of Stalinism. I happened to be visiting in Russia in August 2012 when there were commemorative programs on television about the recently deceased Sergei Kapitza, Petr Kapitza's oldest son. He was a physicist, well known for his long-standing popular science programs on TV. His life and career were detailed in the commemorations. It was mentioned that he was born in 1928 Cambridge, England, and that when he was six years old, his family had *moved* to Moscow. There was, however, no mention that in 1934, on one of his father's visits to his home country, on Stalin's direct order, he was forced to stay in the Soviet Union, and that this is why the family "moved" there (see chapter 4).

The way the Soviet Union treated its people and its scientists changed only incrementally during its existence. Contrary to popular belief, things did not change dramatically when Mikhail Gorbachev came to power in spring 1985. There were still political prisoners in the Soviet Union even though Gorbachev had assured the world that there were none. Sakharov remained in exile in Gorky for *twenty* more months after Gorbachev came to power. Gorbachev knew about Sakharov's plight, and not just because everyone knew about it, but also because Sakharov had written to Gorbachev, and the new Soviet leader did receive his letter. But it took painstaking and lengthy negotiations before Gorbachev finally "invited" the Sakharovs back to Moscow. I mention this to illustrate the point that until the collapse of the Soviet Union, the Soviet regime did not change in a fundamental way.

This does not, however, contradict the notion that the scientists, especially the top scientists in fields vital to the security of Soviet Union, received preferential treatment compared to the rest of the population. They did. But this was not altruism on the part of the regime; rather, it was pure pragmatism in that the regime recognized its own interests. In this, the Soviet leadership was more enlightened than, for example, the British at the time of World War I when they indiscriminately sent scientists to the front, or Hitler, who in 1933 declared that Germany would rather be without science for a few years than tolerate the presence of Jewish scientists. This difference in the Soviet treatment of scientists took time to develop. In the Terror of 1937–1938, scientists were arrested and executed just like anybody else. However, by the time World War II engulfed the Soviet Union, Stalin made sure that scientists were not sent to the front. When the Soviet atomic bomb program began, Stalin stressed the need to create improved working and living conditions for scientists and their families.

This brings us to the second question posed earlier about whether or not the conditions of Soviet life facilitated or hindered the creativity and productivity of scientists. It is generally felt that in the absence of free thought and free speech, science cannot flourish. The fates of cybernetics (informatics) and biology in the Soviet Union, from the mid-1930s on, are examples of the tragedy of science under such circumstances. Physics, in particular nuclear physics, was at the other end of the spectrum. The physicists made the Soviet leadership choose between adherence to rigorous Marxist ideology, on the one hand, and letting the physicists utilize theories that Soviet philosophers had declared to be ideological heresy, on the other. Only the application of the "heretic" theories of relativity and quantum mechanics would enable them to build the world's most powerful bombs. Chemists were in an intermediate position. When they were attacked on ideological grounds, their most outstanding representatives did not let the hard-liners triumph. They let theoretical chemistry be the scapegoat—it did not appear to be too big a sacrifice—and for decades theoretical chemistry was suspect in the Soviet Union.

From the early 1930s to the late 1980s, travel restrictions were detrimental to the work of most Soviet scientists. Some of the top scientists (Zeldovich in chapter 2 and Abrikosov in chapter 7 figured as examples in our narrative) maintained a world map on which they marked all the places from which they had received invitations to visit, and were not allowed to go.

Lacking other attractions, exceptionally gifted people could spend a lifetime on research without too much concern about anything else. They would be given homes, job security, reasonable access to health care facilities, freedom from teaching obligations if they so desired, and even a small allowance in hard currency so they could order Western goods and literature. To the end of his life, Petr Kapitza smoked imported pipe tobacco. Here, we are talking about the uppermost levels of the science hierarchy, but that is the circle we were concerned with in this book. On a lower but still quite privileged level, for example, the top professors had reserved tables and received preferential service in the cafeteria at Moscow State University. It was a scheme, superbly effective, that perks and privileges had many layers in Soviet society. There was always the incentive for someone to move up to the next level. Since we cannot do a control experiment, we must stay undecided on whether and how much the creativity and productivity of the top Soviet scientists suffered under the conditions of Soviet life. Where I have no doubt is that all those I knew personally would have given up their perks and privileges in exchange for personal freedom, including the freedom to travel. The scientists who got out and gained experience in Western Europe and the United States knew only too well that most of what were considered perks and privileges in Soviet society was freely available to members of the middle class in those countries, and considered nothing special.

As for creativity, Rita Levi-Montalcini describes the conditions in Fascist Italy when the racial laws prevented her from continuing her studies and research at the university. She continued them at home with added drive. Professor Giuseppe Levi and a few young people around him, including Levi-Montalcini, acted as if the outside world did not exist. They found refuge in science; it strengthened their resilience, and this helped them not only to survive but also to create.[4]

Resilience served also as protective cushion for the Soviet scientists whose lives and work we have read about in the preceding pages. This resilience manifested itself not only in the prisons and the gulag, but in their resistance to the constant temptations with which the corrupt regime tested its members. Our heroes remained faithful to science. Nor is the question we consider here unique to scientists. Regarding the exceptional poetry of Osip Mandelshtam, Anna Akhmatova, Boris Pasternak, and Marina Tsvetaeva, the question has already been asked: "Would the four have written less remarkable work if they had had an easier time?"[5] Mandelshtam, Akhmatova, Pasternak, and Tsvetaeva were as great in Russian literature as the scientists in this book were in their fields. These poets went through trying and tragic ordeals, and there were some who felt that, yes, those ordeals helped them to create their masterpieces. But for the widow of the martyr Mandelshtam, "There was nothing in the least elevating about her and her husband's sufferings: there was only terror and pain."[6] Mandelshtam perished as a prisoner; others survived, but their fate was not easy. As Akhmatova noted: "Shakespeare's plays—the sensational atrocities, passions, duels—are child's play compared to the life of each one of us. Of the suffering of those executed and sent to concentration camps I dare not speak.... But even our disaster-free biographies are Shakesperian tragedy multiplied by a thousand."[7]

Osip Mandelshtam expresses succinctly what Levi-Montalcini described about the necessity of continuing to create even under the most impossible conditions. The title of his poem is "In nightingale's fever" (other titles have also been used). I quote here the last stanza of this poem in Ilya Shambat's translation:[8]

Because it's helpless here
As the innocent are killed
Heart is in nightingale fever
And remains warm still.

Other translations exist.[9] The resilience of "nightingale fever" characterized both the poets and the scientists who were motivated to create.

As for the third question, about whether we might soon see the extraordinary performance of these Soviet scientists replicated in Russia and the other successor states, my guess is that the answer is in the negative. Of course, by inertia, most of the older scientists have continued their dedicated work in the mostly deteriorating conditions for research in Russia. It is not only that material conditions have worsened, but the prestige of science has also suffered. Thus far, formerly nonexistent or underrecognized professions have gained prominence in Russian society, for example, banking, marketing, and advertising—generally, all kinds of business-related activities as well as communications and communication technologies. These professions have attracted talented young people, as they would in any society, but this seems to have been at the expense of the research professions. Furthermore, many former Soviet scientists sought better working and living conditions, and better futures for their children, abroad. There were top scientists among them; Alexei Abrikosov was merely one example. Tellingly, many of the scientist children of the top Soviet scientists now live and work in Western Europe and the United States. But the migration was even broader. A few years ago, I was invited to be on the jury on a top Middle Eastern international science prize in a country hardly distinguished for its science. I used the occasion to visit a local university and was impressed by the laboratories equipped with the most expensive instrumentation. The researchers were dressed in conventional Western attire, quite different from my hosts' traditional robes, and they made me curious. On impulse, I turned to the researchers in Russian, and they responded in Russian without batting an eye.

On my latest visit to Moscow, in 2011, just as before, I was most favorably impressed by the high level of scientific discussion. But the absence of long-term governmental commitment and the lack of other sources of support does not make one optimistic about the future of science in Russia. There is great difference in how scientific research is administered in Russia compared with the United States, where private foundations and companies provide a considerable share of support for basic science.

None of the heroes in this book was an angel, or a devil. They were outstanding scientists, some even geniuses. As human beings, they had faults, and I did not try to hide them. It is only by finding ourselves under similar conditions that we could know how we would have responded to the challenges they had to face. This comment is not meant to condone any condemnable behaviors, merely to suggest viewing them against the backdrop of the historical conditions. This book was not intended to present Soviet history or the history of Soviet science. The portrayal of scientists in this book is intended to provide a small contribution to the bigger picture. They are scientists of a bygone era, and such a concentration of talent in a relatively small area of human endeavor within one community will probably remain a unique phenomenon for a long time to come.

SOME NOTABLE DATES, 1914–1991

1914	World War I begins; St. Petersburg is renamed Petrograd.
1917	NOVEMBER 7—The "Great October Socialist Revolution"—Vladimir Lenin's communist revolution.
1918	World War I ends; Moscow becomes the capital of the Russian Soviet Socialist Republic (St. Petersburg had been the capital of the Russian Empire since 1712).
1918–1921	Civil War.
1921–1922	Famine along the Volga River and in the Ukraine.
1922	Iosif Stalin becomes general secretary of the Communist Party.
1924	Lenin dies; Petrograd is renamed Leningrad.
1928–1932	The first five-year plan.
1925	through the mid-1930s—Stalin consolidates his power by systematically eliminating his rivals in the communist leadership.
1933	Adolf Hitler and the Nazis come to power in Germany.
1933–1937	Second five-year plan.
1936–1938	The Great Terror.
1938–1939	Lev Landau's arrest and incarceration.
1938	Munich Agreement between the French-British alliance and Germany as a culmination of appeasement policy.
1938	December Lavrentii Beria becomes head of NKVD.
1939	AUGUST—German-Soviet Pact signed in Moscow.
1939–1941	Initial Soviet nuclear research of nuclear explosions follows the discovery of nuclear fission.
1939	SEPTEMBER—World War II begins with simultaneous attacks on Poland by Germany and the Soviet Union.
1939–1940	War between Finland and the Soviet Union.
1941	JUNE—Germany attacks the Soviet Union.
1941	FALL—Scientific institutions evacuate to Kazan.
1942–1943	Battle of Stalingrad.
1943	Soviet research of nuclear explosions resumed.
1944	JUNE—D-Day: The Western Allies open the Second Front.
1945	APRIL—F. D. Roosevelt dies; Harry Truman becomes US president; Hitler commits suicide.
1945	MAY—World War II ends in Europe.
1945	JULY—First atomic bomb tested in US; Potsdam conference of Stalin, Truman, and Winston Churchill/Clement Attlee; Truman informs Stalin about the successful test of US atomic bomb.
1945	AUGUST—US drops atomic bombs on Hiroshima and Nagasaki; Japan surrenders.

1945–1948	Soviet Union builds up the system of satellite states in Eastern Europe.
1946	Soviet administrators and scientists discuss the possibility of thermonuclear weapons.
1949	Communist victory in China.
1949	AUGUST 29—Successful first Soviet test of nuclear device in the Semipalatinsk proving ground (Eastern Kazakhstan).
1950	President Truman's decision about the development of the hydrogen bomb
1950–1953	Korean War.
1952-1953	Peak of anti-Semitic activities in the Soviet Union, "Doctors' plot."
1953	MARCH—Stalin dies.
1953	JUNE—Arrest, then execution of Beria.
1953	AUGUST 12—Soviet nuclear bomb test at Semipalatinsk; it is claimed to be the world's first hydrogen bomb; in reality it is a boosted fission bomb with a minor fusion component—"layered cake" design (*sloika*).
1953	SEPTEMBER—Nikita Khrushchev becomes the first secretary of the Communist Party.
1955	Nikolai Bulganin replaces Georgii Malenkov as prime minister; remains prime minister until March 1958.
1955	Khrushchev and Nikolai Bulganin visit Great Britain, accompanied by Igor Kurchatov, the chief Soviet nuclear scientist.
1955	NOVEMBER 22—First Soviet hydrogen bomb test at Semipalatinsk; utilizes the radiation implosion technology.
1956	FEBRUARY—Twentieth Party Congress, Khrushchev unmasks Stalin's crimes in a secret speech.
1956	First Soviet Nobel Prize, awarded to Nikolai Semenov in Chemistry.
1956	OCTOBER–NOVEMBER—Hungarian Revolution and its suppression by Soviet invasion.
1957	JUNE—Khrushchev consolidates his power by removing most old-timers from the leadership.
1957	OCTOBER—Sputnik.
1958	MARCH—Khrushchev assumes the office of prime minister in addition to party first secretary.
1958	Nobel Prizes for Soviet citizens: Pavel Cherenkov, Ilya Frank, and Igor Tamm in Physics; Boris Pasternak in Literature; Pasternak is blackmailed to decline the award.
1959	Khrushchev visits the US.
1960	Downing of the American U-2 spy plane.
1961	Yurii Gagarin, first human being in space orbit.
1961	OCTOBER—Twenty-Second Party Congress, Stalin's body is removed from the Mausoleum.
1961	OCTOBER 30—Soviet hydrogen bomb test at Novaya Zemlya Archipelago proving ground; highest yield ever (50 megaton), nicknamed "Czar Bomb."
1962	Lev Landau falls victim of tragic automobile accident; receives Nobel Prize.
1962	Cuban missile crisis.
1963	Nuclear test ban treaty signed.

1964 Khrushchev is removed from power; Leonid Brezhnev is the new Soviet leader.

1964 Nikolai Basov and Aleksandr Prokhorov share the Nobel Prize with Charles Townes for their work in quantum electronics leading to the discovery of maser and laser.

1968 JULY—Nuclear Nonproliferation Treaty between the US and the Soviet Union

1968 AUGUST—Soviet and other Warsaw Treaty armies invade Czechoslovakia to abort democratization called the "Prague Spring."

1970 Aleksander Solzhenitsyn is awarded the Nobel Prize in Literature.

1972 Strategic Arms Limitation Treaty (SALT).

1974 Solzhenitsyn is exiled from the Soviet Union.

1975 Andrei Sakharov is awarded the Nobel Peace Prize; Elena Bonner, his second wife, receives it because Sakharov is not allowed out of the country.

1978 Petr Kapitza receives the Nobel Prize.

1979 JUNE—Brezhnev and Jimmy Carter sign SALT II (though later it is not ratified).

1979 DECEMBER—Soviet Union starts Afghanistan War.

1980 JANUARY—Sakharov is exiled to Gorky.

1980 Many democratic countries boycott the Moscow Summer Olympics.

1981 Marital law in Poland; the Solidarity movement is banned.

1982 Brezhnev dies; former KGB chief Yurii Andropov is the new party leader.

1983 The Soviets shoot down a Korean civilian jetliner; President Ronald Reagan calls the Soviet Union the "Evil Empire"; declares the Strategic Defense Initiative.

1984 Andropov dies; Konstantin Chernenko becomes the Soviet party leader.

1985 MARCH—Chernenko dies; Mikhail Gorbachev becomes the Soviet party leader.

1985 NOVEMBER—Gorbachev-Reagan summit takes place in Geneva.

1986 APRIL—Chernobyl nuclear catastrophe.

1986 OCTOBER—Gorbachev-Reagan summit in Reykjavik.

1986 DECEMBER—Sakharov is released from exile; returns to Moscow.

1987 Gorbachev-Reagan summit in Washington, DC; Intermediate-Range Nuclear Forces Treaty is signed.

1989 Soviet Union withdraws from Afghanistan; Congress of People's Deputies in Moscow; East European communist regimes collapse.

1990 Reunification of Germany.

1991 AUGUST—Anti-democracy coup is defeated in Moscow.

1991 Collapse of the Soviet Union; independent Russia and fourteen other independent republics emerge; Leningrad is renamed St. Petersburg (but the surrounding region retains the name Leningradskaya Oblast').

ACKNOWLEDGMENTS

I have received kind and efficient assistance of the most diverse kind from many individuals in the course of the preparation of this book. They included Abrikosov, Alexei (Lemont, IL); Akhrem, Irena (Moscow); Allison, Robert (Chicago); Altshuler, Boris (New York City); Bakh, Abram (Moscow); Balaban, Alexandru (Galveston, TX); Bárány, Anders (Stockholm); Batalka, Krisztina (Budapest); Beck, Mihály (Budapest); Belyakov, Aleksandr (St. Petersburg); Berezanskaya, Valentina (Moscow); Bukatin, Michael (Waltham, MA); Christy, Peter (Los Altos, CA); Curl, Robert (Houston, TX); Dolnik, Milos (Waltham, MA); Dorman, Victoria (Princeton, NJ); Edelsack, Ed and Charlotte (Washington, DC); Efimenko, Lyudmila (St. Petersburg); Epstein, Irving (Waltham, MA); Galpern, Elena (Moscow); Gamow, R. Igor (Boulder, CO); Garwin, Richard (Scarsdale, NY); Gilead, Amihud (Haifa); the late Ginzburg, Vitaly; Gorelik, Gennady (Boston); Gorkov, Lev (Tallahassee, FL); Gorobets, Boris (Moscow); Hargittai, Balazs (Loretto, PA); Hargittai, Eszter (Evanston, IL); Herman, Zelek (Stanford, CA); Hoffmann, Roald (Ithaca, NY); Hurst, Phil (London); Kandror, Jan (Würzburg, Germany); Klein, Georg (Stockholm); Leggett, Anthony (Urbana, IL); Liberman, Mikhail (Uppsala, Sweden); Mackay, Alan (London); the late Mamedov, Khudu; Massa, Lou (New York); Noszticiusz, Zoltán (Budapest); Novakovskaya, Yulia (Moscow); Orbán, Miklós (Budapest); Ovchinnikova, Marina (née Zeldovich) (Moscow); Pickover, Clifford (Yorktown Heights, NY); Ronova, Inga (Moscow); Sakharova, Lyubov (Moscow); Sardanashvily, G. A. (Moscow); Semenov, Alexey (Moscow); the late Semenov, Nikolai; Senechal, Marjorie (Northampton, MA); Shevchenko, Vladimir (St. Petersburg); Shnol, Simon (Pushchino); the late Shoenberg, David; Silberer, Vera (Budapest); Simonovits, András (Budapest); Somorjai, Gabor (Berkeley, CA); Springer, George (Stanford, CA); Stepanov, Nikolai (Moscow); Szűcs, Judit (Budapest); Tamm, Nikita (Moscow); Tsyganova, Tatyana (St. Petersburg); Unna, Issachar (Jerusalem, Israel); Varga, Zoltán (Budapest and Minneapolis, MN); Varshavsky, Alexander (Pasadena, CA); Vernyi, Aleksandr (Moscow); the late Vilkov, Lev; Vilkova, Anna (Moscow); Zasourskaya, Larissa (Moscow); Zeldovich, Olga (Moscow); and the late Zhabotinsky, Anatol.

I owe special thanks to Bob Weintraub, Director of Library, Sami Shamoon College of Engineering (Beersheva and Ashdod, Israel); and Irwin Weintraub, Professor Emeritus of Brooklyn College, CUNY, now living in Beersheva, for tirelessly scrutinizing the entire text of the manuscript and for useful suggestions.

I am grateful to Jeremy Lewis, my editor at Oxford University Press (New York); for his trust and editorial guidance, to the anonymous reviewers for helpful comments, and to the copyeditor, Ginny Faber for care and attention.

As always, my wife, Magdi, has been an inspiration and a constructive critic in this project.

I acknowledge the general support of my work by the Hungarian Academy of Sciences and the Budapest University of Technology and Economics, and the generous specific support for this project by Oxford University Press (New York).

BIOGRAPHICAL NAMES

The following is a list of all the people mentioned in this book. The asterisks following some of the names indicate that there are chapters devoted to them. Abbreviations: CPSU: Communist Party of the Soviet Union; KGB: Komitet Gosudarstvennoi Bezopasnosti (Committee of State Security); NKVD: Narodnii Komissariat Vnutrennikh Del (People's Commissariat of Internal Affairs)

Abrikosov, Alexei I. (1875–1955). Russian-Soviet pathologist; A. A. Abrikosov's father
Abrikosov,* Alexei A. (1928–). Soviet-American physicist; Nobel laureate (2003)
Abrikosov, Dmitrii I. (1876–1950). Russian diplomat and author; A. A. Abrikosov's uncle
Agre, Peter (1949–). American biomedical scientist; Nobel laureate (2003)
Agrest, Mattes M. (1915–2005). Soviet mathematician; later moved to the US
Akhmatova, Anna (1889–1966). Russian-Soviet poet
Akulov, Nikolai S. (1900–1976). Soviet physicist; N. N. Semenov's unprincipled adversary
Aleksandrov, Anatoly P. (1903–1994). Soviet physicist; President of the Soviet Academy of Sciences (1975–1986)
Alekseevskii, Nikolai E. (1912–1993). Soviet physicist
Alikhanov, Abraham I. (1904–1970). Soviet-Armenian physicist
Alikhanov, Gevork (1897–1937). Armenian communist leader; Elena Bonner's father
Alikhanyan, Artem (1908–1978). Soviet-Armenian physicist
Allen, J. F. (1908–2001). Canadian-British physicist; codiscoverer of helium superfluidity
Altshuler, Lev V. (1913–2003). Soviet physicist
Amiton, Ilya P. (1946–). Soviet crystallographer
Andreiev, Nikolai N. (1880–1970). Soviet physicist
Andrianov, Kuzma A. (1904–1978). Soviet chemist
Andropov, Yurii V. (1914–1984). KGB chief (1967–1982); General Secretary of the CPSU (1982–1984)
Arkhipov, R. G. Soviet physicist
Aronov, Yurii E. (1938–1986). Soviet chemist; persecuted for his protest against the invasion of Czechoslovakia in 1968
Arrhenius, Svante (1859–1927). Swedish physical chemist; Nobel laureate (1903)
Artsimovich, Lev A. (1909–1973). Soviet physicist
Balandin, A. A. (1898–1967). Soviet chemist
Baldwin, Stanley (1867–1947). thrice Prime Minister of the UK (1923–1924, 1924–1929, and 1935–1937)
Balfour, Arthur (1848–1930). British politician; Prime Minister (1902–1905); Foreign Secretary (1916–1919)
Bárány, Anders (1942–). Swedish physicist; Secretary of the Nobel Committee for Physics (1989–2003)

Bardeen, John (1908–1991). American physicist; twice Nobel laureate (1956; 1972)
Barton, Derek (1918–1998). British chemist; Nobel laureate (1969)
Bauman, Karl I. (1892–1937). Soviet party official; victim of Stalin's Terror
Bednorz, J. Georg (1950–). Swiss physicist; Nobel laureate (1987)
Belousov,* Boris P. (1893–1970). Soviet chemist
Bergius, Friedrich (1884–1949). German chemist; Nobel laureate (1931)
Beria, Lavrentii P. (1899–1953). Chief of Soviet secret police; Supervisor of the Soviet nuclear program
Bethe, Hans (1906–2005). German-American physicist; Nobel laureate (1967)
Blumenfeld, Lev A. (1921–2002). Soviet biophysicist
Bochvar, Dmitrii A. (1903–1990). Soviet quantum chemist
Bodenstein, Max (1871–1942). German physical chemist
Bohr, Niels (1885–1962). Danish physicist; Nobel laureate (1922)
Bokii, Georgii B. (1909–2001). Soviet crystallographer
Bonhoeffer, Karl F. (1899–1957). German physical chemist
Bonner, Elena (Lusia) (1923–2011). Soviet human rights activist; Andrei Sakharov's second wife
Boreisha, Maria (1879–1923). Russian philologist; Nikolai Semenov's first wife
Born, Max (1882–1970). German-British physicist; Nobel laureate (1954)
Bosch, Carl (1874–1940). German chemical engineer; Nobel laureate (1931)
Botvinnik, Maria M. (1902–1970). Soviet chemist
Bourbaki, Nicolas. invented collective pseudonym of a group of mathematicians
Boyle, Robert (1627–1691). British scientist
Brezhnev, Leonid I. (1906–1982). Soviet politician; General Secretary of the CPSU (1964–1982)
Bronshtein, Matvei P. (1906–1938). Soviet physicist; victim of Stalin's Terror
Bukatin, Michael (Mikhael) A. (1964–). American mathematician and software engineer; Anatol Zhabotinsky's son
Bukatina, Anna E. (1940–). Soviet-American biophysicist; Anatol Zhabotinsky's first wife
Bukharin, Nikolai I. (1888–1938). Soviet politician; victim of Stalin's Terror
Bulganin, Nikolai A. (1895–1975). Soviet politician; Prime Minister (1955–1958)
Bunin, I. A. (1870–1953). Russian poet; Nobel laureate (1933)
Burovskaya, Mirra Ya. (1880–1947). Russian actress; Yulii Khariton's mother
Burtseva (Semenova), Natalia N. (1902–1996). Nikolai Semenov's second wife
Bush, George W. (1946–). US president (2001–2009)
Butlerov, Aleksander M. (1828–1886). Russian chemist
Chadwick, James (1891–1974). British physicist; Nobel laureate (1935)
Chekhov, Anton P. (1860–1904). Russian writer
Chelintsev, G. V.; Soviet chemist; campaigned against the proponents of the theory of resonance
Cherenkov, Pavel A. (1904–1990). Soviet physicist; Nobel laureate (1958)
Chernenko, Konstantin U. (1911–1985). Soviet politician; General Secretary of the CPSU (1984–1985)
Chernomyrdin, Viktor (1938–2010). Prime Minsiter of Russia (1992–1998)
Chernosvitova, Nadezhda. Petr Kapitza's first wife
Chernyakhovskaya, Inna Yu. Yakov Zeldovich's third wife

Chernyshev, Alexei K. (1945–). Soviet physicist
Chirkov, Nikolai M. (1908–1972). Soviet chemist
Choibalsan, Khorloogiin (1895–1952). Communist leader of Mongolia
Chu, Steven (1948–). American physicist; Nobel laureate (1997); US Secretary of Energy (2009–)
Churchill, Winston (1874–1965). British politician; Prime Minister; Nobel laureate (1953)
Cockcroft, John D. (1897–1967). British physicist; Nobel laureate (1951)
Cohen-Tannoudji, Claude (1933–). French physicist; Nobel laureate (1997)
Conan Doyle, Arthur (1859–1930). British author
Cooper, Leon (1930–). American physicist; Nobel laureate (1972)
Crick, Francis (1916–2004). British physicist turned biologist; Nobel laureate (1962)
Dalyell, Tam (now, Lord Dalyell) (1932–). British politician
De Broglie Louis (1892–1987). French physicist; Nobel laureate (1929)
De Kruif, Paul (1890–1971). American biologist turned author of popular science books
Delbrück, Max (1906–1981). German-American scientist; Nobel laureate (1969)
Democritus (460–370 BCE). Greek philosopher
Dirac, Paul A. M. (1902–1984). British physicist; Nobel laureate (1933)
Doroshkevich, A. G. (1936–). Soviet-Russian astrophysicist and cosmologist
Drell, Sidney D. (1926–). American physicist
Drobantseva, Konkordia (Kora) (1908–1984). Soviet chemical engineer; L. D. Landau's wife
Dyatkina, M. E. (1915–1972). Soviet chemist
Dzyaloshinskii, I. E. (1931–). Soviet physicist
Ehrenfest, Paul (1880–1933). Dutch physicist
Einstein, Albert (1879–1955). German Swiss-American physicist; Nobel laureate (1922 for 1921)
Eitington, Max; German psychiatrist; second husband of Yulii Khariton's mother
Elsasser, Walter M. (1904–1991). German-American physicist
Eltenton, George. British chemical engineer; member of the Soviet atomic spy ring
Engelhardt, Vladimir A. (1895–1984). Soviet biochemist
Epstein, Irving R. (1944–). American physical chemist
Erlander, Tage F. (1901–1985). Swedish prime minister (1946–1969)
Ermakova, Nina I. (1922–). Vitaly Ginzburg's second wife
Escher, Maurice C. (1898–1972). Dutch graphic artist
Eyring, Henry (1901–1981). American physical chemist
Ezhov, Nikolai I. (1895–1940). Head of the NKVD during the Great Purge of 1937–1938
Fedin, Erlen I. (1926–2009). Soviet chemist
Fedorov, E. K. (1910–1981). Researcher of polar regions; official of the Soviet Academy of Sciences
Feinberg, E. L. (1912–2005). Soviet physicist
Fermi, Enrico (1901–1954). Italian-American physicist; Nobel laureate (1938)
Fersman, Aleksandr E. (1883–1945). Russian-Soviet geologist
Feynman, Richard (1918–1988). American physicist; Nobel laureate (1965)
Fock, Vladimir A. (1898–1974). Soviet physicist
Fomin, P. F. (1904–1976). Soviet admiral; commander of the Novaya Zemplya proving ground for thermonuclear bomb tests
Frank, Ilya M. (1908–1990). Soviet physicist; Nobel laureate (1958)

Frank-Kamenetskii, David A. (1910–1970). Soviet physicist and chemist
Fredga, Arne (1902–1992). Swedish chemist
Freidlina, Rakhil Kh. (1906–1986). Soviet chemist
Frenkel, Yakov I. (1894–1952). Soviet physicist
Friedman, Aleksandr A. (1888–1925). Russian-Soviet cosmologist
Frumkin, Aleksandr N. (1895–1976). Soviet electrochemist
Fuchs, Klaus (1911–1988). German-British physicist; Soviet atomic spy
Fuller, R. Buckminster (1895–1983). American designer and author
Gagarin, Yurii A. (1934–1968). Soviet pilot; first human in space (1961)
Galpern, Elena (1935–). Soviet chemist
Gambaryan, Natalya P. (1929–). Soviet chemist
Gamow, George (at birth, Georgii A.) (1904–1968). Russian-American physicist and author of popular science books
Gavrilov, Nikolai I. (1892–1966). Soviet chemist
Gershtein, Semyon S. (1929–). Soviet physicist
Gessen, B. M. (1893–1936). Soviet physicist and philosopher; victim of Stalin's Terror
Gilbert, Walter (1932–). American physicist and biologist; Nobel laureate (1980)
Gill, Eric (1882–1940). British sculptor
Ginzburg, Irina V. (1939–). Vitaly Ginzburg's daughter
Ginzburg, Nina I. *See* Ermakova, Nina I.
Ginzburg, Lazar (1863–1942). Russian-Soviet engineer; Vitaly Ginzburg's father
Ginzburg,* Vitaly L. (1916–2009). Soviet physicist; Nobel laureate (2003)
Glashow, Sheldon (1932–). American physicist; Nobel laureate (1979)
Goldanskii, Vitalii I. (1923–2001). Soviet physical chemist; Nikolai Semenov's son-in-law
Gomberg, Moses (1866–1947). American chemist
Gorbachev, Mikhail S. (1931–). Soviet politician; General Secretary of the CPSU (1985–1991); President of the Soviet Union (1990–1991); Nobel laureate (1990)
Gorelik, Gennady E. (1948–). Soviet-American science historian
Gorky, Maxim (1868–1936). Soviet writer
Gorkov, Lev P. (1929–). Soviet-American-Russian physicist
Gorobets, Boris S. (1942–). Soviet-Russian physicist, mineralogist, science historian
Gorskaya, Natalia V. (1941–2008). Soviet crystallographer
Groves, Leslie R. (1896–1970). US Army general; manager of the Manhattan Project
Gurevich, I. I. (1912–1992). Soviet physicist
Hahn, Otto (1879–1968). German chemist; Nobel laureate (awarded in 1945 for the year 1944)
Hartree, Douglas (1897–1958). British physicist
Hecker, Siegfried S. (1943–). American metallurgist; former director of the Los Alamos National Laboratory
Heisenberg, Werner (1901–1976). German physicist; Nobel laureate (1933 for 1932)
Heitler, Walter (1904–1981). German-British physicist
Hinshelwood, Cyril Norman (1897–1967). British physical chemist; Nobel laureate (1956)
Hoffmann, Roald (1937–). American chemist and author; Nobel laureate (1981)
Hulthén, Lamek (1909–1995). Swedish physicist
Ingelstam, Erik (1909–1988). Swedish physicist
Ioffe, Abram F. (1880–1960). Russian-Soviet physicist; mentor of many renowned Soviet physicists

Ipatev, Vladimir N. (1867–1952). Russian-American chemist
Ivanenko, Dmitrii D. (1904–1994). Soviet physicist
Ivanitskii, Genrikh R. (1936–). Soviet biophysicist
John Paul II (1920–2005). Pope of the Catholic Church (1978–2005)
Jordan, Pascual (1902–1980). German physicist
Kabachnik, Martin I. (1908–1997). Soviet chemist
Kabalkina, Sarra S. (1918–1999). Soviet physicist
Kamenev, Lev B. (1883–1936). Soviet politician; victim of Stalin's Terror
Kamerlingh-Onnes, Heike (1853–1926). Dutch physicist; Nobel laureate (1913)
Kanegiesser, Evgeniya; Rudolf Peierls's Russian-born wife
Kapitza, Anna A. (1903–1996). Petr Kapitza's wife
Kapitza, Andrei P. (1931–2011). geographer; Petr Kapitza's son
Kapitza,* Petr L. (1894–1984). Soviet physicist; Nobel laureate (1978)
Kapitza, Sergei P. (1928–2012). Physicist; Petr Kapitza's son
Kargin, Valentin A. (1907–1969). Soviet polymer chemist
Kedrov, Bonifatii M. (1903–1983). Soviet philosopher
Keldysh, Leonid V. (1931–). Soviet-American physicist
Keldysh, Mstislav V. (1911–1978). Soviet mathematician; President of the Soviet Academy of Sciences (1961–1975)
Kendrew, John (1917–1997). British biochemist; Nobel laureate (1962)
Kennedy, John F. (1917–1963). US president (1961–1963)
Khalatnikov, Isaak M. (1919–). Soviet physicist
Khariton, Boris O. (1876–1940). Journalist; Yulii Khariton's father
Khariton, Maria N. (1902–1977). Yulii Khariton's wife
Khariton, Tatyana Yu. (1926–1985). Yulii Khariton's daughter
Khariton,* Yulii B. (1904–1996). Soviet physicist; long-time scientific head of Arzamas-16
Khrushchev, Nikita S. (1894–1971). Supreme leader of the Soviet Union after Stalin's death (1953–1964)
Khvolson, Orest D. (1852–1934). Russian physicist
Kikoin, Isaak K. (1908–1984). Soviet physicist
Kitaigorodskii,* Aleksandr I. (1914–1985). Soviet crystallographer
Kitaigorodskii, Isaac (1888–1965). Soviet chemical engineer; A. I. Kitaigorodskii's father
Klein, Oskar (1894–1977). Swedish physicist
Knunyants, Ivan L. (1906–1990). Soviet chemist
Kochetkov, Nikolai K. (1915–2005). Soviet chemist
Kolmogorov, Andrei N. (1903–1987). Soviet mathematician
Kondratev, Viktor N. (1902–1979). Soviet physical chemist
Konstantinov, Boris P. (1916–1969). Soviet physicist
Konstantinova, Varvara P. (1907–1976). Soviet physicist; Ya. B. Zeldovich's first wife
Korolev, Sergei (1907–1966). General Constructor of the Soviet rocket program
Korshak, Vasilii V. (1909–1988). Soviet chemist
Koshland, Daniel E. Jr. (1920–2007). American biochemist; editor of *Science* (1985–1995).
Kovalev, Sergei (1930–). Soviet biologist and human rights activist
Krasin, Viktor (1929–). Soviet human rights activist; later US citizen
Krinskaya, Albina (1938–). Anatol Zhabotinsky's second wife
Krinskii, Valentin I. (1938–). Soviet biophysicist

Krylov, Aleksey N. (1863–1945). Russian naval engineer

Krylova, Anna A. A. N. Krylov's daughter; *see* Kapitza, Anna A.

Kurchatov, Igor V. (1903–1960). Soviet physicist; supreme leader of the Soviet nuclear program

Kursanov, Dmitrii N. (1899–1983). Soviet chemist

Landau, David L. (1866–1943). Oil engineer; Lev Landau's father

Landau, Igor L. (1944–2011). Soviet-Swiss physicist; Lev Landau's son

Landau,* Lev D. (1908–1968). Soviet physicist; Nobel laureate (1962)

Landau (née Harkavi), Lyubov V. (1876–1941). Lev Landau's mother

Landsberg, Grigory S. (1890–1957). Soviet physicist

Lauterbur, Paul (1929–2007). American chemist; Nobel laureate (2003)

Lavrentiev, Mikhail A. (1900–1980). Soviet mathematician

Lawrence, Ernest O. (1901–1958). American physicist; Nobel laureate (1939)

Lebedev, Petr N. (1866–1912). Russian physicist

Leggett, Anthony J. (1939–). British-American physicist; Nobel laureate (2003)

Leipunskii, Aleksandr I. (1903–1972). Soviet physicist

Leipunskii, Ovsei I. (1909–1990). Soviet physicist

Lenin, Vladimir I. (1870–1924). Russian communist revolutionary; first leader of the Soviet Union

Leontovich, Mikhail A. (1903–1981). Soviet physicist

Letokhov, Vladilen S. (1939–2009). Soviet-Russian physicist

Levi, Saul. Refugee physicist from Germany; Vitaly Ginzburg's physics tutor; later moved to the United States

Levi-Montalcini, Rita (1909–2012). Italian biomedical scientist; Nobel laureate (1986)

Liberman (also, Liberman-Smith), Marina (1968–). Andrei Sakharov's granddaughter

Lifshits Ilya M. (1917–1982). Soviet physicist

Lifshits (née Mazel'), Berta. E. M. and I. M. Lifshits's mother

Lifshits,* Evgenii M. (1915–1985). Soviet physicist

Lindh, Axel E. (1888–1960). Swedish physicist; member of the Nobel Committee for Physics (1935–1960)

Liszt, Franz (Ferenc) (1811–1886). Hungarian composer and virtuoso pianist

Livanova, Anna (1917–2001). Soviet physicist and author of books about scientists; Anatol Zhabotinsky's mother. Livanova was her pen-name; her original surname was Lifshits.

Liverovskii, Alexei; Russian medical doctor; first husband of Maria Boreisha, Nikolai Semenov's first wife

London, Fritz W. (1900–1954). German-American physicist

Lucretius (99–55 BCE). Roman philosopher

Luzhkov, Yury M. (1936–). Mayor of Moscow (1992–2010)

Luzin, Nikolai N. (1883–1950). Russian-Soviet mathematician

Lvov, V. E. Soviet writer, critic of the theory of resonance in chemistry

Lysenko, Trofim D. (1898–1976). Soviet charlatan agronomist; wielded great power and destroyed modern biology in the Soviet Union

MacGillavry, Carolina H. (1904–1993). Dutch crystallographer

MacKinnon, Roderick (1956–). American biologist; Nobel laureate (2003)

Malenkov, Georgii M. (1902–1988). Soviet communist leader; Prime Minister (1953–1955)

Malevich, Kazimir (1879–1935). Russian-Soviet avant-garde artist
Malyshev, Vyacheslav (1902–1957). Soviet official; head of the Ministry of Medium Machine Building (the camouflaged name for the ministry responsible for nuclear matters)
Mamedov, Khudu (1927–1988). Soviet-Azerbaijani crystallographer
Mandelshtam, Leonid I. (1879–1944). Soviet physicist
Mandelshtam, Osip (1891–1938). Russian poet; victim of Stalin's Terror
Mansfield, Peter (1933–). British physicist; Nobel laureate (2003)
Mechnikov, Ilya (in international literature, Élie) (1845–1916). Russian-Ukrainian biologist; Nobel laureate (1908)
Meitner, Lise (1878–1968). Austrian-German physicist; codiscoverer of nuclear fission
Mendel, Gregor (1822–1884). Austrian monk and pioneer geneticist
Mendeleev, Dmitrii I. (1834–1907). Russian chemist
Mezhlauk, Valery I. (1893–1938). Soviet deputy prime minister; head of state planning; victim of Stalin's Terror
Migdal, Alexander (1945–). Soviet-American physicist; Arkadii Migdal's son
Migdal, Arkadii (1911–1991). Soviet physicist
Mikhoels, Solomon (1890–1948). Soviet actor; head of the Jewish Anti-Fascist Committee; assassinated on Stalin's orders
Misener, A. D. British physicist; codiscoverer of helium superfluidity
Mokhov, Viktor N. (193?–2011). Soviet physicist
Molotov, Vyacheslav M. (1890–1986). Leading Soviet politician; Prime Minister (1930–1941); Foreign Minister (1939–1949; 1953–1956)
Mond, Ludwig (1839–1909). German-British chemist-industrialist
Monroe, Marilyn (1926–1962). American actress
Müller, K. Alexander (1927–). Swiss physicist; Nobel laureate (1987)
Myshkis, Anatoli D. (1920–2009). Soviet mathematician
Nametkin, Sergey S. (1876–1950). Russian-Soviet chemist
Nedelin, Mitrofan I. (1902–1960). Soviet military leader
Neizvestny, Ernst I. (1925–). Soviet-American sculptor
Nemchinov, V. S. (1894–1964). Soviet economist
Nernst, Walther (1864–1941). German physical chemist; Nobel laureate (1920)
Nesmeyanov,* Aleksandr N. (1899–1980). Soviet chemist; president of the Soviet Academy of Sciences (1951–1961)
Nesmeyanov, Vasilii N. Aleksandr Nesmeyanov's brother; victim of Stalin's Terror
Neumann, John von (1903–1957). Hungarian-American mathematician
Newton, Isaac (1642–1727). British scientist
Nikolai II of Russia (1868–1918). Last Emperor of Russia
Nixon, Richard M. (1913–1994). US president (1968–1974)
Nobel, Alfred (1833–1896). Swedish chemical engineer and inventor; founded the award known today as the Nobel Prize
Novikov, Igor D. (1935–). Soviet-Danish astrophysicist
Nuzhdin, Nikolai I. (1904–1972). Soviet biologist; close associate of Trofim Lysenko
Obreimov, I. V. (1894–1981). Soviet physicist
Ondra, Annie (1902–1987). German film star
Oparin, Aleksander I. (1894–1980). Soviet biochemist

Oppenheimer, J. Robert (1904–1967). American physicist; first director of the Los Alamos Laboratory
Orbán, Miklós (1939–). Hungarian chemist
Orwell, George (1903–1950). British writer
Osawa, Eiji (1935–). Japanese chemist
Ovchinnikova (née Zeldovich), Marina Ya. (1939–). Soviet physicist, Ya. B. Zeldovich's daughter
Panofsky, Wolfgang (1919–2007). American physicist
Pasternak, Boris (1890–1960). Russian poet, novelist; Nobel laureate (1958; was forced to decline the award)
Pauli, Wolfgang (1900–1958). Austrian-Swiss physicist; Nobel laureate (1945)
Pauling, Linus (1901–1994). American chemist; twice Nobel laureate (1954; 1963 for 1962)
Pavlov, Ivan P. (1849–1936). Russian physiologist; Nobel laureate (1904)
Pavlov, Nikolai I. (1915–1990). KGB general
Peebles, P. J. E. (1935–). American physicist
Peierls, Rudolf E. (1907–1995). German-British physicist
Penzias, Arno (1933–). American astrophysicist; Nobel laureate (1978)
Perutz, Max F. (1914–2002). Austrian-British biochemist
Peter the Great (1672–1725). Emperor of Russia (1682–1725)
Petrovskii, Ivan G. (1901–1973). Soviet mathematician
Phillips, William D. (1948–). American physicist; Nobel laureate (1997)
Pitaevskii, L. P. (1933–). Soviet physicist
Planck, Max (1858–1947). German physicist; Nobel laureate (1919 for 1918)
Pomeranchuk, Isaak Ya. (1913–1966). Soviet physicist
Prigogine, Ilya (1917–2003). Belgian physical chemist; Nobel laureate (1977)
Putin, Vladimir V. (1952–). Russian politician; President (2000–2008; 2012–); Prime Minister (1999–2000; 2008–2012)
Pyatigorskii, Leonid M. (1909–1993). Soviet physicist
Rabi, Isidor I. (1898–1988). American physicist; Nobel laureate (1945 for 1944)
Raman, C. V. (1888–1970). Indian physicist; Nobel laureate (1930)
Rapoport, Iosif A. (1912–1990). Soviet geneticist
Razuvaev, Grigorii A. (1895–1989). Soviet chemist
Reagan, Ronald W. (1911–2004). US president (1981–1989)
Roginskii, Simon Z. (1900–1970). Soviet physical chemist
Romanov, Yurii A. (1926–2010). Soviet physicist
Röntgen, Wilhelm Conrad (1845–1923). German physicist; Nobel laureate (1901)
Rozhdestvenskii, Dmitrii S. (1876–1940). Russian physicist
Russell, Bertrand (1872–1970). British philosopher
Rutherford, Ernest (1871–1937). British physicist; Nobel laureate (1908)
Sagdeev, Roald Z. (1932–). Soviet-American physicist
Sakharov,* Andrei D. (1921–1989). Soviet physicist and human rights activist; Nobel laureate (1975)
Sakharov, Dmitrii I. (1889–1961). Physicist; pedagogue; author; Andrei Sakharov's father
Sakharov, Dmitrii A. (1957). Andrei Sakharov's son
Sakharova, Lyubov (Lyuba) A. (1949–). Teacher; Andrei Sakharov's daughter
Sakharova, Tatyana A. (1945–). Biologist; Andrei Sakharov's daughter

Samoilov, David S. (1920–1990). Soviet poet
Schrieffer, J. Robert (1931–). American physicist; Nobel laureate (1972)
Schrödinger, Erwin (1887–1961). Austrian physicist; Nobel laureate (1933)
Seaborg, Glenn T. (1912–1999). American chemist; Nobel laureate (1951)
Semenov, Alexey Yu. (1951–). Biologist; Nikolai Semenov and Yulii Khariton's grandson
Semenov,* Nikolai N. (1896–1986). Soviet chemical physicist; Nobel laureate (1956)
Semenov, Yurii N. (1925–1995). Philosopher, Nikolai Semenov's son
Semenova (Burtseva), Natalia N. *See* Burtseva, Natalia
Semenova, Ludmilla N. (1928–). Teacher at music school; Nikolai Semenov's daughter
Shabad, Leon M. (1902–1982). Soviet cancer specialist
Shakespeare, William (1564–1616). English poet and playwright
Shalnikov, Aleksandr I. (1905–1986). Soviet physicist
Shcherbakova (Shcherbakova-Semenova), Lidia G. (1926–). Soviet chemist; Nikolai Semenov's third wife
Shemyakin, Mikhail M. (1908–1970). Soviet chemist
Shilov, E. A. (1893–1970). Soviet chemist
Shire, Edward S. (1908–1978). British physicist
Shiryaeva, O. K. (1911–2000). Soviet artist and architect; was incarcerated and exiled; had daughter with Ya. B. Zeldovich
Shnol, Simon (1930–). Soviet-Russian biochemist
Shoenberg, David (1911–2004). British physicist
Shubin, Semyon (1908–1938). Soviet physicist; victim of Stalin's Terror
Shubnikov, Alexey V. (1887–1970). Soviet crystallographer
Shubnikov, Lev V. (1901–1937). Soviet physicist; victim of Stalin's Terror
Shunyaev, Rashid A. (1943–). Soviet-Russian physicist
Sidur, Vadim A. (1924–1986). Soviet avant-garde sculptor
Sigmund Freud (1856–1939). Austrian neurologist
Sillén, Lars Gunnar (1916–1970). Swedish chemist
Simon, Francis (1893–1956). German-British physicist
Sindelevich, Asya; A. I. Kitaigorodskii's mother
Smirnov, Yurii N. (1937–2011). Soviet physicist and historian of science
Sobko, I. D.; Soviet journal editor
Solzhenitsyn, Aleksandr I. (1918–2008). Russian writer; Nobel laureate (1970)
Spinoza, Baruch (1632–1677). Dutch philosopher
Stalin, Iosif V. (1878–1953). Soviet dictator
Stankevich, Ivan V. (1933–2012). Soviet physicist
Steinbeck, John (1902–1968). American writer; Nobel laureate (1962)
Stern, Otto (1888–1969). German-American physicist; Nobel laureate (1943)
Strassmann, Fritz (1902–1980). German chemist; codiscoverer of nuclear fission
Struchkov, Yurii T. (1926–1995). Soviet crystallographer
Syrkin, Yakov K. (1894–1974). Soviet chemist
Szilard, Leo (1898–1964). Hungarian-American scientist
Tamm,* Igor E. (1895–1971). Soviet physicist; Nobel laureate (1958)
Tamm, Leonid E. (1896 or later–1937, 1938 at the latest). Igor Tamm's brother; victim of Stalin's Terror
Tamm, Natalia V. (née Shuiskaya). Igor Tamm's wife

Telegdi, Valentine (1922–2006). Hungarian-American physicist
Teller, Edward (1908–2003). Hungarian-American physicist
Thomson, J. J. (1856–1940). British physicist; Nobel laureate (1906)
Thorne, Kip S. (1940–). American physicist
Timofeev-Resovskii, Nikolai V. (1900–1981). Soviet biologist
Tiselius, Arne (1902–1971). Swedish biochemist; Nobel laureate (1948)
Tisza, Laszlo (1907–2009). Hungarian-American physicist
Todd, Alexander (1907–1997). British chemist; Nobel laureate (1957)
Tolstoy, Lev (in international literature, Leo) N. (1828–1910). Russian writer
Trapeznikova, Olga N. (1901–1997). Soviet physicist; Lev Shubnikov's widow
Trotsky, Lev (in international literature, Leon) (1879–1940). Soviet communist revolutionary; was assassinated in exile on Stalin's order
Truman, Harry S. (1884–1972). US president (1945–1952)
Tsukerman, Veniamin (1913–1993). Soviet physicist
Tsvetaeva, Marina (1892–1941). Russian-Soviet poet
Tupolev, Andrei N. (1888–1972). Soviet aircraft designer
Turgenev, Ivan S. (1818–1883). Russian writer
Ulam Stanislaw (1909–1984). Polish-American mathematician
Vainshtein, Boris K. (1921–1996). Soviet crystallographer
Vainshtein, L. A. (1920–1989). Soviet physicist
Valta, Zinaida (born probably in 1902; year of death unknown). Together with Yulii Khariton investigated the oxidation of phosphorus
Valter, Alexander F. (1898–1941). Soviet physicist
van 't Hoff, Jacobus (1852–1911). Dutch chemist; Nobel laureate (1901)
Vasileva Anzhelika Ya. (? –1985). Yakov Zeldovich's second wife
Vavilov, Nikolai V. (1887–1943). Soviet plant biologist; victim of Stalin's Terror
Vavilov, Sergei V. (1891–1951). Soviet physicist; President of the Soviet Academy of Sciences (1945–1951)
Velikhov, Evgenii P. (1935–). Soviet physicist
Verne, Jules (1828–1905). French writer; pioneer of science fiction
Vernyi, Alexander (1950). Physicist and historian of science; Andrei Sakharov's son-in-law (Lyuba's husband)
Vikhrieva, Klavdia (1919–1969). Chemical engineer; Andrei Sakharov's first wife
Vildauer, Rosa (1891–1948). Vitaly Ginzburg's stepmother
Vildauer-Ginzburg, Augusta (1887–1920). Russian medical doctor; Vitaly Ginzburg's mother
Volpin, Mark E. (1923–1996). Soviet chemist
Watson, James D. (1928–). American biologist; Nobel laureate (1962)
Weinberg, Steven (1933–). American physicist; Nobel laureate (1979)
Weisskopf, Victor (1908–2002). Austrian-American physicist
Weizsäcker, Carl Friedrich von (1912–2007). German physicist
Wells, H. G. (1866–1946). British author
Wheland, George W. (1907–1962). American chemist
Wigner, Eugene P. (1902–1995). Hungarian-American physicist; Nobel laureate (1963)
Wilkins, Maurice (1916–2004). British biophysicist; Nobel laureate (1962)
Wilson, Robert W. (1936–). American physicist; Nobel laureate (1978)
Wöhler, Friedrich (1800–1882). German chemist

Woodward, Robert B. (1917–1979). American chemist; Nobel laureate (1965)
Wul'ff, Fanny D.; Alexei Abrikosov's mother
Yakir, Petr I. (1923–1982). Soviet human rights activist
Yeltsin, Boris N. (1931–2007). President of Russia (1991–1999)
Zababakhin, Evgenii I. (1917–1984). Soviet physicist; director of Chelyabinsk-70
Zaikin, Albert N. (1935–). Soviet biophysicist
Zaks, Yulia B. (1937–). Soviet chemist; was fired from her job for expressing sympathy with persecuted human rights activists; in 1976 immigrated to the United States
Zamsha, Olga I. (1915–). Soviet physicist; Vitaly Ginzburg's first wife
Zavaritskii, Nikolai V. (1925–1997). Soviet physicist
Zavoiskii, Evgenii K. (1907–1976). Soviet physicist; inventor of the electron paramagnetic resonance technique
Zeeman, Pieter (1865–1943). Dutch physicist; Nobel laureate (1902)
Zeldovich, Anna P. (1892–1975). Translator, writer; Ya. B. Zeldovich's mother
Zeldovich, Boris N. (1888–1943). Soviet lawyer; Ya. B. Zeldovich's father
Zeldovich, Boris Ya. (1944–). Soviet-American physicist; Ya. B. Zeldovich's son
Zeldovich, Olga Ya. (1938–). Soviet physicist; Ya. B. Zeldovich's daughter
Zeldovich,* Yakov B. (1914–1987). Soviet physicist
Zelinsky, Nikolai D. (1861–1953). Russian-Soviet chemist
Zhabotinsky,* Anatol M. (1938–2008). Soviet-American biophysicist
Zhabotinsky, Mark E. (1917–2003?). Soviet physicist; A. M. Zhabotinsky's father
Zhdanov, Andrei A. (1896–1948). Soviet Communist Party leader
Zorky, Petr (1933–2005). Soviet crystallographer
Zysin, Yurii A. (1917–1978). Soviet physicist

SELECT BIBLIOGRAPHY

Abrikosov, A. A. *Akademik L. D. Landau: Kratkaya biografiya i obzor nauchnikh rabot* [Academician L. D. Landau: Brief biography and review of scientific works]. Moscow: Nauka, 1965.

Abrikosov, Dmitrii. *Revelations of a Russian Diplomat; The Memoirs of Dmitrii I. Abrikosov.* Edited by George A. Lensen. Seattle, WA: University of Washington Press, 1964.

Akhmatova, Anna. *Pamyati Anny Akhmatova. Stikhi. Pis'ma. L. Chukovskay.* Paris: YMCA, 1974.

Alferov, Zh. Ed., *Ioffe Institute 1918–1998: Development and Research Activities.* St. Petersburg: Ioffe Physico-Technical Institute, 1998.

Altshuler, B. L. Ed. *Andrei Sakharov: Facets of a Life.* Gif-sur-Yvette. France: Edition Frontiers, 1991.

Andreev, A. F., Ed. *Kapitza Tamm Semenov: V ocherkakh i pis'makh* [Kapitza, Tamm, Semenov: In sketches and letters]. Moskva: Vagrius Priroda, 1998.

Applebaum, Anne. *Gulag: A History.* New York: Anchor Books, 2004.

Badash, Lawrence. *Kapitza, Rutherford, and the Kremlin.* New Haven and London: Yale University Press, 1985.

Baev, A. A., Ed. *Vospominaniya o V. A. Engelhardte* [Remembering V. A. Engelhardt]. Moscow: Nauka, 1989.

Baggott, Jim. *Atomic: The First War of Physics and the Secret History of the Atom Bomb: 1939–1949.* London: Icon Books, 2009.

Bailey, George. *The Making of Andrei Sakharov.* London: Allen Lane the Penguin Press, 1989.

Bergman, Jay. *Meeting the Demands of Reason: The Life and Thought of Andrei Sakharov.* Ithaca, NY, and London: Cornell University Press, 2009.

Birstein, Vadim J. *The Perversion of Knowledge: The True Story of Soviet Science.* Cambridge, MA: Westview Press, 2001.

Blokh, A. M. *Sovetskii Soyuz v interyere nobelevskikh premii: Fakti. Dokumenti. Razmyshleniya. Kommentarii* [The Soviet Union in the interior of the Nobel Prizes: Facts; documents; reflections; commentaries]. Second edition. Moscow: Fizmatlit, 2005.

Boag, J. W., P. E. Rubinin, and D. Shoenberg. Compilers and Editors. *Kapitza in Cambridge and Moscow: Life and Letters of a Russian Physicist.* Amsterdam: North-Holland, 1990.

Bongard-Levin, G. M., and Zakharov, V. E., Eds. *Rossiiskaya nauchnaya emigratsiya: Dvadtsat portretov* [Scientific Emigration from Russia: Twenty portraits]. Moscow: URSS, 2001.

Bonner, Elena. *Alone Together.* New York: Vintage Books, 1988.

Born, Max. *My Life: Recollections of a Nobel Laureate.* New York: Charles Scribner's Sons, 1978.

Campbell, John. *Rutherford: Scientist Supreme.* Christchurch, New Zealand: AAS Publications, 1999.

DeWolf Smyth, H. J. *Atomic Energy for Military Purposes: The Official Report on the Development of the Atomic Bomb under the Auspices of the United States Government, 1940–1945*. Princeton, NJ: Princeton University Press, 1945.

Editorial board (E. P. Velikhov et al.). *Physics of the 20th Century: History and Outlook*. Moscow: MIR, 1987.

Farmelo, Graham. *The Strangest Man: The Hidden Life of Paul Dirac, Quantum Genius*. London: Faber and Faber, 2009.

Fedin, Erlen. *Izbrannoe* [Selected works]. Krasnoyarsk: Polikom, 2008.

Feinberg, E. L. *Physicists: Epoch and Personalities*. New Jersey, London, Singapore: World Scientific, 2011.

Felshtinskii, Yu. G. *Razgovori s Bukharinym* [Conversations with Bukharin]. New York: Telex, 1991.

Gamow, George. *My World Line: An Informal Autobiography*. New York: Viking Press, 1970.

Ginzburg, Vitaly L. *O fizike i astrofizike* [About physics and astrophysics]. Third revised edition. Moscow: Buro Quantum, 1995.

Ginzburg, Vitaly L. *About Science, Myself and Others*. Bristol and Philadelphia: Institute of Physics Publishing, 2005.

Gol'danskii, Vitalii I. *Essays of a Soviet Scientist: A Revealing Portrait of a Life in Science and Politics*. Woodbury, NY: American Institute of Physics, 1997.

Goldanskii, V. I., Ed. *Yulii Borisovich Khariton: Put' dlinoyu v vek* [Century-long journey], Moscow: Nauka, 2005.

Gorelik, Gennady E., and Victor Ya. Frenkel, *Matvei Petrovich Bronshtein and Soviet Theoretical Physics in the Thirties*. Basel: Birhäuser, 1994.

Gorelik, Gennady, with Antonina W. Bouis. *The World of Andrei Sakharov: A Russian Physicist's Path to Freedom*. New York: Oxford University Press, 2005.

Gorobets, Boris S. *Krug Landau i Lifshitsa* [Landau's and Lifshits's circle]. Moscow: URSS, 2008.

Gorobets, Boris S. *Krug Landau: Zhizn geniya* [Landau's circle: The life of a genius]. Second corrected and augmented edition. Moscow: URSS, 2008.

Gorobets, Boris S. *Krug Landau: Fizika voini i mira* [Landau's circle: Physics of war and peace]. Moscow: URSS, 2009.

Gorobets, Boris S. *Sekretnie fiziki iz atomnogo proekta SSSR: Semya Leipunskikh* [Classified physicists from the atomic project of the Soviet Union: The Leipunskii family]. Second edition. Moscow: URSS, 2009.

Gorobets, Boris S. *Sovetskie fiziki shutyat...khotya bivalo ne do shutok* [Soviet physicists joking...although it was not joyful]. Moscow: URSS, 2010.

Goudsmit, Samuel A. *Alsos*. New York: Henry Schuman, 1947.

Hargittai, Balazs, and István Hargittai. *Candid Science V: Conversations with Famous Scientists*. London: Imperial College Press, 2005.

Hargittai, I., Ed. *Symmetry: Unifying Human Understanding*. Oxford, UK: Pergamon Press, 1986.

Hargittai, I., and Kálmán, A. Guest editors. "A. I. Kitaigorodskii Memorial Issue on Molecular Crystal Chemistry," Part I. *Acta Chimica Hungarica: Models in Chemistry*. Budapest: Akadémiai Kiadó, 1993.

Hargittai, Istvan. *Candid Science: Conversations with Famous Chemists*. Edited by Magdolna Hargittai. London: Imperial College Press, 2000.

Hargittai, Istvan, and Magdolna Hargittai. *In Our Own Image: Personal Symmetry in Discovery.* New York: Kluwer Academic/Plenum Publishers, 2000.

Hargittai, Istvan. *Candid Sciernce II: Conversations with Famous Biomedical Scientists.* Edited by Magdolna Hargittai. London: Imperial College Press, 2002.

Hargittai, Istvan. *Candid Science III: More Conversations with Famous Chemists.* Edited by Magdolna Hargittai. London: Imperial College Press, 2003.

Hargittai, Istvan. *Our Lives: Encounters of a Scientist.* Budapest: Akademiai Kiado, 2004.

Hargittai, Istvan. *Martians of Science: Five Physicists Who Changed the Twentieth Century.* New York: Oxford University Press, 2006.

Hargittai, Istvan, and Magdolna Hargittai. *Candid Science VI: More Conversations with Famous Scientists.* London: Imperial College Press, 2006.

Hargittai, Istvan. *Judging Edward Teller: A Closer Look at One of the Most Influential Scientists of the Twentieth Century.* Amherst, NY: Prometheus Books, 2010.

Hargittai, Istvan. *Drive and Curiosity: What Fuels the Passion for Science.* Amherst, NY: Prometheus, 2011.

Hargittai, Magdolna, and Istvan Hargittai. *Candid Science IV: Conversations with Famous Physicists.* London: Imperial College Press, 2004.

Hartshorne, E. Y. *The German Universities and National Socialism.* Cambridge, MA: Harvard University Press, 1937.

Hingley, Ronald. *Nightingale Fever: Russian Poets in Revolution.* New York: Alfred A. Knopf, 1981.

Holloway, David. *Stalin and the Bomb: The Soviet Union and Atomic Energy 1939–1956.* New Haven and London: Yale University Press, 1994.

Hyde, H. Montgomery. *Stalin: The History of a Dictator.* New York: Popular Library, 1971.

Isaacs, Jeremy, and Downing, Taylor. *Cold War: The Book of the Ground-Breaking TV Series.* London: Bantam Press, 1998.

Ivanenko, D. D., Ed. *W. Heisenberg, E. Shrödinger, P. A. M. Dirac: Sovremennaya kvantovaya mekhanika: Tri nobelevskikh doklada* [W. Heisenberg, E. Shrödinger, P. A. M. Dirac: Modern quantum mechanics: Three Nobel reports]. Leningrad and Moscow: Technico-Theoretical State Press, 1934.

Ivanov, V. T., Ed. *Yurii Anatolevich Ovchinnikov: Zhizn i nauchnaya deyatelnost* [Yurii Anatolevich Ovchinnikov: Life and scientific oeuvre]. Moscow: Nauka, 1991.

Josephson, Paul R. *Red Atom: Russia's Nuclear Power Program from Stalin to Today.* New York: W. H. Freeman, 2000.

Josephson, Paul R. *Lenin's Laureate: Zhores Alferov's Life in Communist Science.* Cambridge, MA, and London: MIT Press, 2010.

Kabachnik, M. I., Ed. *Aleksandr Nikolaevich Nesmeyanov: Uchonii i Chelovek* [Aleksandr Nikolaevich Nesmeyanov: Scientist and Human Being]. Moscow: Nauka, 1988.

Kapitza, P. L. *Experiment, Theory, Practice.* Dordrecht, Boston, London: D. Reidel, 1980.

Khalatnikov, I. M., Ed. *Vospominaniya o L. D. Landau* [Reminiscences about L. D. Landau]. Moscow: Nauka, 1988.

Khalatnikov, I. M. *Dau, Kentavr i drugie. (Top nonsecret)* [Dau, Centaur and others (top non-secret)]. Moscow: Fizmatlit, 2008.

Khrushchev, N. S. *Khrushchev Remembers: The Last Testament.* Little Brown, 1974.

Knight, Amy. *Beria: Stalin's First Lieutenant.* Princeton, NJ: Princeton University Press, 1993.

Kojevnikov, Alexei B. *Stalin's Great Science: The Times and Adventures of Soviet Physicists.* London: Imperial College Press, 2004.

Kumanev, V. A. *Tragicheskie sudbi: Repressirovannie uchonie Akademii nauk SSSR* [Tragic fates: Suppressed scientists of the Soviet Academy of Sciences]. Moscow: Nauka, 1995.

Kuznetsova, N. I., Ed. *Chelovek, kotorii ne umel bit ravnodushnim: Yurii Timofeevich Struchkov v nauke i zhizhn.* [The man who was unable to be indifferent: Yurii Timofeevich Struchkov in science and in life]. Moscow: Russian Academy of Sciences, 2005.

Landau-Drobantseva, Kora. *Akademik Landau: Kak mi zhili* [Academician Landau: How we lived]. Moscow: Zakharov-AST, 1999.

Leonova, E. B., compiler, *A. I. Kitaigorodskii: Uchonii, uchitel, drug* [A. I. Kitaigorodskii: Scientist, teacher, friend]. Moscow: Moskvovedenie, 2011.

Levi-Montalcini, Rita. *In Praise of Imperfection.* Basic Books, 1989.

Levinshtein, Michael. *The Spirit of Russian Science.* New Jersey, London, Singapore: World Scientific, 2002.

Lobikov, E. A. *Sovremennaya Fizika i Atomnii Proekt* [Modern physics and atomic project]. Moscow and Izhevsk: Institut Komputernikh Issledovanii, 2002.

Lourie, Richard. *Sakharov: A Biography.* Hanover, NH, and London: Brandeis University Press, 2002.

Lozansky, Edward D., Ed. *Andrei Sakharov and Peace.* New York: Avon, 1985.

Lucretius. *The Nature of Things* [*De rerum natura*]. First edition. Translated by F. O. Copley. New York: W. W. Norton, 1977.

MacGillavry, C. H. *Symmetry Aspects of M. C. Escher's Periodic Drawings.* Utrecht: Bohn, Scheltem and Holkema, 1976.

Mackay, Alan L. *A Dictionary of Scientific Quotations.* Bristol: Adam Hilger, 1991.

Montefiori, Simon Sebag. *Stalin: The Court of the Red Tsar.* New York: Alfred A. Knopf, 2004.

Moss, Walter G. *A History of Russia, Volume II: Since 1855.* New York: McGraw-Hill, 1997.

Nesmeyanov, A. N. *Na kachelyakh XX veka* [Sitting on the swings of the twentieth century]. Moscow: Nauka, 1999.

Lundqvist, Stig, Ed. *Nobel Lectures: Physics 1971–1980.* Singapore: World Scientific, 1992.

Ozkan, Svetlana. *Novodevichy Necropolis in Moscow.* Moscow: Ritual, 2007.

Paloczi-Horvath, George. *The Facts Rebel: The Future of Russia and the West.* London: Secker & Warburg, 1964.

Pauling, Linus. *The Nature of the Chemical Bond.* Third edition. Ithaca, NY: Cornell University Press, 1961.

Pollock, Ethan. *Stalin and the Soviet Science Wars.* Princeton, NJ: Princeton University Press, 2006.

Popovsky, Mark. *Science in Chains: The Crisis of Science and Scientists in the Soviet Union Today.* London: Collins and Harvill Press, 1980.

Popovsky, Mark. *The Vavilov Affair.* Hamden, CT: Archon Books, 1984.

Pringle, Peter. *The Murder of Nikolai Vavilov: The Story of Stalin's Persecution of One of the Great Scientists of the Twentieth Century.* New York: Simon & Schuster, 2008.

Redlich, Shimon. *War, Holocaust and Stalinism: A Documented History of the Jewish Anti-Fascist Committee in the USSR.* Luxembourg: Harwood Academic Publishers, 1995.

Rhodes, Richard. *Dark Sun: The Making of the Hydrogen Bomb.* New York: Touchstone, 1996.

Roll-Hansen, Nils. *The Lysenko Effect: The Politics of Science.* Amherst, NY: Humanity Books, 2005.

Ryabev, L. D., Ed. *Atomnii Proekt SSSR I 1938–1945*. Moscow: Nauka-Fizmatlit, 1999; Ryabev, L. D., Ed. *Atomnii Proekt SSSR II 1945–1954*. Moscow-Sarov: Nauka-Fizmatlit, 1999.

Sagdeev, Roald Z. *The Making of a Soviet Scientist: My Adventures in Nuclear Fusion and Space from Stalin to Star Wars*. New York: John Wiley & Sons, 1994.

Sakharov, Andrei. *Memoirs*. Translated by Richard Lourie. New York: Alfred A. Knopf, 1990.

Sakharov, Andrei. *Moscow and Beyond 1986–1989*. New York: Alfred A. Knopf, 1991.

Sardanashvily, Genaddii A. *Dmitrii Ivanenko: Superzvezda sovetskoi fiziki: Nenapisennie memuari* [Dmitrii Ivanenko: Superstar of Soviet physics: Unwritten memoirs]. Moscow: URSS, 2010.

Semenov, Nikolai N. *Tsepnie Reaktsii* [Chain reactions]. Leningrad: Goskhimizdat, 1934; in English translation, Oxford: Oxford University Press, 1935.

Shilov, A. E., Ed. *Vospominaniya ob akademike Nikolae Nikolaeviche Semenove* [Reminiscences about academician Nikolai Nikolaevich Semenov]. Moscow: Nauka, 1993.

Shnol, Simon E., *Geroi, zlodei, konformisti rossiiskoi nauki* [Heroes, villains, conformists of Russian science]. Third edition. Moscow: URSS, 2009.

Smolegovskii, A. M. *I. I. Kitaigorodskii i ego trudi v oblasti khimii i khimicheskoi tekhnologioi stekla, keramiki i sitallov* [I. I. Kitaigorodskii and his works in chemistry and chemical technology of glass, ceramics, and sitals]. Perm, Russia: Bazaltovie Tekhnologii, 2005.

Snow, C. P. *Variety of Men*. New York: Charles Scribner's Sons, 1966.

Sonin, A. S. *"Fizicheskii idealism": Istoriya odnoi ideologicheskoi kampanii* ["Physical idealism": The history of an ideological campaign]. Moscow: Fiziko-Matematicheskaya Literatura, 1994.

Sostoyanie teorii khimicheskogo stroeniya v organicheskoi khimii [State of theory of chemical structure in organic chemistry]. All-Union Conference June 11–14, 1951, stenographic minutes. Moscow: Publishing House of the Academy of Sciences of the USSR, 1952.

Spufford, Francis. *Red Plenty*. London: Faber and Faber, 2010.

Sukhomlinov, Andrei. *Kto Vi, Lavrentii Beria? Neizvestnie stranitsi uglovogo dela* [Who are you, Lavrentii Beria? Unknown pages of a criminal case]. Moscow: Detektiv-Press, 2003.

Sunyaev, Rashid A., Ed. *Zeldovich: Reminiscences*. Boca Raton, LA: CRC Press, 2005.

Takeuchi, Y. *Global Dynamical Properties of Lotka-Volterra Systems*. Singapore: World Scientific, 1996.

Tauber, Alfred I., and Chernyak, Leon. *Metchnikoff and the Origin of Immunology*. New York and Oxford: Oxford University Press, 1991.

Teller, Edward, with Allen Brown. *The Legacy of Hiroshima*. Garden City, NY: Doubleday, 1962.

Wheland, George. *Theory of Resonance and Its Applications to Organic Chemistry*. New York: Wiley, 1944.

Zeldovich, Yakov B. *My Universe: Selected Reviews*. London: Routledge, 1992.

Zeldovich, Ya. B., *Selected works of Yakov Borisovich Zeldovich. Volume I. Chemical Physics and Hydrodynamics*. Princeton, NJ: Princeton University Press, 1992; Zeldovich, Ya. B., *Selected works of Yakov Borisovich Zeldovich. Volume II. Particles, Nuclei, and the Universe*. Princeton, NJ: Princeton University Press, 1993.

Zhabotinsky, A. M. *Kontsentratsionnie avtokolebaniya* [Concentrational oscillations]. Moscow: Nauka, 1974.

NOTES

Preface

1. G. A. Mesyats, "P. N. Lebedev Physical Institute RAS: past, present, and future." *Physics—Uspekhi* 2009, 52, 1084–1097; actual quote, p. 1091.

2. Svetlana Ozkan, ed., *Novodevichy Necropolis in Moscow* (Moscow: Ritual, 2007). The Novodevichy Cemetery was opened in 1898, and soon it was surrounded by a brick wall. In 1932, it became a burial place of the Soviet elite, but it also includes many of the greats of pre-Soviet times.

3. At the top of his power, Khrushchev publicly criticized and humiliated Neizvestny for his modern art. After Khrushchev's death, his family asked Neizvestny to create a tombstone for Khrushchev's grave. Today, Ernst Neizvestny lives in New York City.

4. Istvan Hargittai, "Limits of Perfection," in *Symmetry: Unifying Human Understanding*, ed. I. Hargittai (Oxford, UK: Pergamon Press, 1986), pp. 1–17.

Introduction

1. Andrei Sakharov, *Memoirs,* trans. Richard Lourie (New York: Alfred A. Knopf, 1990), pp. 123–124.

2. The fact that there was a token Jewish student from time to time in such institutions did not change the situation.

Chapter 1

1. I. I. Tamm [I. E. Tamm's daughter], "Family Chronicle," in *Kapitsa Tamm Semenov: v ocherkakh i pis'makh*, general ed., A. F. Andreev (Moskva: Vagrius Priroda, 1998), pp. 359–382; actual reference, p. 253.

2. I. E. Tamm, letter to Natalia Vasilevna, May 22, 1917; Andreev, *Kapitza Tamm Semenov*, p. 257.

3. Graham Farmelo, *The Strangest Man: The Hidden Life of Paul Dirac, Quantum Genius* (London: Faber and Faber, 2009), pp. 149–150.

4. I. I. Tamm in Andreev, *Kapitza Tamm Semenov*, pp. 364–365

5. Ibid., pp. 367–368.

6. Ibid., pp. 359–382.

7. Andreev, *Kapitsa Tamm Semenov*, p. 309.

8. E. Y. Hartshorne, *The German Universities and National Socialism* (Cambridge, MA: Harvard University Press, 1937), p. 112.

9. Andrei Sakharov, *Moscow and Beyond 1986–1989* (New York: Alfred A. Knopf, 1991), p. 64.

10. Istvan Hargittai, *Judging Edward Teller: A Closer Look at One of the Most Influential Scientists of the Twentieth Century* (Amherst, NY: Prometheus Books, 2010), p. 355.

11. http://www.nobelprize.org/nobel_prizes/physics/laureates/1958/tamm-speech.html.
12. Andreev, *Kapitza Tamm Semenov*, p. 314.
13. Ibid., p. 397.
14. Ibid., p. 298.
15. Ibid., p. 335.
16. Ibid., p. 338.
17. Ibid., p. 341.
18. Ibid., pp. 342–343.
19. Ibid., pp. 343–345.
20. Istvan Hargittai, "A Curious Case of Soviet Nobel Aspirations," *Chemical Intelligencer* 1999, 5(3), 61–64; actual quote, p. 61. The Russian original of the quote was in A. Blokh, *Poisk,* nos. 31–32, July 25–August 7, 1998, p. 12. See also A. M. Blokh, *Sovetskii Soyuz v interyere nobelevskikh premii: Fakti. Dokumenti. Razmyshleniya. Kommentarii* [The Soviet Union in the interior of the Nobel Prizes: facts; documents; reflections; commentaries], 2nd ed. (Moscow: Fizmatlit, 2005).
21. Hargittai, "A Curious Case."
22. Blokh, *Sovetskii Soyuz*, p. 486.
23. Ibid., p. 819.
24. Private communication from Sheldon Glashow via e-mail, June 2012.
25. As for Glashow, he was at the CERN at the time of his waiting for his Soviet visa. His colleagues observed him reading whatever he could find about the Soviet Union in the CERN Library. Private communication by Richard Garwin, via e-mail, June 13, 2012.
26. I am grateful to Anders Bárány for having conducted this research in March 2012.
27. The four leading scientists involved in the antiproton discovery were Owen Chamberlain, Emilio Segrè, Clyde Wiegand, and Thomas Ypsilantis. In 1959, two of the four, Chamberlain and Segrè, were awarded the Nobel Prize in Physics for the discovery of the antiproton.
28. Andreev, *Kapitza Tamm Semenov*, p. 384 [by Leonid Vernskii, Tamm's grandson].
29. Ibid., p. 394.
30. Blokh, *Sovetskii Soyuz*, pp. 488–489.
31. Ibid., p. 490.
32. George Paloczi-Horvath, *The Facts Rebel: The Future of Russia and the West* (London: Secker & Warburg, 1964), pp. 82–83.
33. Andreev, *Kapitza Tamm Semenov*, p. 346 [after *Moskovskaya Pravda* No. 137, June 19, 1994].
34. Ibid., pp. 350–358.
35. Ibid., p. 356.
36. Andreev, *Kapitza Tamm Semenov*, p. 384 [by Leonid Vernskii].
37. Ibid., p. 385.
38. Ibid., p. 386
39. Ibid., p. 388.

Chapter 2

1. Perhaps the best source for learning more about Ya. B. Zeldovich's life and oeuvre is Rashid A. Sunyaev, ed., *Zeldovich: Reminiscences* (Boca Raton, FL: CRC Press, 2005).

2. R. A. Sunyaev, "When We Were Young," in Sunyaev, *Zeldovich: Reminiscences*, pp. 233–240.

3. Istvan Hargittai, "Walter Gilbert," in *Candid Science II: Conversations with Famous Biomedical Scientists*, ed. Magdolna Hargittai (London: Imperial College Press, 2002), pp. 98–113.

4. M. Ya. Ovchinnikova, "Enchanted with the World," in Sunyaev, *Zeldovich: Reminiscences*, pp. 61–68; actual quote, p. 61.

5. http://www.the-tls.co.uk/tls/public/article1235922.ece (as of April 10, 2013).

6. "From the Eulogy of A. D. Sakharov," in Sunyaev, *Zeldovich: Reminiscences*, pp. 97–98; actual quote, p. 98.

7. Ya. B. Zeldovich, "Autobiographical Afterword," in Sunyaev, *Zeldovich: Reminiscences*, pp. 338–348; actual quote, p. 342.

8. Boris S. Gorobets, *Krug Landau i Lifshitsa* [Landau's and Lifshits's circle] (Moscow: URSS, 2008), p. 99.

9. Ibid., p. 100.

10. See, Boris S. Gorobets, *Sekretnie fiziki iz atomnogo proekta SSSR: Semya Leipunskikh* [Classified physicists from the atomic project of the Soviet Union: The Leipunskii family], 2nd ed. (Moscow: URSS, 2009), pp. 151–157.

11. Yu. B. Khariton, in *Znakomii neznakomii Zeldovich* (Moscow 1993), p. 107

12. N. A. Konstantinova, " 'Don't forget to write about me,' " in *Zeldovich: Reminiscences*, pp. 26–28; actual quote, p. 27.

13. Ovchinnikova, "Enchanted with the World," p. 62.

14. Ibid., p. 63.

15. Ibid.

16. Olga Zeldovich, Marina Ovchinnikova, and Boris Zeldovich, private communication of February 20, 2012, by e-mail.

17. Boris Gorobets, private communication of February 25, 2012, by e-mail.

18. Ovchinnikova, "Enchanted with the World," p. 63.

19. Olga Zeldovich, Marina Ovchinnikova, and Boris Zeldovich, private communication of February 20, 2012, by e-mail.

20. Ibid.

21. Yu. N. Smirnov, "A Knight of Science," in Sunyaev, *Zeldovich: Reminiscences*, pp. 105–120; actual quote, p. 119.

22. Ibid., p. 116.

23. Kip S. Thorne, "Zeldovich Predicts That Black Holes Radiate," in Sunyaev, *Zeldovich: Reminiscences*, pp. 323–329; actual reference, p. 323.

24. V. Ya. Frenkel, "I Can See Him in Front of Me, as if Still Alive," in Sunyaev, *Zeldovich: Reminiscences*, pp. 14–25; actual quote, p. 19.

25. V. S. Pinaev, " 'For Me, They Were the Happy Years,' " in Sunyaev, *Zeldovich: Reminiscences*, pp. 131–138; actual quote, p. 138.

26. V. E. Zakharov, "My Reminiscences about Ya. B. Zeldovich," in Sunyaev, *Zeldovich: Reminiscences*, pp. 203–207; actual reference, p. 204.

27. Gorobets, *Krug Landau i Lifshitsa*, p. 108.

28. V. I. Goldanskii, "Forty-Five Years," in Sunyaev, *Zeldovich: Reminiscences*, pp. 30–45; actual reference, p. 40.

29. Ovchinnikova, "Enchanted with the World."

30. Shimon Redlich, *War, Holocaust and Stalinism: A Documented History of the Jewish Anti-Fascist Committee in the USSR* (Luxembourg: Harwood Academic Publishers, 1995), p. 131.

31. Gorobets, *Krug Landau i Lifshitsa*, p. 108.

32. Richard Lourie, *Sakharov: A Biography* (Hanover, NH, and London: Brandeis University Press, 2002), p. 131.

33. Andrei Sakharov, *Memoirs,* trans. Richard Lourie (New York: Alfred A. Knopf, 1990), p. 112.

34. "Letter from Ya. B. Zeldovich to M. M. Agrest," June 14, 1981, in Sunyaev, *Zeldovich: Reminiscences*, pp. 69–70.

35. "Letter from M. M. Agrest to the Editors of the Journal *Khimiya i Zhizn* [Chemistry and Life]," February 13, 1992, in Sunyaev, *Zeldovich: Reminiscences*, p. 71.

36. A. D. Myshkis, "The 'Humanizing' of Mathematics," in Sunyaev, *Zeldovich: Reminiscences*, pp. 208–219; actual quote, p. 218.

37. V. E. Zakharov, "My Reminiscences about Ya. B. Zeldovoch," p. 206.

38. Andrei D. Sakharov, "A Man of Universal Interests," in Sunyaev, *Zeldovich: Reminiscences*, pp. 125–128; actual reference, p. 128.

39. Ibid.

40. L. V. Altshuler, "The Beginning of the Physics of Extreme States," in Sunyaev, *Zeldovich: Reminiscences*, pp. 99–104; actual quote, p. 99.

41. "From the Eulogy of A. D. Sakharov," in Sunyaev, *Zeldovich: Reminiscences*, pp. 97–98; actual quote, p. 97.

42. Smirnov, "A Knight of Science," in Sunyaev, *Zeldovich: Reminiscences*, pp. 105–120.

43. See, e.g., Yakov B. Zeldovich, *My Universe: Selected Reviews* (Routledge, 1992), p. 3.

44. Arno Penzias, "The Origin of the Elements," in *Nobel Lectures: Physics 1971–1980* (Singapore: World Scientific, 1992), pp. 444–457; actual quote, p. 454, referring to Doroshkevich and Novikov.

45. A. G. Doroshkevich and I. D. Novikov, "Mean Density of Radiation in the Metagalaxy and Certain Problems in Relativistic Cosmology," *Soviet Physics-Dokl.* 1964, 9, 111; Russian original, *Doklady Akademii Nauk* SSSR 1964, *154*, 809.

46. Penzias, *Nobel Lectures*, p. 455.

47. R. A. Sunyaev, "When We Were Young," in Sunyaev, *Zeldovich: Reminiscences*, pp. 233–243; actual quote, p. 237.

48. Thorne, "Zeldovich Predicts That Black Holes Radiate," in Sunyaev, *Zeldovich: Reminiscences*, p. 329.

49. P. J. E., "Zeldovich and Modern Cosmology," in Sunyaev, *Zeldovich: Reminiscences*, pp. 281–293; actual quote, p. 281.

50. Sakharov, *Memoirs*, p. 259.

51. Smirnov, "A Knight of Science," in Sunyaev, *Zeldovich: Reminiscences*, p. 105.

52. Ibid., p. 107.

53. I. D. Novikov, "The Beginning of Work in Astrophysics," Sunyaev, in *Zeldovich: Reminiscences*, pp. 224–228; actual reference, p. 227.

54. Smirnov, "A Knight of Science," in Sunyaev, *Zeldovich: Reminiscences*, p. 113.

55. On the comparison of Fermi and Szilard, see, I. Hargittai, *The Martians of Science* (New York: Oxford University Press, 2006; 2007), pp. 188–195.

56. Gorobets, *Krug Landau i Lifshitsa.*

57. I appreciate Roald Hoffmann's kindness in translating the poem, January 2012.

Chapter 3

1. In collecting biographical information, I relied on a variety of sources, especially on Andrei Sakharov, *Memoirs,* trans. Richard Lourie (New York: Alfred A. Knopf, 1990).

2. Ibid., p. 4.

3. Ibid., p. 74.

4. Ibid., p. 47.

5. Ibid., p. 74.

6. Ibid., p. 95.

7. Ibid., p. 117.

8. Ibid., pp. 139–147.

9. Tokamak, toroidal chamber with magnetic coils (in Russian, toroidalnaya kamera s aksialnym magnitnym polem).

10. Sakharov, *Memoirs*, pp. 149–155.

11. Ibid., p. 94.

12. Ibid., p. 97.

13. Istvan Hargittai, *The Martians of Science: Five Physicists Who Changed the Twentieth Century* (New York: Oxford University Press, 2006; 2007), p. 165.

14. Ibid., p. 166.

15. Sakharov, *Memoirs*, p. 100.

16. Istvan Hargittai, *Judging Edward Teller: A Closer Look at One of the Most Influential Scientists of the Twentieth Century* (Amherst, NY: Prometheus Books, 2010).

17. David Holloway, "New Light on Early Soviet Bomb Secrets," *Physics Today* 1996, November, 26–27.

18. Hargittai, *Judging Edward Teller*, p. 221, and subsequent mentions.

19. Ibid., p. 239.

20. German A. Goncharov, "Thermonuclear Milestones," *Physics Today* 1996 November, 44; (1) "The American Effort," ibid., 45–48; (2) "Beginnings of the Soviet H-Bomb program," ibid., 50–54; (3) "The Race Accelerates," ibid., 56–61.

21. Sakharov, *Memoirs*, p. 124.

22. Ibid., p. 133.

23. Ibid., p. 146.

24. Ibid., p. 164.

25. Ibid., p. 171.

26. Ibid., p. 184.

27. Ibid., p. 193.

28. Ibid., p. 194.

29. Ibid.

30. Ibid., p. 217.

31. Andrei Sakharov, *Moscow and Beyond 1986–1989* (New York: Alfred A. Knopf, 1991), p. 24.

32. Sakharov, *Memoirs*, p. 160.

33. Ibid.

34. Hargittai, *Judging Edward Teller*, pp. 319–337. Teller's comments accepting the harmful consequences of testing quoted here demonstrate one of his two alternate approaches to this issue. The other was belittling any possible danger from testing. Apparently it did not matter to him that his two alternate approaches were in direct contradiction.

35. Edward Teller with Allen Brown, *The Legacy of Hiroshima* (Garden City, NY: Doubleday, 1962), pp. 180–181. Daughters of the American Revolution is a genealogical society; any women eighteen years or older may become a member if she can prove lineal descent from a patriot of the American Revolution, in which toward the end of the eighteenth century, thirteen colonies of North America decided to break free from the British Empire.

36. Sakharov, *Memoirs*, p. 201.

37. Ibid., p. 221.

38. Ibid.

39. E-mail exchange with Alexander Vernyi, February–March 2012.

40. Gennady Gorelik with Antonina W. Bouis, *The World of Andrei Sakharov: A Russian Physicist's Path to Freedom* (New York: Oxford University Press, 2005), p. 226.

41. Sakharov, *Moscow and Beyond*, p. 42.

42. Sakharov, *Memoirs*, pp. 267–268.

43. Ibid., p. 282

44. Ibid., p. 289.

45. Ibid., p. 386.

46. Vitaly L. Ginzburg, *O fizike i astrofizike* [About physics and astrophysics]. 3rd rev. ed. (Moscow: Byuro Quantum, 1995), pp. 467–468.

47. Boris Ya. Zeldovich, private communication in February 2012, by e-mail.

48. Richard Lourie, *Sakharov: A Biography* (Hanover, NH: University Press of New England, 2002), p. 219.

49. Sidney D. Drell, "Andrei Sakharov and the Nuclear Danger," *Physics Today* 2000, May, 37–41.

50. http://www.aps.org/units/fps/awards/ (as of March 31, 2012).

51. Sakharov, *Memoirs*, p. 663.

52. Elena Bonner, *Alone Together* (New York: Vintage Books, 1988), pp. 271–280.

53. Ginzburg, *O fizike*, pp. 469–476.

54. Ibid., p. 469.

55. Ibid., p. 476.

56. Even as late as November 1988, almost one year after Sakharov's return from exile to Moscow and three and a half years after Gorbachev's having become the supreme leader of the Soviet Union, Sakharov still had to demand freedom for political priosoners, long after Gorachev had announced that there were no longer any. C. M., "Sakharov Visits US to Launch International Group," *Nature* 1988, *336*, (November 17), 191.

57. Sakharov, *Moscow and Beyond*, pp. 24–25.

58. Magdolna Hargittai and Istvan Hargittai, *Candid Science IV: Conversations with Famous Physicists* (London: Imperial College Press, 2004), pp. 272–285

59. Arno A. Penzias, "Sakharov and SDI," in *Andrei Sakharov: Facets of a Life*, edited by B. L. Altshuler (Gif-sur-Yvette, France: Edition Frontiers, 1991), pp. 507–516.

60. Ibid., p. 507.

61. "Ostensibly," because the principal tool of SDI was to be the X-ray laser whose application necessitated triggering an explosion of a nuclear device, even though the purpose of SDI, was to make nuclear weapons "impotent and obsolete" (in President Reagan's words).

62. Penzias, "Sakharov and SDI," pp. 511–515.

63. Hargittai, *Candid Science IV*, p. 279.

64. Edward Kline, "Foreword" to Sakharov, *Moscow and Beyond*, pp. vii–xvii; actual quote, pp. viii–ix.

65. Ibid., p. xvi.

66. Ibid., p. xii.

67. Hargittai, *Martians of Science*, p. 74

68. Daniel E. Koshland Jr., "Andrei Sakharov, 1921–1989," *Science* 1990, 247, 265.

Chapter 4

1. Magdolna Hargittai and Istvan Hargittai, *Candid Science IV: Conversations with Famous Physicists* (London: Imperial College Press, 2004), "David Shoenberg," pp. 688–697. I have also learned a lot about Kapitza from the following sources: A. F. Andreev, general editor, *Kapitza Tamm Semenov: v ocherkakh i pismakh* [Kapitza Tamm Semenov: In sketches and letters] (Moscow: Vagrius Priroda, 1998), "Petr Leonidovich Kapitza," pp. 13–218; J. W. Boag, P. E. Rubinin, and D. Shoenberg, compilers and editors, *Kapitza in Cambridge and Moscow: Life and Letters of a Russian Physicist* (Amsterdam: North-Holland, 1990); Lawrence Badash, *Kapitza, Rutherford, and the Kremlin* (New Haven and London: Yale University Press, 1985).

2. C. P. Snow, *Variety of Men* (New York: Charles Scribner's Sons, 1966), p. 17.

3. Boag, Rubinin, and Shoenberg, *Kapitza*, pp. 40–45.

4. Ibid.

5. Trotsky was murdered by the Soviet secret police while he lived in exile, but this was admitted only in 1989, under Mikhail S. Gorbachev; Kamenev was rehabilitated in 1988, during Gorbachev's reign.

6. Boag, Rubinin, and Shoenberg, *Kapitza*, p. 691.

7. Ibid., p. 29.

8. Ibid., p. 11.

9. I. M. Khalatnikov, *Dau, Kentavr i drugie (Top nonsecret)* [Dau, Centaur and others (top nonsecret)] (Moscow: Fizmatlit, 2008), pp. 40–41. The expression in English, "top nonsecret," is part of the Russian title.

10. Boag, Rubinin, and Shoenberg, *Kapitza*, p. 31.

11. Hargittai and Hargittai, *Candid Science IV*.

12. Ibid., p. 692.

13. Ibid., p. 693.

14. George Gamow, *My World Line: An Informal Autobiography* (New York: Viking Press, 1970), p. 131.

15. Letter of P. L. Kapitza to L. B. Kamenev, February 2, 1929, in Boag, Rubinin, and Shoenberg, *Kapitza*, pp. 313–314.

16. B. S. Gorobets, *Istoriya nauki i tekhniki* 2010, no. 3, 19–32.

17. Ibid., p. 20.

18. Letter of P. L. Kapitza to Niels Bohr, November 15, 1933, in Boag, Rubinin, and Shoenberg, *Kapitza*, pp. 315–316.

19. Ibid.

20. Letter of P. L. Kapitza to Max Born, February 26, 1936, ibid., p. 324.

21. Max Born, *My Life: Recollections of a Nobel Laureate* (New York: Charles Scribner's Sons, 1978), p. 281.

22. Today, there is an S. V. Vavilov Institute of History of Science and Technology of the Russian Academy of Sciences. Sergei Vavilov had a brother, Nikolai Vavilov, who was a world renowned botanist. His disagreement with the charlatan Lysenko—whom initially Nikolai had supported—led to Nikolai's murder in 1943 by malnutrition in prison at the age of fifty-five. He was rehabilitated in 1955.

23. D. Shoenberg, "Piotr Leonidovich Kapitza July 9, 1894–April 8, 1984," *Biographical Memoirs of Fellows of the Royal Society* 1985, 31, 326–374; actual quote, p. 349.

24. Ibid.

25. Alan L. Mackay, *A Dictionary of Scientific Quotations* (Bristol: Adam Hilger, 1991), p. 150.

26. B. S. Gorobets, *Krug Landau i Lifshitsa* [Landau's and Lifshits's circle] (Moscow: URSS, 2008), chap. 10, "P. L. Kapitza: 'Kentaur,'" pp. 180–225.

27. Letter of P. L. Kapitza to I. V. Stalin, December 1, 1935, in Andreev, *Kapitza Tamm Semenov*, pp. 159–164.

28. Letter of I. V. Stalin to P. L. Kapitza, April 4, 1946, ibid., p. 359 (the facsimile of Stalin's letter) and p. 378 (the translation of the letter).

29. Letter of P. L. Kapitza to V. M. Molotov, July 6, 1936, in Boag, Rubinin, and Shoenberg, *Kapitza*, pp. 331–333.

30. Letter of P. L. Kapitza to V. M. Molotov, July 6, 1936, ibid., pp. 331–333.

31. Ibid., p. 332.

32. Ibid., p. 339.

33. A. S. Sonin, *"Fizicheskii idealism" Istoriya odnoi ideologicheskoi kampanii* ["Physical idealism": The history of an ideological campaign] (Moscow: Fiziko-Matematicheskaya Literatura, 1994), p. 57.

34. John Campbell, *Rutherford: Scientist Supreme.* (Christchurch, New Zealand: AAS Publications, 1999), p. 474

35. Sébastian Balibar, "Looking Back at Superfluid Helium," *Séminare Poincaré* 2003, 1, 11–20.

36. P. L. Kapitza, "Plasma and the Controlled Thermonuclear Reaction," *Nobel Lectures Physics 1971–1980* (Singapore: World Scientific, 1992), pp. 424–436.

37. P. L. Kapitza, *Nature* 1938, 141, 74; J. F. Allen, A. D. Misener, *Nature* 1938, 141:75.

38. Boris S. Gorobets, "Piotr L. Kapitza Summoned to 'Lubianka' (Ministry of Interior Affairs—NKVD): Worn Myth and a First Non-Contradictory Version of Landau's Liberation in 1939," *Istoria nauki i tekhniki*, 2011, no. 10, pp. 24–34; and private communications from Boris Gorobets by e-mail in October 2011.

39. Anne Applebaum, *Gulag: A History* (New York: Anchor Books, 2004).

40. H. J. DeWolf Smyth, *Atomic Energy for Military Purposes: The Official Report on the Development of the Atomic Bomb under the Auspices of the United States Government, 1940–1945* (Princeton, NJ: Princeton University Press, 1945).

41. Boag, Rubinin, and Shoenberg, *Kapitza*, pp. 372–378.

42. Ibid., p. 378.

43. Ibid., p. 389.

44. Andreev, *Kapitza Tamm Semenov*, p. 44.

45. Ibid., p. 28.

46. Alden Whitman, "A Brilliant Scientist," *New York Times*, April 3, 1968, pp. 1 and 47; the article was mainly about Lev Landau.

47. Boag, Rubinin, and Shoenberg, *Kapitza*, p. 390.

48. Ibid., p. 399. Indeed, for example, even the rigorous dress codes for the Nobel Prize award ceremony prescribing white ties for males allow wearing a national dress instead.

49. Letter of P. L. Kapitza to N. S. Khrushchev, September 22, 1955, ibid., pp. 403–404.

50. Letter of P. L. Kapitza to N. S. Khrushchev, December 15, 1955, ibid., pp. 404–408.

51. N. S. Khrushchev, *Khrushchev Remembers: The Last Testament* (Boston, 1976), pp. 67–73.

52. Ibid.

53. P. L. Kapitza, *Experiment, Theory, Practice* (Dordrecht, Boston, London: D. Reidel, 1980). This reference is not to the Italian edition mentioned in the text.

54. Letter from P. L. Kapitza to Yu. V. Andropov, April 22, 1980, in Boag, Rubinin, and Shoenberg, *Kapitza*, pp. 413–416.

55. Andrei Sakharov, *Memoirs* (New York: Alfred A. Knopf, 1990), pp. 303–304; Andrei Sakharov, *Moscow and Beyond 1986–1989* (New York: Alfred A. Knopf, 1991), pp. 11–12.

56. Snow, *Variety*, p. 19.

Chapter 5

1. George Gamow, *My World Line: An Informal Autobiography* (New York: Viking Press, 1970), p. 50.

2. Boris Gessen was the author of the entry. Gessen became a victim of Stalin's Terror in 1937–1938. Note that Gessen's name often figures as Hessen in the literature elsewhere.

3. Gamow, *My World Line*, p. 96.

4. The original Russian article was translated into English and published: G. Gamow, D. Ivanenko, and L. Landau, "World Constants and Limiting Transitions," *Physics of Atomic Nuclei* 2002, 65, 1373–1375.

5. A. S. Sonin, *"Fizicheskii idealism" Istoriya odnoi ideologicheskoi kampanii* ["Physical idealism": The history of an ideological campaign] (Moscow: Fiziko-Matematicheskaya Literatura, 1994), p. 129.

6. I. Hargittai and M. Hargittai, "Lev D. Landau (1908–1968): in Memoriam," *Structural Chemistry*, 2008, 19, 181–184.

7. Istvan Hargittai, *Judging Edward Teller: A Closer Look at One of the Most Influential Scientists of the Twentieth Century* (Amherst, NY: Prometheus Books, 2010).

8. Gorobets, *Landau I.*, p. 181. Note that in his Landau trilogy, being referred to here as *Landau I*, *Landau II*, and *Landau III*, Gorobets made use of other books about Landau in addition to his own research. The sources are meticulously documented in Gorobets's volumes.

9. Ibid.

10. The text of the leaflet is reproduced in Gorobets, *Landau I*, p. 284.

11. Feinberg, E. L., *Physicists: Epoch and Personalities* (New Jersey, London, Singapore: World Scientific, 2011), p. 396.

12. Boris S. Gorobets, *Krug Landau: Zhizn geniya* [Landau's circle: The life of a genius], 2nd corrected and augmented ed. (Moscow: URSS, 2007), p. 298, n.1.

13. Ibid., p. 299.

14. Ibid., p. 301.

15. Ibid.

16. Ibid.

17. Ibid., pp. 304–305.

18. Ibid., p. 304
19. Gorobets, *Landau I*, p. 37.
20. Magdolna Hargittai and Istvan Hargittai, *Candid Science IV: Conversations with Famous Physicists* (London: Imperial College Press, 2004), "Laszlo Tisza."
21. Gorobets, *Landau I*, p. 113.
22. Hargittai and Hargittai, *Candid Science IV*, "David Shoenberg," pp. 688–697; actual quote, p. 695.
23. Gorobets, *Landau I*, p. 271; see also G. E. Gorelik, *Priroda*, 1991, no. 11, 93–104.
24. Ibid., p. 121, after Landau-Drobantseva, 1999, p. 82.
25. Hargittai and Hargittai, *Candid Science IV*, p. 696.
26. Ibid., p. 697.
27. Ibid., p. 295.
28. A. A. Abrikosov, *Akademik L. D. Landau: Kratkaya biografiya i obzor nauchnikh rabot* [Academician L. D. Landau: Brief biography and review of scientific works] (Moscow: Nauka, 1965).
29. I. M. Khalatnikov, ed., *Vospominaniya o L. D. Landau* [Reminiscences about L. D. Landau] (Moscow: Nauka, 1988).
30. E. M. Lifshits, ibid., pp. 7–31
31. N. E. Alekseevskii, ibid., pp. 40–42.
32. I. M. Khalatnikov, *Dau, Kentaur i drugie (Top nonsecret)* [Dau, Cantaur and others (top nonsecret)] (Moscow: Fizmatlit, 2008), p. 58.
33. Feinberg, *Physicists: Epoch and Personalities,* p. 387.
34. Ibid., p. 388.
35. Gorobets *Landau I*, appendix 3, doc. no.12, pp. 296–297.
36. *Atomnii Proekt SSSR I 1938–1945* (Moscow: Nauka-Fizmatlit, 1999); *Atomnii Proekt SSSR II 1945–1954* (Moscow-Sarov: Nauka-Fizmatlit, 1999).
37. Gorobets *Landau II*, p. 51.
38. Ibid., p. 56.
39. Ibid., pp. 61–63.
40. Gorobets, *Landau II*, p. 108.
41. Ibid., p. 109.
42. Feinberg, *Physicists: Epoch and Personalities*, p. 379.
43. Gorobets *Landau II*, p. 114.
44. Feinberg, *Physicists: Epoch and Personalities*, p. 382.
45. Ibid., pp. 114–123.
46. Ibid., pp. 123–126.
47. Gorobets *Landau I*, p. 185.
48. Ibid., pp. 185–188.
49. Ibid., p. 189.
50. Ibid., p. 190.
51. Ibid.
52. Ibid., pp. 192–194.
53. A. M. Blokh, *Sovietskii Soyuz v interyere nobelevskikh premii: Fakti, dokumenti, razmyshlenie, kommentarii* [The Soviet Union in the interior of the Nobel Prizes: Facts; documents; reflections; commentaries] (Moscow: Fizmatlit, 2005), p. 501.
54. Ibid., p. 502.

55. Ibid., p. 556.

56. Alden Whitman, "A Brilliant Scientist," *New York Times,* April 3, 1968, pp. 1 and 47; actual quote, p. 47.

57. Ibid.

58. Gorobets, *Landau I*, p. 254.

59. Ya. B. Zeldovich and M. I. Kaganov, trans. J. B. Sykes, "Evgenii Mikhailovich Lifshitz February 21, 1915–October 29, 1985," *Biographical Memoirs of Fellows of the Royal Society* 1990, 36, 336–357; actual quote, p. 345.

60. Ibid., p. 340.

61. A. A. Roukhadze, M. A. Liberman, and B. S. Gorobets, "Academician E. M. Lifshits: Outstanding Physicist and Scientific Writer. Part II. The Scale of a Scientist," *Istoriya nauki i tekhniki*, 2011, no. 3, 44–53.

62. Ibid., p. 45.

63. B. S. Gorobets, "Academician E. M. Lifshits: An Outstanding Physicist and Scientific Writer. Part I: The Scale of Personality," *Istoriya nauki i tekhniki,* 2011, no. 2. The quotation is from D. E. Khmelnitskii, 1994.

64. Ibid.

65. Zeldovich and Kaganov, *Memoirs*, p. 343.

66. Roukhadze, Liberman, and Gorobets, "Academician E. M. Lifshits," p. 50.

67. Gorobets, *Landau III*, p. 41.

68. Ibid.

69. *Berezka* means "birch tree" in a diminutive version.

70. Gorobets, *Landau III*, p. 37.

71. Ibid, p. 24.

72. Ibid., p. 45.

Chapter 6

1. Istvan Hargittai and Magdolna Hargittai, *Candid Science VI: More Conversations with Famous Scientists* (London: Imperial College Press, 2006), "Vitaly L. Ginzburg," pp. 808–837; actual quote, p. 836.

2. Vitaly L. Ginzburg, "Notes of an Amateur Astrophysicist," *Annual Review of Astronomy and Astrophysics*, 1990, 28, 1–36; actual quote, p. 1.

3. V. L. Ginzburg, "Ob ottse i nashei seme" ["About my father and my family"], *Uspekhi fizicheskikh nauk,* 2010, *180*, 1217–1230; actual reference, p. 1226.

4. Ibid.; see also note 6, p. 1226.

5. Vitaly L. Ginzburg, "Autobiography," Nobel Foundation website.

6. Vitaly L. Ginzburg, *About Science, Myself and Others* (Bristol and Philadelphia: Institute of Physics Publishing, 2005), p. 373.

7. Ginzburg, "Notes of an Amateur Astrophysicist," p. 4.

8. Ginzburg, Nobel autobiography.

9. Ginzburg, *About Science,* pp. 375–376.

10. Ginzburg, "Ob ottse i nashei seme," p. 1222.

11. Ginzburg, *About Science,* p. 378.

12. Ibid., p. 382.

13. Ibid.

14. Ibid.

15. Genaddii A. Sardanashvily, *Dmitrii Ivanenko—Superzvezda sovetskoi fiziki: Nenapisennie memuari* [Dmitrii Ivanenko—superstar of Soviet physics: Unwritten memoirs], (Moscow: URSS, 2010).

16. Ginzburg, *About Science,* p. 220 and p. 384.

17. Ibid., p. 220.

18. Hargittai and Hargittai, *Candid Science VI*, p. 821.

19. On Heisenberg's attitude during and after World War II, see, e.g., Istvan Hargittai, *Judging Edward Teller: A Closer Look at One of the Most Influential Scientists of the Twentieth Century* (Amherst, NY: Prometheus Books, 2010), pp. 81–90.

20. Hargittai and Hargittai, *Candid Science VI*, p. 821.

21. Ibid., p. 822.

22. Ibid.

23. Ginzburg, "Notes of an Amateur Astrophysicist," p. 12.

24. Vitaly L. Ginzburg, *O fizike i astrofizike* [About physics and astrophysics], 3rd rev. ed. (Moscow: Buro Quantum, 1995), pp. 467–468.

25. Ibid., p. 812.

26. Vitaly Ginzburg, Nobel Lecture, December 8, 2003, Nobel Foundation website, pp. 96–127.

27. Hargittai and Hargittai, *Candid Science VI*, p. 821.

28. Ginzburg, Nobel Lecture, p. 104.

29. Yu. I. Solov'ev, *Herald of the Russian Academy of Science* 1997, 7, 627.

30. Hargittai and Hargittai, *Candid Science VI*, pp. 815–818.

31. Ginzburg, Nobel Lecture, p. 121.

32. Ginzburg, *O fizike i astrofizike,* pp. 312–349 "Experience of Scientific Autobiography".

33. Ginzburg, "Notes of an Amateur Astrophysicist," p. 14.

34. These two volumes are available in English translation, *Selected Works of Yakov Borisovich Zeldovich. Volume I. Chemical Physics and Hydrodynamics* (Princeton, NJ: Princeton University Press, 1992); *Selected Works of Yakov Borisovich Zeldovich. Volume II. Particles, Nuclei, and the Universe* (Princeton, NJ; Princeton University Press, 1993).

35. V. L. Ginzburg, *Biographical Memoirs of Fellows of the Royal Society*, vol. 40, (London: The Royal Society, 1994), pp. 431–441.

36. Hargittai and Hargittai, *Candid Science VI*, p. 831.

37. Ginzburg, *O fizike i astrofizike*, pp. 312–349.

38. Hargittai and Hargittai, *Candid Science VI*, p. 810.

39. Ginzburg, "Notes of an Amateur Astrophysicist," p. 21.

Chapter 7

1. Balazs Hargittai and István Hargittai, *Candid Science V: Conversations with Famous Scientists* (London: Imperial College Press, 2005), "Alexei A. Abrikosov," pp. 176–197.

2. E-mail message from Alexei A. Abrikosov to the author, January 3, 2012.

3. Dmitrii Abrikosov, *Revelations of a Russian Diplomat; The Memoirs of Dmitrii I. Abrikosov*, ed. George A. Lensen (Seattle: University of Washington Press, 1964).

4. Isaak M. Khalatnikov, *Dau, Kentavr i drugie (Top nonsecret)* [Dau, Centaur and others (top nonsecret)] (Moscow: Fizmatlit, 2008), pp. 26–27.

5. Boris S. Gorobets, *Krug Landau i Lifshitsa* [Landau's and Lifshits's circle] (Moscow: URSS, 2008), p. 136

6. Hargittai and Hargittai, *Candid Science V*, pp. 187–188.

7. Ibid.

8. Ibid., p. 188.

9. Ibid., p. 189.

10. Ibid.

11. Gorobets, *Krug Landau i Lifshitsa*, p. 131.

12. Sergei Leskov, op-ed article, "America's Soviet Scientists," *New York Times*, July 15, 1993. The English text here is the translation from the original Russian statement, e.g., Gorobets, *Krug Landau i Lifshitsa*, p. 142.

13. Ibid., p. 145; *Pravda*, December 17, 1993, no. 211.

14. Alexander Migdal, letter in the *New York Times*, July 25, 1993.

15. Alexei A. Abrikosov, Nobel Lecture, Nobel Foundation website, pp. 59–67; actual quote, p. 65.

16. The two papers appeared in 1952, in the same issue of *Doklady Akademii Nauk SSSR*; Abrikosov, p. 489; Zavaritskii, p. 501.

17. Hargittai and Hargittai, *Candid Science V*, p. 184.

18. Ibid., p. 181.

19. Gorobets, *Krug Landau i Lifshitsa*, p. 140.

20. Hargittai and Hargittai, *Candid Science V*, p. 183.

21. Abrikosov, Nobel Lecture, pp. 65–66.

22. István Hargittai and Magdolna Hargittai, *Candid Science VI: More Conversations with Famous Scientists* (London: Imperial College Press, 2006), "Vitaly L. Ginzburg," pp. 808–837; actual quote, p. 816.

23. Hargittai and Hargittai, *Candid Science V*, p. 195.

24. Ibid., p. 196.

25. Ibid.

Chapter 8

1. Istvan Hargittai, *Candid Science: Conversations with Famous Chemists*, ed. Magdolna Hargittai (London: Imperial College Press, 2000), "Nikolai N. Semenov," 466–475. When the interview was broadcast and then also printed in Hungarian, I provided its Hungarian translation. For the *Candid Science* series, the conversation was translated into English. When the Russian translation of the first volume of *Candid Science* was being prepared, the English translation was translated back into Russian. When I learned about this, I sent a copy of the original tape to the editor of the Russian translation of *Candid Science*, Professor Petr M. Zorkii of Moscow State University. He found the two texts in excellent agreement.

2. *Ioffe Institute 1918–1998: Development and Research Activities,* editor-in-chief, Zh. Alferov (St. Petersburg: Ioffe Institute, 1998), p. 245.

3. I. V. Obreimov, "Molodie godi" ["Young Years"], in *Vospominaniya ob akademike Nikolae Nikolaeviche Semenove* [Reminiscences about academician Nikolai Nikolaevich Semenov], ed. A. E. Shilov (Moscow: Nauka, 1993), pp. 29–30.

4. Olga Liverovskaya, "Istoriya odnoi semi" ["The history of a family"], *Neva* 2005, no. 7. I am grateful to Alexey Semenov for a copy of this article. Alexey Semenov is the grandson of Nikolai Semenov and Yulii Khariton.

5. Hargittai, *Candid Science*, pp. 468–469.

6. Yu. B. Khariton, "Nachalo" ["The Beginning"], in Shilov, *Vospominaniya*, pp. 30–42.

7. Letter of Yu. B. Khariton to N. N. Semenov, March 13, 1927, in A. F. Andreev, general editor, *Kapitza Tamm Semenov: v ocherkakh i pismakh* [Kapitza Tamm Semenov: In sketches and letters] (Moscow: Vagrius Priroda, 1998), "Nikolai Nikolaevich Semenov," pp. 401–575; actual reference, pp. 439–441.

8. Yu. B. Khariton, "U istokov yadernogo dela" ["At the sources of things nuclear"), in Andreev, *Kapitza Tamm Semenov,* Khariton's chapter, pp. 432–441.

9. Istvan Hargittai, *The Martians of Science: Five Physicists Who Changed the Twentieth Century* (New York: Oxford University Press, 2006; 2007), p. 76.

10. Nikolai N. Semenov, *Tsepnie Reaktsii* [Chain reactions] (Leningrad: Goskhimizdat, 1934; in English translation, Oxford: Oxford University Press, 1935).

11. Khariton in Andreev, *Kapitza Tamm Semenov*, p. 437.

12. Based on private communication from Alexey Semenov in March 2012, via e-mail.

13. Andreev, *Kapitza Tamm Semenov*, p. 557.

14. Ibid., p. 556.

15. Khariton in Andreev, *Kapitza Tamm Semenov*, p. 433.

16. A. Yu. Semenov in Andreev, *Kapitza Tamm Semenov*, p. 569.

17. There is a book about Bronshtein: Gennady E. Gorelik and Victor Ya. Frenkel, *Matvei Petrovich Bronshtein and Soviet Theoretical Physics in the Thirties* (Birhauser, 1994).

18. A. S. Sonin, *"Fizicheskii idealism" Istoriya odnoi ideologicheskoi kampanii* ["Physical idealism" The history of an ideological campaign] (Moscow: Fiziko-Matematicheskaya Literatura, 1994), p. 55.

19. Nikolai S. Akulov (1900–1976) was a sad figure. In the civil war, he fought in the Red Army. Afterward, he studied physics at Moscow State University and worked there, rising to Professor of Physics. His research covered ferromagnetism and also chain reactions.

20. Yu. I. Solovev in Andreev, *Kapitza Tamm Semenov*, pp. 442–445.

21. A. M. Blokh, *Sovietskii Soyuz v interyere nobelevskikh premii: Fakti, dokumenti, razmyshlenie, kommentarii* [The Soviet Union in the interior of the Nobel Prizes: Facts; documents; reflections; commentaries] (Moscow: Fizmatlit, 2005), p. 326.

22. The label "cosmopolitan" later became a euphemism for Jews, but before the war it did not yet have this meaning. It simply meant, for example, bowing to the West.

23. Andreev, *Kapitza Tamm Semenov*, pp. 444–445.

24. Ibid., p. 445.

25. The title of Akulov's book was *Teoria tsepnikh processov* [Theory of Chain Processes].

26. F. I. Dubovitskii in Andreev, *Kapitza Tamm Semenov*, pp. 446–453.

27. At the time it was called the Presidium of the Central Committee; at other times, it was called the Politburo of the Central Committee.

28. A. Yu. Semenov, in Andreev, *Kapitza Tamm Semenov*, pp. 573–574.

29. S. G. Entelis, in Shilov, *Vospominaniya*, p. 83.

30. Osypyan, in Andreev, *Kapitza Tamm Semenov*, p. 558.

31. Photo insert in Andreev, *Kapitza Tamm Semenov*, preceding page 449.

32. Khariton, in Shilov, *Vospominaniya*, p. 42.

33. L. G. Shcherbakova, "My Great Man and Friend," in Shilov, *Vospominaniya*, pp. 217–222, p. 222

34. Shilov, in Andreev, *Kapitza Tamm Semenov*, pp. 561–562.

35. Entelis, in Shilov, *Vospominaniya*, p. 85.

36. Private communication from Alexey Semenov in March 2012, via e-mail.

37. *Nauka i Zhizn* 1965, no 4, pp. 38–43

38. A. Yu. Semenov, in Andreev, *Kapitza Tamm Semenov*, pp. 570–571.

39. Shilov, in Andreev, *Kapitza Tamm Semenov*, p. 563.

40. Entelis, in Shilov, *Vospominaniya*, p. 86.

41. Blokh, *Sovietskii Soyuz*, p. 326.

42. Julius Glaser, "About Lars Gunnar Sillén," *Chemical Intelligencer* 1999, 5(4), 4–5.

43. Istvan Hargittai, "A Curious Case of Soviet Nobel Aspirations," *Chemical Intelligencer* 1999, 5(3), 61–64; actual quote, p. 62.

44. Blokh, *Sovietskii Soyuz*, p. 429.

45. Ibid.

46. Ibid., p. 478.

47. Conversations with A. Yu. Semenov, June 2011, Moscow.

48. Vitalii I. Goldanskii, *Essays of a Soviet Scientist* (Springer, 2009), p. 53.

49. Solovev, in Andreev, *Kapitza Tamm Semenov*, p. 445.

50. Sonin, "*Fizicheskii idealism*," p. 133

51. Entelis, Shilov, *Vospominaniya*, p. 83.

52. Letter of N. N. Semenov to P. L. Kapitza, March 25, 1922, in Andreev, *Kapitza Tamm Semenov*, pp. 482–486; actual quote, pp. 484–485; see also, A. Yu. Semenov, in Andreev, *Kapitza Tamm Semenov*, pp. 553–554.

53. Hargittai, *Candid Science*, pp. 472–474.

Chapter 9

1. An excellent book with many contributions by Yulii B. Khariton and many about him as scientist and human being is available in Russian, *Yulii Borisovich Khariton: Put' dlinoyu v vek* [Yulii Borisovich Khariton: Century-long journey] (Moscow: Nauka, 2005).

2. J. Chariton and Z. Walta, "Oxydation von Phosphordämpfen bei niedrigen Drucken," *Z. Phys.* 1926, 39(7–8), 547–556. Khariton's and Valta's names were spelled according to the German transliteration in this German-language article.

3. *Yulii Borisovich Khariton*, 2005, p. 37.

4. V. A. Tsukerman, "Kriterii Kharitona" ["Khariton criteria"], ibid., pp. 277–280.

5. P. E. Rubinin, "Khariton i Kapitza" ["Khariton and Kapitza"], ibid., pp. 253–277.

6. Samuel A. Goudsmit, *Alsos* (New York: Henry Schuman, 1947).

7. *Yulii Borisovich Khariton*, 2005, pp. 125–127.

8. Ibid.

9. L. V. Altshuler, "'Zateryannii mir' Kharitona" [Khariton's lost world], ibid., pp. 286–287.

10. The Presidential Commission found that there had been internal warnings about the faulty seal design, but they were ignored. Taking a risk seemed acceptable, "because they 'got away with it last time.' As Commissioner [Richard] Feynman observed, the decision making was 'a kind of Russian roulette,'" http://history.nasa.gov/rogersrep/v1ch6.htm.

11. Boris S. Gorobets, *Krug Landau i Lifshitsa* [Landau's and Lifshits's circle] Moscow: URSS, 2008), p. 103.

12. S. V. Vasilchenko, "Tysyacha trista slov" [One thousand three hundred words], *Khariton*, pp. 392–394. According to Khariton's grandson, in the early 1990s Khariton served as consultant for a British documentary for which he was paid about fifteen hundred

to two thousand US Dollars, which he kept in his safe before giving the money to his grandson. Alexey Semenov, private communication in March 2012 via e-mail.

13. V. E. Fortov, "...chtoby stremyas' k luchshemy, ne natvorit' khudshego" (...in striving for the best, we should avoid the worst), *Yulii Borisovich Khariton*, pp. 288–292.

14. A. K. Chernyshev, "Rol' Yuliya Borisovicha Kharitona v obespechenii yadernogo pariteta v 70–80 gody" [Yulii Borisovich Khariton's role in maintaining nuclear parity in the 1970s and 1980s], ibid., pp. 381–385; actual quote, p. 384.

15. Yu. B. Khariton, "Obrashchenie k chitatelyam" ["Appeal to the readers"], ibid., pp. 363–364.

16. Yuli Khariton and Yuri Smirnov, "The Khariton Version," *Bulletin of the Atomic Scientists* 1993 May, 20–31.

17. Ibid., p. 26.

18. Alexey Semenov, private communication in March 2012 via e-mail.

19. Khariton and Smirnov, "Khariton Version," p. 29.

20. John von Neumann said this with reference to J. Robert Oppenheimer's quoting the Hindu scripture, "Now I am become Death, the destroyer of worlds." Istvan Hargittai, *Martians of Science* (New York: Oxford University Press, 2006), p. 125.

21. Edward Teller's letter of January 17, 1995, to the Honorable Hazel O'Leary, secretary of energy, U.S. Department of Energy, Washington, DC, 20545. I am grateful to Alexey Semenov, Moscow, for a copy of this letter. See more about it in Istvan Hargittai, *Judging Edward Teller: A Closer Look at One of the Most Influential Scientists of the Twentieth Century* (Amherst, NY: Prometheus Books, 2010), p. 434.

22. Hargittai, *Judging Edward Teller*, p. 433.

23. Edward Teller, letter of February 9, 1999, to Siegfried S. Hecker, Los Alamos National Laboratory. Hoover Archives, Stanford University.

24. Tam Dalyell, "Obituary: Yuli Khariton," *The Independent*, Monday, December 23, 1996.

25. *Khariton*, 2005, p. 428.

Chapter 10

1. The citation of Prigogine's Nobel Prize was "for his contributions to nonequilibrium thermodynamics, particularly the theory of dissipative structures." The dissipative forces usually destroy structures and thus they act in the direction of entropy increase. Usually, we are used to thinking in terms of destroying structures and thus violating spatial symmetry in space. The oscillating reactions are considered dissipative structures violating symmetry in time.

2. Irving R. Epstein, "Anatol Zhabotinsky (1938–2008)," *Nature* 2008, October 23, 455, 1053.

3. Istvan Hargittai, *Candid Science III: More Conversations with Famous Chemists*, ed. Magdolna Hargittai (London: Imperial College Press, 2003), "Ilya Prigogine," pp. 422–431; actual quote, p. 426.

4. Simon E. Shnol, *Geroi, zlodei, konformisti rossiiskoi nauki* [Heroes, villains, conformists of Russian science] (Moscow: Kron-Press, 2001; my copy is the second edition, but there is a third edition (Moscow: URSS, 2009).

5. A brief historical introduction is given in Anatol M. Zhabotinsky, "A History of Chemical Oscillations and Waves," *CHAOS*, 1991, *1*(4), 379–386.

6. I. Prigogine and R. Balescu, *Bull. Acad. R. Belg.*, 1955, *41*, 917; 1956, *42*, 256.

7. The Krebs cycle was named after the German-British biochemist Hans Krebs. It is part of the chemical reactions describing the process of aerobic respiration in the organism. Sometimes it is mentioned as the citric acid cycle—citric acid being the first product in the cycle, which then reappears at the end of the cycle.

8. Private communications in February 2012 from Simon Shnol, by e-mail.

9. B. P. Belousov, "Periodicheski deistvuyushchaya reaktsiya i ee mekhanizm" ["Periodically working reaction and its mechanism"], in *The Collected Abstracts of Radiation Medicine in 1958* (Moscow: Medgiz, 1959), pp. 145–147.

10. Shnol, *Heroes, Villains*, p. 296.

11. Ibid., p. 434.

12. Hargittai, *Candid Science III*, p.

13. I am grateful to Zoltán Nosztíciusz for this example.

14. The rabbit/fox population example is usually referred to as an illustration for the so-called Lotka-Volterra model in mathematical ecology; Y. Takeuchi, *Global Dynamical Properties of Lotka-Volterra Systems* (Singapore: World Scientific, 1996).

15. Private communication in March 2012 from Michael Bukatin, by e-mail.

16. A. M. Zhabotinsky, *Biofizika* 1964, *9*, 306.

17. A. M. Zhabotinsky, *Proc. Akad. Sci.* USSR 1964, *157*, 392.

18. V. M. Vitvitsky, D. P. Kharakoz, T. A. Tverdislov, L. A. Piruziyan, F. I. Ataullakhanov, G. R. Ivanitsky, and E. E. Fesenko, "Anatol M. Zhabotinsky (1938–2008)," *Biophysics*, 2009, *54*, 549–550. *Biophysics* is the English translation of the Russian journal *Biofizika* in which Zhabotinsky's first paper appeared in 1964.

19. S. E. Shnol, "In Memoriam A. M. Zhabotinsky," *Biophysics*. 2009, 54, 551–553.

20. The English translation of its title was "Study of Homogeneous Chemical Auto-Oscillatory Systems."

21. A. M. Zhabotinsky, *Kontsentratsionnie avtokolebaniya* [Concentrational Oscillations] (Moscow: Nauka, 1974).

22. A. N. Zaikin and A. M. Zhabotinsky, *Nature*, 1970, 225, 535.

23. I am grateful to Michael Bukatin for a discussion of this approach in April 2012, by e-mail. See also, Vitvitsky et al., "Anatol M. Zhabotinsky (1938–2008)," p. 550.

24. Shnol, "In Memoriam A. M. Zhabotinsky," p. 552.

25. Private communication from Michael Bukatin, April 2012, by e-mail.

26. Istvan Hargittai, *Our Lives: Encounters of a Scientist* (Budapest: Akadémiai Kiadó, 2004), pp. 79–81.

27. Zhabotinsky's former wife and their son, Michael Bukatin, also moved to the United States, in 1989. Michael eventually earned his PhD degree from Brandeis University. Zhabotinsky kept up his connection with his son, his only child: Private communication from Michael Bukatin, April 2012, by e-mail.

28. I am grateful to Miklós Orbán for sharing his impressions of Zhabotinsy's Brandeis experience with me.

29. Private communications in March 2012 from Simon Shnol, by e-mail.

30. Hargittai, *Candid Science III*, p. 427.

Chapter 11

1. Erlen Fedin, *Izbrannoe* [Selected works] (Krasnoyarsk: Polikom, 2008), p. 83.

2. Istvan Hargittai, *Drive and Curiosity: What Fuels the Passion for Science* (Amherst, NY: Prometheus, 2011), pp. 95–96.

3. L. Pauling, M. Delbrück, "The Nature of the Intermolecular Forces Operative in Biological Processes," *Science*, 1940, 92, 77–79.

4. A. I. Kitaigorodskii, "The Close-Packing of Molecules in Crystals of Organic Compounds," *Journal of Physics* (USSR) 1945, 9, 351–352.

5. Istvan Hargittai and Magdolna Hargittai, *In Our Own Image: Personal Symmetry in Discovery* (New York: Kluwer Academic/Plenum, 2000), chap. 6, "Aleksandr Kitaigorodskii," pp. 112–142.

6. Ibid., pp. 130–132.

7. C. H. MacGillavry, *Symmetry Aspects of M. C. Escher's Periodic Drawings* (Utrecht: Bohn, Scheltem and Holkema, 1976).

8. See, e.g., with references therein, Istvan Hargittai, "Crystal Structures and Culture: In Memoriam Khudu Mamedov (1927–1988)," *Structural Chemistry*, 2007, *18*, 535–536.

9. Lucretius, *The Nature of Things* [*De rerum natura*], 1st ed., trans. F. O. Copley (New York: W. W. Norton, 1977). The quoted passage is from Book VI, p. 72, lines 1084–1086.

10. N. I. Kuznetsova, ed., *Chelovek, kotorii ne umel bit ravnodyshnim: Yurii Timofeevich Struchkov v nauke i zhizhni* [The man who was unable to be indifferent: Yurii Timofeevich Struchkov in science and in life] (Moscow: Russian Academy of Sciences, 2005), pp. 121–123.

11. Tatyana Mastryukova, "Years of Studying," in Kuznetsova, *Chelovek*, pp. 79–82; actual quote, p. 82.

12. Tatyana Khotsyanova, "How It All Started," in Kuznetsova, *Chelovek*, pp. 83–97; actual quote, p. 85.

13. Ibid., p. 87.

14. Ibid., p. 90.

15. Kuznetsova, *Chelovek*, p. 165.

16. Kuznetsova, *Chelovek*, p. 137.

17. P. M. Zorky, "The Development of Organic Crystal Chemistry at the Moscow State University," *ACH Models in Chemistry*, 1993, 130, 173–181; actual story, p. 174.

18. Hargittai and Hargittai, *In Our Own Image*, pp. 120–121.

19. Istvan Hargittai, *Our Lives: Encounters of a Scientist* (Budapest: Akadémiai Kiadó, 2004), p. 100.

20. Jack D. Dunitz, "A. I. Kitaigorodsky: Personal Reminiscences," *ACH Models in Chemistry*, 1993, 130, 153–157.

21. Zorky, pp. 177–178.

22. Ibid.

23. N. N. Petropavlov, in *A. I. Kitaigorodskii: Uchonii, uchitel, drug* [A. I. Kitaigorodskii: Scientist, teacher, friend] (Moscow: Moskvovedenie, 2011), pp. 96–113.

24. Fedin, pp. 83–86.

Chapter 12

1. Emiliya G. Perevalova, "Aleksandr N. Nesmeyanov," *Chemical Intelligencer*, 2000, 6(2), 32–36.

2. Ibid., p. 33.

3. A. Todd, "O Nesmeyanove," in M. I. Kabachnik, ed., *Aleksandr Nikolaevich Nesmeyanov: Uchonii i chelovek* [Aleksandr Nikolaevich Nesmeyanov: Scientist and Human Being] (Moscow: Nauka, 1988), pp. 49–50.

4. Perevalova, "Aleksandr N. Nesmeyanov," pp. 35–36.

5. Yu. A. Ovchinnikov. "Neskolko slov ob Engelhardte" ["A few words about Engelhardt"]. In *Vospominaniya o V. A. Engelhardte* [Remembering V. A. Engelhardt] ed. A. A. Baev (Moscow: Nauka, 1989), pp. 104–108.

6. Ibid., p. 106.

7. W. A. Engelhardt, "Life and Science," *Annual Review of Biochemistry* 1982, 51, 1–19.

8. Ibid., p. 36.

9. N. I. Kuznetsova, ed., *Chelovek, kotorii ne umel bit ravnodyshnim: Yurii Timofeevich Struchkov v nauke i zhizhni* [The man who was unable to be indifferent: Yurii Timofeevich Struchkov in science and in life], (Moscow: Russian Academy of Sciences, 2005), p. 126.

10. Ibid., p. 165.

11. Erlen Fedin, *Izbrannoe* [Selected works] (Krasnoyarsk: Polikom, 2008).

12. Kuznetsova, *Chelovek*, p. 294.

13. Ibid., pp. 125–126.

14. Jan Kandror, private communication in February 2012, by e-mail.

15. Kuznetsova, *Chelovek*, pp. 125–126.

16. Ibid. p. 128

17. A. N. Nesmeyanov, *Na kachelyakh XX veka* [Sitting on the swings of the twentieth century] (Moscow: Nauka, 1999).

18. Kuznetsova, *Chelovek*, pp. 129–130.

19. *Sostoyanie teorii khimicheskogo stroeniya v organicheskoi khimii* [State of theory of chemical structure in organic chemistry], All-Union Conference June 11–14, 1951, stenographic minutes (Moscow: Publishing House of the Academy of Dciences of the USSR, 1952).

20. Linus Pauling, *The Nature of the Chemical Bond* (Ithaca, NY: Cornell University Press). The Russian translation of its second edition appeared in 1947.

21. George Wheland, *Theory of Resonance and Its Applications to Organic Chemistry* (New York: Wiley, 1944). Its Russian translation appeared in 1948.

22. The English translation of the report has appeared: D. N. Kursanov, M. G. Gonikberg, B. M. Dubinin, M. I. Kabachnik, E. D. Kaverzneva, E. N. Prilezhaeva, N. D. Sokolov, and R. Kh. Freidlina, "The Present State of the Chemical Structural Theory," *Journal of Chemical Education*, 1952, (1), 2–13

23. V. M. Tatevskii and M. I. Shakhparonov, "About a Machistic Theory in Chemistry and Its Propagandists," *Journal of Chemical Education*, 1952, (1), 13–15. These are extracts from the original Russian article by these authors in *Voprosi Filosofii* [Problems in philosophy] 1949, 3, 176.

24. Istvan Hargittai, *Candid Science: Conversations with Famous Chemists*, ed. Magdolna Hargittai (London: Imperial College Press, 2000).

25. J. Chatt and M. I. Rybinskaya, "Aleksandr Nikolaevich Nesmeyanov September 9, 1899–January 17, 1980," *Biographical Memoirs of Fellows of the Royal Society*, 1983, 29, 399–480.

26. Hargittai, *Candid Science*, "Elena G. Galpern and Ivan V. Stankevich," pp. 322–331; actual quote, p. 327. As is known, each of the two teams has eleven players. Nowadays not only men, but women also play soccer.

27. D. A. Bochvar and E. G. Galpern, *Doklady Akademii Nauk*, 1973, 209, 610–612.

28. H. W. Kroto, J. R. Heath, S. C. O'Brien, R. F. Curl, and R. E. Smalley, *Nature* 1985, 318, 162–163.

29. W. Krätschmer, L. D. Lamb, K. Fostiropoulos, and D. R. Huffman, *Nature* 1990, 347, 354–358.

Epilogue

1. The expression of "Iron Curtain" was used in Winston Churchill's speech on March 5, 1946, in Fulton, Missouri: "From Stettin in the Baltic to Trieste in the Adriatic, an iron curtain has descended across the Continent."

2. Istvan Hargittai, *Martians of Science: Five Physicists Who Changed the Twentieth Century* (New York: Oxford University Press, 2006), pp. 156–157.

3. Mark Popovsky, *The Vavilov Affair* (Hamden, CT: Archon Books, 1984), pp. 197–199.

4. Rita Levi-Montalcini, *In Praise of Imperfection* (New York: Basic Books, 1989).

5. Ronald Hingley, *Nightingale Fever: Russian Poets in Revolution* (New York: Alfred A. Knopf, 1981), p. xii.

6. Ibid., p. xiii.

7. Anna Akhmatova, *Pamyati Anny Akhmatova. Stikhi. Pis'ma. L. Chukovskaya* (Paris, YMCA, 1974) p. 188. Quoted in Hingley, *Nightingale Fever*, p. xiii.

8. Translation by Ilya Shambat: http://lib.udm.ru/lib/POEZIQ/MANDELSHTAM/tristia_engl.txt_Piece40.01.

9. For references, see, Hingley, *Nightingale Fever*, p. 73.

INDEX

Asterisks indicate the names of the fourteen principal heroes of the book and the pages of their respective chapters are given in bold. Academy refers to the Soviet Academy of Sciences and MSU to Moscow State University.